Agenda Setting in a 2.0 World

Agenda Setting in a 2.0 World explores agenda-setting theory in light of changes in the media environment in the twenty-first century. In the decades since the original Chapel Hill study that launched agenda-setting research, the theory has attracted the interest of scholars worldwide.

This volume features the work of a new generation of scholars. The research provided by these young scholars reflects two broad contemporary trends in agenda setting: A centrifugal trend of research in the expanding media landscape and in domains beyond the original focus on public affairs, and a centripetal trend further explicating agenda setting's core concepts.

Thomas J. Johnson is the Amon G. Carter Jr. Centennial Professor in the School of Journalism at the University of Texas at Austin. His most recent co-edited book, *International Media Communication in a Global Age* (Routledge, 2009), examines key issues regarding global communication.

New Agendas in Communication

A Series from Routledge and the College of Communication at the University of Texas at Austin

Roderick Hart and Stephen Reese, Series Editors

This series brings together groups of emerging scholars to tackle important interdisciplinary themes that demand new scholarly attention and reach broadly across the communication field's existing courses. Each volume stakes out a key area, presents original findings, and considers the long-range implications of its "new agenda."

Global Communication
edited by Karin Wilkins, Joe Straubhaar, and Shanti Kumar

Identity and Communication
edited by Dominic L. Lasorsa and America Rodriguez

Political Emotions
edited by Janet Staiger, Ann Cvetkovich, and Ann Reynolds

Media Literacy
edited by Kathleen Tyner

Communicating Science
edited by LeeAnn Kahlor and Patricia Stout

Journalism and Citizenship
edited by Zizi Papacharissi

The Interplay of Truth and Deception
edited by Matthew S. McGlone and Mark L. Knapp

Agenda Setting in a 2.0 World

New Agendas in Communication

A Tribute to Maxwell McCombs

Edited by Thomas J. Johnson

Routledge
Taylor & Francis Group
NEW YORK AND LONDON

First published 2014
by Routledge
711 Third Avenue, New York, NY 10017

Simultaneously published in the UK
by Routledge
2 Park Square, Milton Park, Abingdon, Oxon OX14 4RN

Routledge is an imprint of the Taylor & Francis Group, an informa business

Library of Congress Cataloging in Publication Data

Agenda setting in a 2.0 world : new agendas in communication /
edited by Thomas J. Johnson.
pages cm
1. Mass media—Political aspects. 2. Digital media—Political aspects. 3. Mass media and public opinion. 4. Mass media—Research. I. Johnson, Thomas J.
P95.8.A34 2013
302.23'1—dc23
2013022059

ISBN: 978-0-415-83701-9 (hbk)
ISBN: 978-0-415-83700-2 (pbk)
ISBN: 978-0-203-73587-9 (ebk)

Typeset in Sabon
by Apex CoVantage, LLC

Contents

Contributors

Lindita Camaj (Ph.D., Indiana University at Bloomington) is an assistant professor at the Jack J. Valenti School of Communication, University of Houston. Her dissertation investigates media effects on citizens' political trust and participation in Southeastern Europe. Her broad research and teaching interests include the media role in democratization, agenda-setting and priming effects, the interaction between journalism and culture, and freedom of information legislation. Her scholarly work has been awarded by the AEJMC and published in *International Communication Gazette*, *Global Media Journal*, and *Central European Journal of Communication.*

Lei Guo is a doctoral student in the School of Journalism at the University of Texas at Austin. She has an M.A. in journalism from UT Austin (2010). Her research focuses mainly on media and marginalized groups, the development of agenda-setting theory, and international communications. Her dissertation will explore alternative media practices, social movement of Asian communities in the United States, and technology and empowerment.

Vanessa de Macedo Higgins Joyce (Ph.D., University of Texas at Austin) is an assistant professor in the Division of Journalism at Southern Methodist University in Dallas, TX. Her research interests include international and Latin American communication, political communication and public opinion, journalists' roles, and, of course, agenda-setting theory. Her dissertation focused on the consensus-building function of agenda setting, analyzing national and transnational news media in Europe. Her work has been published in the journals *Journalism*, *Brazilian Journalism Research*, *Revista Logos: Comunicação e Universidade*, and *Media Tenor Forschungsbericht*. She has also contributed chapters in the book *Women, Men, and News: Divided and Disconnected in the News Media Landscape* and the first and second editions of *Research Methods in Communication.*

Yonghwan Kim (Ph.D., University of Texas at Austin) is an assistant professor at the University of Alabama. His research interests include new communication technologies, media psychology, media effects, and public opinion research. His research has been recognized by the International Communication Association and his work has been published in *Computers in Human Behavior, Journal of Communication and Cyberpsychology, Behavior and Social Networking, Asian Journal of Communication, International Journal of Strategic Communication, International Journal of Public Opinion Research,* and *Mass Communication & Society.*

Jennifer Kowalewski (Ph.D., University of North Carolina at Chapel Hill) is an assistant professor of multimedia journalism in the Department of Communication Arts at Georgia Southern University. Her research interests lie in political communication, with special focus on agenda-setting and priming effects of different news media. Her academic work has been presented at a number of conferences such as the Association for Education in Journalism and Mass Communication, International Communication Association, and the Midwest Association of Public Opinion Research. She also has published her academic work.

Jae Kook Lee (Ph.D., University of Texas at Austin) is an assistant professor at the Indiana University at Bloomington. His research focuses on a wide range of issues concerning public opinion and media effects. Lee's latest question is about how the new media environment influences public opinion and political communication processes.

Jason A. Martin (Ph.D., Indiana University at Bloomington) is an assistant professor of journalism in the College of Communication at DePaul University. He received his Ph.D. in mass communication from Indiana University at Bloomington in May 2011 under the direction of David Weaver. His teaching and research interests include political communication during election campaigns, media effects on civic participation, and issues of media law regarding public affairs journalism and newsgathering. He has published articles in journals such as *Communication Law & Policy, Journal of Media Law & Ethics,* and *Newspaper Research Journal.*

Sharon Meraz (Ph.D., University of Texas at Austin) is an assistant professor at the University of Illinois at Chicago. Her interdisciplinary research interests center on the political impact of emergent media such as blogs, social networking sites, and mobile technologies on political communication and citizen political engagement. Her interests also include studying the impact of social influence and social networks on the evolution of traditional mass media industry and mass media effect theories. Her work has appeared in *Journalism and Mass Communication Quarterly*, *Journalism*, and *Journal of Computer Mediated Communication*. She is the coauthor of *Women, Men and News: Divided and Disconnected in the News Media Landscape*. Meraz has published several book chapters on social media, citizen media, and mass media evolution, and comes to academia after a career in the information technology industry, where she developed software and database applications.

Soo Jung Moon (Ph.D., University of Texas at Austin) is an assistant professor in the Department of Mass Communications at the University of West Georgia. Prior to her academic career, she worked as a professional journalist for 12 years in Seoul, Korea. Her research interests include media effects and new communication technologies. Her work has been published in *Communication Research* and *Journal of Broadcasting & Electronic Media*.

Matthew W. Ragas (Ph.D., University of Florida) is an assistant professor of public relations in the College of Communication at DePaul University in Chicago. Spiro Kiousis directed his dissertation, which received the Nafziger-White-Salwen award from AEJMC for outstanding dissertation. His agenda-setting research focuses on corporate communication and business news, and political communication. His work has been published in *Journalism & Mass Communication Quarterly*, *Mass Communication & Society*, *Public Relations Review*, *International Journal of Strategic Communication*, *Journal of Health & Mass Communication*, and *Web Journal of Mass Communication Research*. He is also the author of two trade books and several book chapters.

Hai Tran (Ph.D., University of North Carolina at Chapel Hill) is an assistant professor of journalism at DePaul University. As a doctoral student at North Carolina, he assisted Prof. Donald Shaw in agendamelding research. Hai studies media effects across platforms, including agenda setting and multimedia news. His work examines theoretical issues in the context of political and international communication.

Sebastián Valenzuela (Ph.D., University of Texas at Austin) is an assistant professor in the School of Communications at Pontificia Universidad Católica de Chile, where he conducts research on political communication, public opinion, social media, and journalism. His work has been published in *Journal of Communication*, *Communication Research*, *International Journal of Public Opinion Research*,

International Journal of Press/Politics, *Journalism & Mass Communication Quarterly*, *Journal of Computer-Mediated Communication*, *International Communication Gazette*, *Journalism*, *CyberPsychology & Behavior*, and *Cuadernos de Información*. His research has been awarded the Naomi C. Turner prize from WAPOR and the Chaffee-McLeod student paper award from AEJMC.

Foreword

Agenda Setting in a 2.0 World

David H. Weaver
Distinguished Professor Emeritus
School of Journalism, Indiana University

I'm very pleased to be invited to write a forward to this book honoring Professor Maxwell McCombs, as well as Donald Shaw, upon their retirement from unusually productive and focused careers of 45 years of university teaching and research.

As one of their earlier doctoral students from 1971–74 at the University of North Carolina, I not only learned how to *do* research from these two young scholars, but also how to become *excited* about doing it. They both have a rare ability to inspire and focus others on certain kinds of research, especially studies dealing with newspaper readership, journalism history, and the agenda-setting function of the news media.

As we celebrate 40 years since the original agenda-setting article by Maxwell McCombs and Donald Shaw was published in *Public Opinion Quarterly* in the summer of 1972, it is truly inspiring to look back at what that simple little study comparing a few media issue agendas with the issue agendas of 100 undecided Chapel Hill voters has led to.

For one thing, that original publication has been one of the most-cited articles in our field, and the original concept of agenda setting has been elaborated and expanded over the past three decades by a steady flow of studies by McCombs and Shaw and their students, as well as many outside their original circle of influence. The latest estimate is around 500 published articles on agenda setting, not to mention hundreds of conference papers and numerous books and book chapters.

But more than that, the entire area of agenda-setting theory and research has been expanded and refined to include not only (a) basic first-level issue or topic agenda-setting effects, but also (b) the study of attribute agenda setting, (c) the development of contingent conditions affecting the strength of agenda setting, (d) the study of the consequences of agenda setting (including cognitive, affective, and behavioral effects), (e) the study of the origins of media agendas, and (f) the relatively new approach of agendamelding developed by Donald Shaw and his associates.

In addition, the *domain or focus* of agenda-setting research has moved beyond politics and public affairs to the worlds of business, sports, the economy, education, entertainment, and religion; and the *methods* used in such studies have gone far beyond the simple rank-order correlations between media and public agendas to sophisticated time-series analyses and studies of curvilinear relationships. Another recent trend is research involving the newer social media such as Facebook, Twitter, blogs, and other Web-based services and sites.

If we consider the chapters included in this book, which is based on the New Agendas conference in Austin, Texas, in late September and early October 2011, we see examples of different agenda-setting domains, settings, and concepts, including some that are focused on

1. social media and websites, such as those by Sharon Meraz, Jason Martin, and Hai Tran;
2. contingent conditions, such as those by Yonghwan Kim and Jae Kook Lee;
3. agenda-setting research beyond the traditional domain of political news, such as that by Jennifer Kowalewski and Matt Ragas;
4. different levels of agenda setting such as those by Lindita Camaj, Lei Guo, and Vanessa Higgins Joyce;
5. the key concept of issue salience by Sebastián Valenzuela; and
6. the behavioral consequences of agenda setting by Soo Jung Moon.

These chapters—and recent journal articles on different ways of measuring the public agenda using Internet search trends instead of surveys, different measures of need for orientation and other contingent conditions, second- and third-level agenda setting, attitudinal and behavioral consequences of agenda setting, various influences on media agendas (including new studies of intermedia agenda setting using websites and social media), new long-term studies of policy agenda setting, and new studies of the societal consequences of agenda setting and agenda melding—all attest to the continuing interest in, and usefulness of, the agenda-setting approach.

And these recent studies also illustrate vividly the advantage of building a cumulative pattern of research over time. I think it is safe to say that Maxwell McCombs and Donald Shaw are among the very few people in our field who have had such a major impact on mass communication research with their conceptualization, elaboration, and empirical testing of the agenda-setting idea during the past 44 years. And one important reason for this impact is that they have kept coming back to this fruitful idea, elaborating and refining it, and making the implicit explicit. Another reason is that they both write so clearly and convincingly about this approach that it is difficult *not* to be interested in it!

In addition, their interest and enthusiasm are contagious, making them natural research leaders as well as producers, something that few of us can claim even if our own personal research productivity is high. I feel very fortunate to have been one of their early doctoral students and one of their friends and colleagues for my entire career.

As we look ahead to the future of agenda-setting theory and research in this brave new media world of the twenty-first century, I think we should keep in mind some of the areas where we need to know more, such as

1. the impact of different cultures and media systems on agenda-setting processes and effects;
2. the conceptual and operational definitions of salience, which is THE key concept of all agenda-sctting research;
3. the impact of an increasingly complex media environment on public and policy agenda setting (is there really as much fragmentation of media and public agendas as many are claiming?);
4. the relationship of framing research and theory to agenda setting;
5. the multiple influences on media agendas and how these vary across time and space; and
6. the consequences of agenda setting not only for individuals but for the larger society and its policies.

As Toshio Takeshita (2006, p. 291) has written, "Agenda-setting theory is rare in the history of journalism and mass communication research. . . . [It] is an indigenous theory born from this field of research, not from other disciplines such as sociology, psychology, or political science."

Takeshita goes on to point out that agenda setting has helped various research approaches to connect with one another, including gatekeeping, uses and gratifications, the study of message sources, framing, priming, stereotyping, and image making.

He ends by writing that "agenda setting has played an important role in theory building in mass communication research, and it continues to do so. This research tradition is still worth passing along." (p. 291)

I think we can all agree with that. Agenda setting is not only worth passing along, it *is* being passed along and adapted to our new media and social environment, and it continues to be refined, elaborated, and expanded by many promising younger scholars such as those represented in this book.

I don't think that there is any higher tribute that could be paid to the two founders of this approach, Professors Maxwell McCombs and Donald Shaw.

References

McCombs, M.E., & Shaw, D.L. (1972). The agenda-setting function of mass media. *Public Opinion Quarterly*, *36*, 176–187.

Takeshita, T. (2006). Current critical problems in agenda-setting research. *International Journal of Public Opinion Research*, *18*(3), 275–296.

Introduction

Agenda Setting in a 2.0 World

Thomas J. Johnson

McCombs and Reynolds' 2002 chapter "News Influence on Our Pictures of the World" may have been written to help mass communication students navigate the literature on agenda setting, but it also served as an unofficial marker of how far the theory had come in its first 30 years. The single McCombs and Shaw (1972) Chapel Hill study had blossomed into more than 500 empirical works (Weaver, Foreword, this volume). More importantly, the accumulated evidence showed strong support for agenda-setting theory's central claim that the media focus the public's attention and influence the public's perceptions of what are the important issues of the day. The McCombs and Reynolds agenda-setting review demonstrated that the theory had been supported from different geographic and historical settings worldwide in studies that examined a wide range of media and public issues using different methodological approaches.

But while some scholars were celebrating agenda setting at its thirtieth anniversary for being one of the dominant theories in the mass communication field, others questioned whether the theory could remain viable until its fortieth anniversary (Chaffee & Metzger, 2001; Takeshita, 2006). A fundamental assumption of the theory, the researchers argued, was that people got their news from a finite number of news outlets run by professional gatekeepers with shared news values. But as the numbers of news outlets increased dramatically and the audience for them shrank, the assumption of a unified media agenda has become problematic. Further, as the audience gains more control over what news sources it will use, the audience will likely seek sources that support their point of view (Iyengar & Hahn, 2009; Stroud, 2010, 2011) or avoid news altogether (Atre & Katz, 2005; Prior, 2005, 2007), obliterating the sense of a public consensus.

But, to paraphrase Mark Twain, reports of agenda setting's impending death have been greatly exaggerated. More than 40 years after McCombs and Shaw's seminal study, agenda setting remains a vibrant theory that continues to illuminate what issues we perceive as important and how we perceive them.

So how has agenda setting survived in a 2.0 world?

First, critics may have overemphasized the role of the Internet in affecting the public agenda. Though certain components of the Internet such as Twitter (Kushin, 2010) and blogs (Project for Excellence in Journalism, 2010), may advance different agendas than the traditional media, the online public still gets most of its news from online versions of traditional news sources and news aggregators such as Google and Yahoo! that rely heavily on mainstream media for their information (Smith, 2011). Consequently, news sites on the Internet and off-line forms share a similar agenda (Hamilton, 2004; Lee, 2009).

Second, the authors seem to harbor misunderstandings about agenda setting. First, the notion that people get news from a small set of national sources has never been a fundamental assumption of agenda setting. After all, the original McCombs and Shaw agenda-setting study focused on four local sources in addition to the *New York Times, Time, Newsweek*, NBC and CBS evening news. More importantly, critics still treated agenda setting as a simple transfer of salience from the media to the public—the first stage of agenda setting. However, agenda setting has undergone four additional stages that have deepened the theoretical understanding of agenda setting: attribute agenda setting, the psychology of agenda-setting effects, sources of the media agenda, and consequences of media effects (Coleman et al., 2009; McCombs, 2004, 2005; McCombs, Shaw & Weaver, 2007; Valenzuela & McCombs, 2008. Further, agenda setting has expanded beyond its initial focus on public affairs news to explore other areas such as professional sports (Fortunato, 2001), classroom teaching (Rodriguez Diaz, 2004), environmental issues (Chan, 1999; Salwen, 1988); and health issues, including HIV/AIDS (Pratt, Ha, & Pratt, 2002) and smoking (Sato, 2003). In particular it has become an important theory used in corporate studies (Carroll, 2009, 2010, 2011; & McCombs, 2003; Ragas, 2010, Chapter 12, this volume; Ragas, Kim & Kiousis, 2011). Agenda setting also has not been limited to national issues but applied to local ones as well (Palmgreen & Clarke, 1977; Smith, 1987).

I argue that agenda setting has now entered a sixth stage of development: agenda setting in an online world. In this stage researchers have adapted agenda-setting research to the new media landscape as well as addressed new areas created by this changing landscape.

The Internet, along with cable news and talk radio, has introduced fundamental changes in both the amount and the nature of online information as well as between audience and technology. The Internet provides such an abundance of information that makes selective exposure not only possible but a necessity (Hart, Albarracin, Eagly, Brechan, Lindberg, & Merill, 2009; Smith Fabrigar, & Norris, 2008). The Internet, cable news, and radio talk shows not only provide more partisan content but also increase people's ability to customize and control information, making it easier to find sources that support one's predispositions (Bennett & Iyenger, 2008; Garrett, 2009; Knobloch-Westerwick, & Meng, 2011; Johnson, Bichard, & Zhang, 20009; Sunstein, 2001). This could lead to a more polarized and fragmented electorate (Iyengar & Hahn, 2009; Stroud, 2008, 2010, 2011) that may be drawn to different agendas instead of being shaped by one common one, although there is some disagreement about how fragmented the public actually is (Garrett, 2009, Hart et al., 2009; Johnson, Zhang, & Bichard, 2010; Webster & Ksiazek, 2012). Also, netizens are not simply consumers of news, but are what Bruns (2008) calls "produsers," individuals who actively search for information as well as those who create content in a networked, participatory environment.

Agenda-setting researchers have taken advantage of these changing audience relationships in how they have measured salience. For instance, the Yahoo! Buzz Index and number of hits on Google News have both been employed as measures of public salience (Hester & Gibson, 2007) and online search terms have been used as indicators of media salience (Hester & Gibson, 2007; Weeks & Southwell, 2010).

As Hai Tran (Chapter 10, this volume) notes in his review of online agenda setting at least 26 refereed journal articles have explored online agenda setting. Most of these applied what we have known about agenda setting to the online world. For instance, Tran found that almost 70% of the studies explored intermedia agenda setting, examining the relationship between traditional news sites and their online counterparts (Groshek, 2008; Lim, 2011) as well as traditional news sources and social media (Maier, 2010), blogs (Meraz, 2009, 2011a, 2011b), candidate websites (Sweetser, Golan & Wanta, 2008), YouTube (Ragas & Kiousis, 2010; Sayre et al., 2010), and online discussion boards (Lee, Lancendorfer, & Lee, 2005; Roberts, Wanta, & Dzwo, 2002). Other studies have examined the consequences of online agenda setting from increases in charitable giving online during the 2004 Asian tsunami (Waters & Tindall, 2011) to increased support for efforts to reduce global warming (Lee, 2010). Several studies have also explored how the rise of 24-hour media has affected the time span for agenda-setting effects to occur. Studies have found agenda-setting effects can decline after one day (Roberts, Wanta, & Dzwo, 2002) or even during the course of a single day (Weeks & Southwell, 2010).

In light of the changes in the media landscape and fundamental changes in the relationship between the audience and the media, the "Agenda Setting in a 2.0 World: New Agendas in Communication" conference was held Sept. 29-Oct. 1, 2011, in the School of Journalism at the University of Texas at Austin. The purpose of the conference was twofold. First, it was designed to honor retired UT Austin professor Max McCombs for his numerous contributions to the University of Texas, and the mass communication field, particularly his work in agenda setting. Second, it was to invite 12 young scholars (graduate students and assistant professors) to present drafts of their papers for this book in a collaborative atmosphere to try to make sense of what role agenda setting plays in the new media environment, to explore theoretical issues of agenda setting in a 2.0 world, as well as to chart how the theory has been expanded to topics beyond traditional public affairs to areas such as business.

The papers in this volume address agenda setting in a 2.0 world through all five stages of the agenda-setting process from the original agenda-setting idea to consequences of agenda setting. While all the papers shed light on the agenda-setting process, a few make more groundbreaking discoveries. For instance, Lei Guo, a doctoral student in the School of Journalism at UT Austin, proposes a third level of agenda setting: a network agenda-setting model. Relying on network analysis theory and employing a new survey approach—mind-mapping—that allows respondents to indicate what connections they perceive between objects or attributes, Guo develops a Network Agenda Setting Model that suggests that media help transfer the salience of the networked relationships among objects and attributes to the public. More specifically, the model asserts that the media can help to connect new information to old information in the audience's associative network memory. Media are able to strengthen the connections by activating pairs of constructs in tandem. The model suggests that the news media not only succeed in telling us what to think about (first-level agenda setting), and how to think (attribute agenda setting) but are also capable of telling us what and how to associate objects or attributes (third–level agenda setting).

Sharon Meraz, assistant professor at the University of Illinois at Chicago, advances the new notion of agenda supporting. Meraz's chapter examines the role of sources in setting the discussion agenda in the Tea Party Patriots' Facebook Group. She found that while traditional media and blogs helped set the group's discussion agendas, citizen media such as YouTube and Facebook played an agenda-supporting role. That is, Tea Party members searched these sources for information that supported or confirmed their pre-existing agendas.

Different chapters in this volume provide insight into the five stages of agenda-setting research.

Matt Ragas, assistant professor at DePaul University, primarily reviews the literature on traditional agenda-setting effects, but in the corporate world rather than in public affairs news. He focuses on three main areas: agenda setting and corporate reputation, agenda setting and the economy, and agenda setting in the stock market. For instance, in terms of corporate reputation, Ragas notes that past studies find that media play a role in increasing brand recognition and top-of-mind awareness as well as how the public judges a corporation, but that corporations use public relations to manage corporate reputations. The relationship for agenda setting and the economy is more complex. Ragas notes that most studies find that shifts in news coverage lead to changes in perceptions of the economy rather than vice versa, but real-world economic conditions also influence public perceptions of the economy and the White House can affect both the public and media agendas. In studies of the stock market, the media may not only influence how people think but also how they act. For instance, amount of media coverage in the year before a company starts selling stock on the public market suggests amount of media coverage can affect trading volume and capital raised on the first day of trading. Ragas may indicate early on in the article that corporate research focuses on the traditional transfer of salience from the media to the public, but his review shows a more complex process involving attribute agenda setting, contingent conditions, questions of who influences the media agenda, and what effects the media have.

Lindita Camaj, assistant professor at the University of Houston, and Vanessa de Macedo Higgins Joyce, assistant professor at Southern Methodist University, explored attribute agenda setting in very different ways. Camaj employed an attribute agenda-setting index that measured the difference between media emphasis on institutional positive attributes and citizens' positive evaluations of institutional traits and found that it led to increased trust of the Kosovo government. Furthermore, those who relied on partisan, pro-government media were more likely to adopt the media agenda on positive institutional attributes when judging institutional performance. Higgins Joyce studied the potential of the media to build consensus on how people in Europe thought about terrorism (a substantive dimension) and how they felt about Muslim and Arabic communities (an affective dimension). She found that national television built consensus more than national newspapers and discovered more of an increase for consensus on the substantive than the affective dimension.

Studies that have examined contingent conditions of agenda-setting effects often focus on the need for orientation (NFO), the idea that people have an innate curiosity about the world that is defined by how personally relevant one perceives an issue and how uncertain one is about information on the topic (Weaver, 1980). Need for orientation is identified as the most important contingent condition in agenda setting (McCombs, 2004). Three of the studies in this volume examine need for orientation and all find it increases agenda-setting effects. But all of the studies go beyond the NFO/agenda setting connection. Yonghwan Kim, assistant professor at the University of Alabama, explores the relationships between NFO, issue obtrusiveness, and attitude strength. Kim finds that in candidate ads featuring both obtrusive (job) and unobtrusive (real estate) issues, relevance but not uncertainty affected issue salience. Although NFO did not influence attitude strength, it interacted with the tone of the ad so that low NFO was linked to negative ad messages in influencing issue salience. Sebastián Valenzuela, assistant professor at Catholic University of Chile, hypothesized that materialist values (security and survival goals that can be gained by material means) and postmaterialist values (autonomy and quality of life aspects) may interact with NFO to influence agenda setting. That is, values moderate the process by which people determine the most important issues of the day that need to be addressed by public officials. Camaj discovered that NFO influenced her attribute agenda-setting index of the distance between public and media positive attitudes toward the Kosovo government through media use. The relationship, though, was different for partisan and independent media. Need for orientation interacted with independent newspapers, but not with partisan television and radio content, to increase positive assessments of the Kosovo government. Camaj argues that in a partially polarized media environment such as Kosovo, level of political uncertainty (as measured by partisan strength) might prove to be more important than relevance in determining media sources. Therefore, differences in attribute agenda-setting effects occur when differentiating between exposure to independent and partisan-oriented media.

But as the studies above suggest, while need for orientation may be the most important predictor of agenda setting, it is not the only one. Jae Kook Lee, assistant professor at Indiana University, examined the effects of age, specifically life-cycle effects, on the *New York Times* agenda from 1980 to 2004. Life-cycle effects assume that as people age their needs and interests change and therefore which issues they view as important should also change. In terms of agenda setting, researchers assume that people's different situations in their life span affect their media use, which leads to different perceptions of issue importance by people of different ages. However, Lee found that except for Baby Boomers, there was little evidence of a life-cycle effect. People had similar issue agendas throughout their life spans.

Jennifer Kowalewski, assistant professor at Georgia Southern University, examined the effects of information vs. entertainment content on agenda setting. When Kowalewski conducted an experiment exposing people to a radio show as a traditional news program and as an entertainment program, she found that the method of presentation did not affect whether respondents judged the issues discussed as the most important issue in the nation, suggesting individuals accept political information from nontraditional sources and that they want to be entertained as well as informed.

As noted earlier, about 70% of the journal articles that have explored agenda setting in an online atmosphere have focused on the intermedia agenda-setting relationship between new and old media (Tran, in this volume). But as the public has become part of the news production process and gained more control over what information it will consume, there has also been renewed interest in agenda building, an old concept that can be said to predate agenda setting (Cobb & Elder, 1971). Meraz examines the public agenda of the Tea Party Patriots group and finds that for its discussion group it prefers citizen media such as YouTube and social network sites such as Facebook as sources over traditional media sites, political blogs, political establishment sites, or online news-only/aggregator sites. On the other hand, traditional media and political blogs helped set the agendas of the Tea Party discussants, while social media sources were employed primarily to support existing agendas. Jason Martin, assistant professor at DePaul University, conducted two studies to look at agenda-building effects during local election campaigns. In the first study he found that while those issues covered in the online edition of a local online paper influenced the issues discussed on that paper's discussion board, the public discussion also had some influence on the online newspaper's agenda. In the second study, the public responded to the agendas of both the newspapers and the candidates (through the press releases they issued), but the media agenda had a stronger influence. Together, these studies showed the continued importance of traditional media in setting the public agenda, but in the 2.0 world the public has a role to play in what issues the media view as salient.

Most agenda-setting studies examine agenda setting as an end state. However, a group of studies have explored the influence of agenda setting on a host of political attitudes and behaviors. Soo Jung Moon, assistant professor at the University of West Georgia, explored the direct and indirect effects of agenda setting on a host of political participation and civic engagement measures. Specifically, the agendas of the *New York Times* and NBC directly predicted voting, political discussion, attending political meetings, donating to political causes, community involvement, and membership in civic organizations, both directly and indirectly through its relationship with strength of opinion on a host of issues.

Earlier, this author identified two primary reasons why agenda setting has survived in the 2.0 world despite the proliferation of media and the decline of a unified public agenda. There is at least one more reason. The three primary architects of agenda setting, Max McCombs, Don Shaw, and David Weaver, though recently retired, are still actively researching and expanding the boundaries of agenda-setting research. More importantly, they trained a second generation of active agenda-setting researchers, several of whom participated in our conference, including Wayne Wanta, Spiro Kiousis, Salma Ghanem, Lars Wilnat, and Craig Carroll. Both of these generations trained a third generation of agenda-setting researchers, many of whom were presenting at this Agenda Setting in the 2.0 World conference. As you will see from the research in this volume, the future of agenda setting appears in very secure hands.

References

Atre, J., & Katz, E. (2005). *What's killing television news? Experimentally assessing the effects of multiple channels on media choice.* Paper presented at the International Communication Association Annual Conference, New York, NY.

Bennett, W. L., & Iyengar, S. (2008). A new era of minima effects: The changing foundation of political communication. *Journal of Communication, 58,* 707–731.

Bruns, A. (2008). *Blogs, Wikipedia, Second Life, and beyond: From production to produsage.* New York: Peter Lang.

Carroll, C.E. (2009). The relationship between firms' media favorability and public esteem. *Public Relations Journal, 3*(4).

Carroll, C.E. (2010). Should firms circumvent or work through the news media? *Public Relations Review, 36*(3), 278–280.

Carroll, C.E. (Ed.). (2011). *Corporate reputation and the news media: Agenda-setting within business news coverage in developed, emerging, and frontier markets.* New York: Routledge.

Carroll, C.E., & McCombs, M. (2003). Agenda-setting effects of business news on the public's images and opinions about major corporations. *Corporate Reputation Review, 6*(1), 36–46.

Chaffee, S.H., & Metzger, M.J. (2001). The end of mass communication? *Mass Communication & Society, 4*(4),365–379.

Chan, K. (1999). The media and environmental issues in Hong Kong 1983–95. *International Journal of Public Opinion Research, 11*(2),135–151.

Cobb, R.W., & Elder, C. (1971). The politics of agenda-building: An alternative perspective for modern democratic theory. *Journal of Politics, 33,* 892–915.

Coleman, R., McCombs, M., Shaw, D., & Weaver, D. (2009). Agenda setting. In K. Wahl-Jorgensen & T. Hanitzsch (Eds.), *The handbook of journalism studies* (pp. 147–160). New York: Routledge.

Fortunato, J. (2001). *The ultimate assist: the relationship and broadcasting strategies of the NBA and television networks.* Cresskill, NJ: Hampton Press.

Garrett, R.K. (2009). Politically motivated reinforcement seeking: Reframing the selective exposure debate. *Journal of Communication, 58,* 676–699.

Groshek, J. (2008). Homogenous agendas, disparate frames: CNN and CNN International coverage online. *Journal of Broadcasting & Electronic Media, 52*(1), 52–68.

Hamilton, J.T. (2004). *All the news that's fit to sell: How the market transforms information into news.* Princeton, NJ: Princeton University Press.

Hart, W., Albarracin, D., Eagly, A.H., Brechan, I., Lindberg, M.J., & Merill, L. (2009). Feeling validated versus being correct: A meta-analysis of selective exposure to information. *Psychological Bulletin, 135*, 555–588.

Hester, J.B., & Gibson, R. (2007). The agenda-setting function of national versus local media: A time-series analysis for the issue of same-sex marriage. *Mass Communication & Society, 10*(3), 299–317.

Iyengar, S., & Hahn, K.Y. (2009). Red media, blue media: Evidence of ideological selectivity in media use. *Journal of Communication, 57*, 19–39.

Johnson, T.J., Bichard, S.L., & Zhang, W. (2009). Communication communities or 'cyberghettos'?: A path analysis model examining factors that explain selective exposure to blogs. *Journal of Computer-Mediated Communication 15*(1), 60–82.

Johnson, T.J., Zhang, W., & Bichard, S.L. (2011). Voices of convergence or conflict? A path analysis investigation of selective exposure to political Web sites. *Social Science Computer Review, 29*(4), 449–469.

Knobloch-Westerwick, S., & Meng, J. (2011). Reinforcement of the political self through selective exposure to political messages. *Journal of Communication, 61*, 349–368.

Kushin, M.J. (2010). *Tweeting the issues in the age of social media? Intermedia agenda setting between the New York Times and Twitter.* Unpublished dissertation. Washington State University.

Lee, B., Lancendorfer, K., & Lee, K.J. (2005). Agenda-setting and the Internet: The intermedia influence of Internet bulletin boards on newspaper coverage of the 2000 general election in South Korea. *Asian Journal of Communication, 15*(1), 57–71.

Lee, G. (2010). Who let priming out? Analysis of first- and second-level agenda setting effects on priming. *International Communication Gazette, 72*(8), 759–776.

Lee, J.K. (2009). *Incidental exposure to news: Limiting fragmentation in the new media environment.* Doctoral dissertation. University of Texas at Austin. Retrieved from http://repositories.lib.utexas.edu/

Lim, J. (2011). First-level and second-level intermedia agenda-setting among major news websites. *Asian Journal of Communication, 21*(2), 167–185.

Maier, S. (2010). All the news fit to post? Comparing news content on the Web to newspapers, television, and radio. *Journalism & Mass Communication Quarterly, 87*(3/4), 548–562.

McCombs, M.E. (2004). *Setting the agenda: The mass media and public opinion.* Cambridge, UK: Polity.

McCombs, M.E. (2005). A look at agenda-setting: Past, present and future. *Journalism Studies, 6*(4), 543–557.

McCombs, M., & Shaw, D. (1972). The agenda-setting function of mass media. *Public Opinion Quarterly, 36*, 176–187.

McCombs, M.E. Shaw, D.L., & Weaver, D.H. (2007). *Communication and democracy: exploring the intellectual frontiers in agenda setting theory*. Mahwah, NJ: Lawrence Erlbaum Associates.

Meraz, S. (2009). Is there an elite hold? Traditional media to social media agenda setting influence in blog networks. *Journal of Computer-Mediated Communication, 14*(3), 682–707.

Meraz, S. (2011a). Using time series analysis to measure intermedia agenda-setting influence in traditional media and political blog networks. *Journalism & Mass Communication Quarterly, 88*(1), 176–194.

Meraz, S. (2011b). The fight for "how to think": Traditional media, social networks, and issue interpretation. *Journalism, 12*(1), 107–127.

Palmgreen, P., & Clarke, P. (1977). Agenda setting with local and national issues. *Communication Research,4,* 435–452.

Pratt, C.B., Ha, L., & Pratt, C.A. (2002). Setting the public health agenda on major diseases in Sub-Saharan Africa: African popular magazines and medical journals, 1981–1997. *Journal of Communication,52*(4),889–905.

Prior, M. (2005). News vs. entertainment: How increasing media choice widens gaps in political knowledge and turnout. *American Journal of Political Science, 49,* 577–592.

Prior, M. (2007). *Post-broadcast democracy: How media choice increases inequality in political involvement and polarizes elections*. Cambridge, UK: Cambridge University Press.

Project for Excellence in Journalism. (2010). *New media, old media: How blogs and social media agendas relate and differ from the traditional press*. Retrieved from http://www.journalism.org/analysis_report/new_media_old_media

Ragas, M.W. (2010). *Agenda-building and agenda-setting in corporate proxy contests: Exploring influence among public relations efforts, financial media coverage and investor opinion*. Unpublished doctoral dissertation, University of Florida.

Ragas, M.W., Kim, J., & Kiousis, S. (2011). Agenda-building in the corporate sphere: Analyzing influence in the 2008 Yahoo!-Icahn proxy contest. *Public Relations Review, 37*(3), 257–265.

Ragas, M.W., & Kiousis, S. (2010). Intermedia agenda-setting and political activism: MoveOn.org and the 2008 presidential election. *Mass Communication & Society, 13*(5), 560–583.

Roberts, M., Wanta, W., & Dzwo, T.-H. (2002). Agenda setting and issue salience online. *Communication Research, 29*(4), 452–465.

Rodríguez Diaz, R. (2004). *Teoría de la agenda-setting aplicación a la enseñanza universitaria*. Observatorio Europeo de Tendencias Sociales.

Salwen, M. (1988). Effects of accumulation of coverage on issue salience in agenda setting. *Journalism Quarterly, 65,* 100–106, 130.

Sato, H. (2003). Agenda setting for smoking control in Japan, 1945–1990: Influence of the mass media on national health policy making. *Journal of Health Communication, 8*(1), 23–41.

Sayre, B., Bode, L., Shah, D., Wilcox, D., & Shah, C. (2010). Agenda setting in a digital age: Tracking attention to California Proposition 8 in social media, online news and conventional news. *Policy & Internet,* 2(2), 7.

Smith, A. (2011). The Internet and campaign 2010. *Pew Internet and American Life Project.* Retrieved from http://pewinternet.org/Reports/2011/The-Internet-and Campaign-2010.aspx

Smith, K. (1987). Newspaper coverage and public concern about community issues. *Journalism Monographs, 101,* 1–32.

Smith, S.M., Fabrigar, L.R., & Norris, M.E. (2008). Reflecting on six decades of selective exposure research: Progress, challenges and opportunities. *Social and Personality Psychology Compass*, *2*(1), 464–493.

Stroud, N.J. (2008). Media use and political predispositions: Revisiting the concept of selective exposure. *Political Behavior, 30*, 341–366.

Stroud, N.J. (2010). Polarization and partisan selective exposure. *Journal of Communication, 60, 556–576.*

Stroud, N.J. (2011). *Niche news: The politics of news choice*: Oxford, UK: Oxford University Press.

Sunstein, C. (2001). *Republic.com.* Princeton, NJ: Princeton University Press.

Sweetser, K.D., Golan, G.J., & Wanta, W. (2008). Intermedia agenda setting in television, advertising and blogs during the 2004 election. *Mass Communication & Society*, *11*(2), 197–216.

Takeshita, T. (2006). Current critical problems in agenda-setting research. *International Journal of Public Opinion Research, 18*(3), 275–296.

Valenzuela, S. & McCombs, M. (2008). The agenda-setting role of the news media. In D. W. Stacks & M. B. Salwen (Eds.), An integrated approach to communication theory and Research (pp. 90–105). Mahwah, NJ: Lawrence Erlbaum Associates.

Waters, R.D., & Tindall, N.T.J. (2011). Exploring the impact of American news coverage on crisis fundraising: Using media theory to explicate a new model of fundraising communication. *Journal of Nonprofit & Public Sector Marketing*, *23*, 20–40.

Weaver, D.H. (1980). Audience need for orientation and media effects. *Communication Research, 3*,361–376.

Webster, J., & Ksiazek, T. (2012), The dynamics of audience fragmentation: public attention in an age of digital media.*Journal of Communication,* 62, 39–56.

Weeks, B., & Southwell, B. (2010). The symbiosis of news coverage and aggregate online search behavior: Obama, rumors, and presidential politics. *Mass Communication & Society*, *13*(4), 341–360.

Chapter 1

Media Agenda Setting in a Competitive and Hostile Environment

The Role of Sources in Setting Versus Supporting Topical Discussant Agendas in the Tea Party Patriots' Facebook Group

Sharon Meraz

The theory of agenda setting was conceptualized before the fragmentation now evidenced in the twenty-first century new media climate. The strength of the agenda-setting theory has been captured by its continued prolific usage, not only within the field of political communication (Bryant & Miron, 2004), but across other varied disciplines (McCombs, 2004). The expansion of the theory to include contingent conditions for agenda-setting effectiveness (McCombs & Weaver, 1973) and intermedia agenda setting (Meraz, 2009, 2011a, 2011b; Roberts & McCombs, 1994; Siu, 2010; Sweetser, Golan & Wanta, 2008) has further bolstered the strength of the theory's applicability in relation to the changing media marketplace (Anderson, 2010; Benkler, 2006) and to changing generational news-reading habits (Coleman & McCombs, 2007). The media's role in fostering democratic citizenship to the public (Page, 1996; Schudson, 1995) through highlighting significant issues and issue attributes remains a central role for traditional media entities in a twenty-first century news climate in which long-tail media options abound on the Internet (Anderson, 2006; Shirky, 2008).

Yet, competing with the agenda-setting theory are social influence theories such as the two-step flow theory (Berelson, Lazarsfeld, & McPhee, 1954; Coleman, Katz, and Mendel, 1957, 1966; Katz, & Lazarsfeld, 1955; Merton, 1949) and social network science theory (Adamic & Huberman, 2001; Barabasi & Albert, 1999; Barabasi, Albert, & Jeong, 2000). The two-step flow theory contends that news and information from traditional media are absorbed by opinion leaders, who then inform followers of this information (Lang & Lang, 2006). Unlike the two-step flow theory, social network science theory frees itself from hierarchical channels by suggesting multi-step pathway flows to information diffusion based on an individual's membership and location in dyads, triads, cliques, and large networks (Monge &

Contractor, 2003). With both social influence models, media's agenda-setting influence is diffused through the social process of sharing, networking, and connectivity, a phenomenon that can either bolster or weaken traditional media's agenda-setting influence, depending first on the type of media shared and second on the network's stance toward traditional media.

Within this environment, where Web 2.0 technologies have empowered like-minded political groups to coalesce in homophilous networks, it is questionable to what extent traditional media's influence can be powerfully maintained. Groups with predetermined causes and agendas can now sample the Internet's long tail of media, selecting whatever source supports their perspective. Access to backdated content is also more readily available than in the past through improved Web access to media archives via blog tools and search engines. The general assumption that all political groups depend on traditional media for the talking points of the day is complicated by the growth of political groups that publicly articulate the hostile media effect, the tendency for partisans to view neutral media content as biased against their position on issues they care about (Giner-Sorolla & Chaiken, 1994; Gunther & Chai., 2001 Vallone et.al., 1985;). This paper examines one such group, the Tea Party Patriots, a wing of the Tea Party Movement.

Background

The Tea Party Movement traced its beginning to CNBC's Rick Santelli's rant on February 19, 2009, on the grounds of the Chicago Mercantile Exchange, against the Obama administration's mortgage bailout plan (Armey & Kibbe, 2010; Lepore, 2010; Paul, 2011; Zernike, 2010). This arguably leaderless party (Cohen, 2010; Fitzgibbons, 2010; Good, 2009) picked up 5 of 10 Senate seats and won 40 of 130 House races in the 2010 midterm elections. The majority of the polls conducted on supporters of the Tea Party Movement took place in 2010, months after the movement's inception. One poll suggested that those who identify with the movement are fairly mainstream (Saad, 2010). A *New York Times* poll highlighted the demographic composition of Tea Party supporters as older, overwhelmingly white, male, married, conservative, wealthy, and educated (Zernike & Thee-Brenan, 2010). Exit polls for the 2010 midterm elections revealed that 47% of those over 60 years of age supported the Tea Party, with seniors accounting for 24% of the general 2010 voting block (Rasmussen & Schoen, 2010).

The Tea Party Movement's effects on both U.S. politics and the GOP/conservative movement remained tangible in the 2012 U.S. electoral cycle. In July 2011, the Tea Party movement's influence on raising the

U.S. debt ceiling was palpable. Their unrelenting desire for deeper spending cuts resulted in tempestuous negotiations between the two mainstay political parties (Nocera, 2011), a process referred to by one writer as a "Beltway form of theatre" (Tanenhaus, 2011). The desire by 2012 Republican presidential contenders to capture the Tea Party vote occurred alongside diminishing public support for the Tea Party Movement among the U.S. public in the wake of the midterm elections (Zernike, 2011).

The Tea Party has publicly documented their hostile traditional media perception. Tea Party Movement's leaders accuse traditional media elites of persistent misrepresentation, claiming that these entities have chosen to focus on the extremist behavior of the fringe members of their movement rather than on the issues that are important to the movement (Rasmussen & Schoen, 2010). With such a hostile public stance toward traditional media (Armey & Kibbe, 2010; Rasmussen & Schoen, 2010), Tea Party elites credit Web 2.0 technologies such as social networking sites (for example, Facebook and Ning), Twitter, blogs, and other insurgent/emergent media with growing the movement and facilitating social information sharing (Bai, 2012; Rasmussen & Schoen, 2010). Their open hostility toward traditional media makes this group a perfect test case for examining the extent to which traditional media is capable of setting source agendas for daily, political conversation in a social media space.

This chapter examines the sources of media used for conversation on the Tea Party Patriots' Facebook group. Given their publicly articulated distrust of traditional media, this chapter questions the success of traditional media in relation to source agenda setting within a community that has a hostile media perception, as well as how traditional media source agenda setting operates within communities such as the Tea Party Patriots that are seeking agenda support for a strong, predetermined, and pre-articulated agenda. Finally, this chapter examines whether traditional media source agenda setting impacts the nature and quality of conversation as it relates to more factual (substantive) versus more emotive (affective) political dialogue.

This chapter randomly samples discussant post data from the Tea Party Patriots' Facebook group during the timeframe of October 15, 2010, through June 15, 2011, in an effort to explore traditional media's source agenda-setting influence in a Facebook political discussion group arguably hostile to its influence. Through examining the initial, orienting post within post topics, sources utilized for framing discussant agendas were located and coded for their media site affiliation. Sources were also examined in situ for their impact on the character of the political discussion.

Literature Review

Agenda Setting Versus Agenda Supporting in an Environment of Hostile Media Perception

Unlike the twentieth-century media climate,in which media largely was limited to traditional media entities, twenty-first-century media options are endless, captured by the economic metaphor of a long tail (Brynjolfsson, Hu, & Simester, 2007, 2011). Citizens can opt to be creators and/or distributers. In relation to this era of never-ending media supply, citizens now curate the Web in large numbers for interesting content, sharing links and other newsworthy information outside the confines of newsroom home pages and within Web 2.0 technologies such as blogs, Twitter, and social networking sites (Meraz, 2012)

Evidence suggests that in this environment, traditional media is no longer capable of leveraging complete media agenda-setting influence (Meraz, 2009; Tran, Chapter 10, this volume). Studies reveal a sharing of agenda-setting power between citizen media and traditional media. Within citizen media environments, traditional media has been shown to leverage a "superstar effect," maintaining dominant influence in the short head of the long tail, but not singular influence when compared to the combined, smaller linkages to citizen media entities (Elberse & Oberholzer-Gee, 2008; Meraz, 2009). In sites such as Digg and Reddit, where group members share news items from the Web and vote on the story, prior studies have found nontraditional media sources to be significantly more likely to be promoted to popular or top pages when compared with traditional media (Meraz, 2012, 2009). Many of these social news communities seek niche-related news in such genres as science, technology, and entertainment/humor/odd news. These studies reveal that traditional media's ability to set a community's agenda is qualified by the news and information that is both interesting to the community and supporting of its agenda.

The degree to which specific traditional media entities are selected is clearly mediated by partisanship. In partisan blog communities with strong ideological perspectives, blogs are significantly more likely to also link to media (traditional and citizen) that share their ideological perspective (Adamic & Glance, 2005; Meraz, 2009; Tremayne, Zheng, Lee, & Jeong, 2006). This ideological cocooning has led critics to express concern regarding the dangers of partisan discourse to a more tolerant form of democratic citizenship (Stroud, 2011; Sunstein, 2000, 2001, 2002, 2009). In one of the only studies to examine second-level agenda-setting influence in blog networks and traditional media entities, Meraz (2011b) found that echo chamber or cocooning of partisan perspectives extends to the attributes of issues. Left-leaning blogs share issue attributes perspectives with other left-leaning blogs more closely than with traditional media's attribute frames. In the same study, it was found that through three disparate issues, right-leaning blogs resisted the attribute agenda-setting influence of traditional media entities. Such evidence confirms that partisan blogs seek out news and information in a selective manner, linking and focusing on issues that relate to their issue-specific agenda.

Given that partisan blogs focus on issues that are relevant to their ideological agenda, the question could be posed: What role does traditional media agenda setting play in an interpersonal scenario where the group has a very strong, predetermined agenda? Is it possible that traditional media can leverage strong agenda-setting power in environments where members are seeking agenda support for a predetermined issue agenda? Tea Party elites have developed issue manifestos, said to bind this distributed movement together. These core issues are said to revolve around individual freedom, fiscal responsibility, and states' rights (Armey & Kibbe, 2010; Rasmussen & Schoen, 2010). Given such a strong prearticulated agenda, it is possible that this group is less in need of its agenda being set and more in pursuit of news and information that supports its agenda. In environments such as Facebook, where Web publics can link to media options that best support their point of view, it is arguable that traditional media agenda setting may be more difficult to sustain. The increased freedom that Web publics have to locate and choose media options that satisfy topical agendas within social media sites (Tran, Chapter 10, this volume) makes the notion of agenda supporting a feasible counter-force to the operation of traditional media agenda setting.

The idea of agenda supporting exists outside other mitigating factors that may regulate the strength of traditional media's agenda-setting effect, such as the need for orientation (Kim, Chapter 4, this volume; McCombs, 2004; McCombs & Weaver, 1973) or value resonance (Sebastian, Chapter 3, this volume. The notion that media is selected in service to the support of a predetermined, pre-articulated topical agenda can be evidenced through many separate criteria. This study examines time as a variable in the selection of media for discussion among Web news publics. The ubiquitous media supply present on the Internet's unlimited shelf space provides expanded choice across time for discerning and self-selecting Web publics. In this study, agenda supporting is operationalized as curation of the Internet's long tail of media outside the day-to-day news offerings of traditional media. Active curation of the Internet's media supply across time periods can provide evidence of a Web public searching for news and information not readily available from the home pages or RSS feeds of traditional media news entities. If evidence is found that Web publics within the Tea Party Patriots' Facebook group page bypass the daily, traditional media news offerings for older time-stamped news and information, one can reasonably assume that this group is actively searching for items that support or confirm their topical agenda. Locating backdated news and information outside the offerings of traditional media entities can provide evidence of a Web public utilizing media for supporting an agenda already set and in need of confirmation.

Tea Party elites have publicly articulated their mistrust of media. The hostile media effect was first documented in 1985 through a case study of the Beirut massacre (Vallone, Ross, & Lepper, 1985), and it was initially defined as the tendency by partisans to view neutral media content as unfavorable to their point of view (Christen, Kannaovakun, & Gunther, 2002; Gunther & Schmitt, 2004; Schmitt, Gunther, & Liebhart, 2004). Relative media hostile perception was later identified as partisans' perception of bias in media content that was non-neutral (Gunther & Chai, 2001; Gunther & Christen, 2002) Though effects range in some studies from small to large, a meta-analytic study confirms the moderate existence of the hostile media effect (Hansen & Kim, 2011). The hostile media effect has been widely documented in political contexts (Schmitt, Gunther, & Liebhart, 2004h), such as in the Middle East conflict (Giner-Sorolla & Chaiken, 1994; Jee Oh, Park, & Wanta, 2011; Perloff, 1989; Tsfati, 2007; Tsfati & Cohen, 2005) and in presidential election contexts (Dalton, Beck, & Huckfeldt, 1998).

Effect sizes for the hostile media effect also tend to be larger for highly involved issue publics (Choi, Soon Park, & Chang, 2011; Hansen & Kim, 2011; Giner-Sorolla & Chaiken, 1994; Vallone, Ross, & Lepper, 1985). Many psychological mechanisms have been advanced to explain why hostile media perception occurs. Strong support has been found for individuals' fear of media's widespread reach (Gunther & Liebhart, 2006). Evidence also supports the operation of a process similar to the third-person effect, where partisan audiences negatively perceive the effect that potentially biased media coverage can have on audiences paying attention to news media (Gunther, Christen, Liebhart, & Chia, 2001). Though there is a negative perception of the effect of this biased coverage on news publics, little actual evidence has been found for the hostile media effect on perceived public opinion (Huge & Glynn, 2010).This chapter proposes to probe for evidence of the hostile media effect through frequency of hyperlink sourcing to traditional media outlets among members of the Tea Party Patriots' Facebook group. Tea Party elites argue that their general coverage in traditional media's reports is biased (Armey & Kibbe, 2010; Rasmussen & Schoen, 2010). Operationalizing the issue of high involvement as general coverage of the Tea Party in traditional media, this chapter examines whether Tea Party members avoid dedicating hyperlinks to traditional media within their Facebook discussion group in an effort to register their discontent with the nature of Tea Party coverage in these reports.

Agenda Setting on the Nature of Political Conversation

With the growth of socially networked communities such as the blogosphere, Facebook, and Twitter, it is possible that different media forms inspire or result in different agenda-setting effects. Theories such as the two-step flow (Lang & Lang, 2006) and social network science theory (Monge & Contractor, 2003; Neuman, 2003) find strong evidence for the power of social influence in relation to persuasion, attitude formation, and behavioral change. In sharp contrast, traditional media have greater documented effects at the cognitive or informational level (McCombs, 2004). Traditional news organizations depend on professional norms and adhere to shared media sociology norms in the identification of news items worthy of the public's attention (Shoemaker & Reese, 1996). Due to their lack of professional ties, the majority of blogs function as opinion platforms, spinning the news of the day in a form that is congenial to their like-minded publics (Meraz, 2011a). With the ready availability of both traditional media and blog outlets, this chapter also questions whether the media sourced on the Tea Party Patriots' Facebook discussion wall inspire differences in the nature of the ensuing political conversation. More specifically, this chapter questions whether the citation of traditional media sources inspires more substantive conversation when compared with blogs and other social media platforms. As opposed to the objective and factual nature of traditional media news reports, the largely opinionated nature of blogs and social media, spun from news of the day as identified from traditional media, may result in less substantive and more affective elaboration when members of the Tea Party Patriots' Facebook group depend on these outlets to frame their political discussion topics.

Social Networks, Political Conversation, and Political Activity

Since the strong growth of the blogosphere in the post-9/11 climate, Web 2.0 technologies have advanced to include microblogging applications like Twitter and social networking platforms like Facebook and MySpace. During the 2008 presidential election season, social media emerged as a defining force in American politics, enabling a visible resurgence in online activism by both political candidates and citizens campaigning on behalf of their candidates. President Barack Obama's usage of social networking became the standard for growing an online base of ardent followers and supporters (Budak, 2010; Fraser & Dutta, 2008). Facebook's involvement in facilitating political activity was enhanced by the opening of its application programming interface (API), allowing third-party developers to host applications on the social networking site. In addition to this technological signal that the platform was supporting political activism, Facebook developed an iPhone app, partnered with ABC news to host political debates (ABC News, 2007), supported candidate political advertisements (Kaye, 2009), and enabled online town halls between citizens and political candidates. The popularity of social networking applications in supporting political activity is now palpable and tangible among the U.S. public (Boerma, 2011; Mehta, 2011), with almost one in four individuals utilizing these sites for political activity (Smith, 2011b).

Outside official political candidate Facebook activity, citizen activism, specifically progressive/liberal organizing, was visible on Facebook during the 2008 presidential election. Facebook groups such as One Million Strong for Barack Obama, and One Million Strong against Hillary emerged,allowing political Web publics to voice their perspective on liberal/progressive agendas. Though progressive/liberal campaigning outpaced conservative activity in the 2008 presidential election (Meraz, 2008a), observers now point to an uptick of social media use by conservatives online (Bai, 2012; Preston, 2011). Republicans were identified as more effective users of Twitter than Democrats in the 2010 midterm elections (Livne, Simmons, Adar, & Adamic, 2011). As a subgroup of the conservative movement, the Tea Party movement has used social media technologies, namely Facebook, Twitter, and Ning, as common channels for their distributed Web publics to coalesce into single, conversational spaces (Eggen & Bacon, Jr., 2009; Kuhn, 2010).

Candidates are increasingly using Facebook to spur political involvement. Williams and Gulati (2009) found an increase in congressional candidates' usage of Facebook over the period 2006 to 2008, with Democratic candidates capturing more supporters, fan photos, and wall photos. The authors also found that Facebook pages were significantly more likely to be updated if the candidate were Democrat, well-funded, a non-incumbent, or a candidate running in a competitive race. In relation to electoral success, Williams and Gulati (2008) found that Facebook support was an important additional indicator of electoral success that is independent of traditional measures such as expenditures, media coverage and organizing activities as represented by campaign events.

Web publics who use social networking sites also share and obtain increasing amounts of political information through their friendship connections (Purcell, Rainie, Mitchell, Rosenstiel & Olmstead, 2010; Rainie & Smith, 2008). Though some studies have found no connection between general usage of the site and political participation or changes in political attitudes and behavior (Postelnicu & Cozma, 2008; Zhang, Johnson, Seltzer & Bichard, 2010), political usage of social networks has resulted in increased voting totals (Kim & Geidner, 2008) and increased civic participation (Zhang, Johnson, Seltzer, & Bichard, 2010). Political usage of social networks has also predicted offline and online political behavior (Vitak, Zube, Smock, Carr, Ellison & Lampe, 2010, with political activity affected by the political activity of an individual. Other Web publics utilize these social networking sites to connect with other like-minded supporters of political candidates (Ancu & Cozma, 2009; Kaye 2010).

Given the heated debate surrounding the significance of Internet political conversation (Brundidge, 2006; Davis, 1999; Holt, 2004), it is surprising that little research exists to date on content generated in Facebook groups. In one of the few studies to examine political conversation on Facebook, Kushin and Kitchener (2009) found that group political conversation involved greater agreement toward the group's stated objective; however, disagreement was met by an absence of flaming. A related area, the role of media as an agenda-setting source in supporting/influencing discursive conversation in Facebook groups, has also not been examined.

This current chapter provides a deeper analysis of audience political activity within Facebook political groups through the examination of one political Web public, the Tea Party Patriots' Facebook group. Specifically, it examines the types of media selected by political discussants to frame discussion topics, and the usage of media in inspiring differing forms of political conversation within this group.

Research Questions

This chapter examines source agenda-setting effects on the Tea Party Patriots' Facebook group. Leaders of this group have accused traditional media of biased, unfair coverage (Rasmussen & Schoen, 2010), and have suggested that their reliance on social media forms was designed to circumvent the influence of traditional media who have portrayed them as a fringe group. This chapter questions whether the movement's hostile perception toward traditional media may predict less reliance on these outlets for determining the movement's agenda. This chapter advances the following research question:

> RQ1: To what extent do traditional media entities exert overall source agenda-setting influence in discussion topics within the Tea Party Patriots' Facebook group when compared with nontraditional media entities?

News articles from traditional news organizations are generally factual and objective (unless it is an opinion column piece). The professional training of journalists (Shoemaker & Reese, 1996) ensures shared news values and norms across disparate news organizations. An intermedia agenda-setting process has been well documented, as news flows from elite to non-elite news organizations (for review, see Meraz, 2011a). As part of their routine news gathering practices, newsrooms also rely on bureaucratic sources for information (Sigal, 1973).

Given the nature of traditional media news coverage, it can be argued that depending on traditional media as a source may result in a discursive post that is weighted toward substantive/factual/informational characteristics, as opposed to purely opinionated/affective/nonfactual characteristics. This study questions whether it is possible to discern different agenda-setting outcomes from posts that depend on traditional media as a source to frame their post topic. This chapter advances the following question:

> RQ2: To what extent do traditional media entities generate more substantive commentary within discussion posts in comparison to nontraditional media entities on the Tea Party Patriots' Facebook group?

With the strong pre-articulated agenda of this movement, and the unlimited shelf space for hosting media on the Internet, members of the Tea Party Patriots' Facebook group can elect to source both backdated, archived media content and recent/current media content. Agenda-setting studies have found optimal effects within a time lag of one month (for review, see McCombs, 2004). This study proposes that if significant linkages are forged to backdated/archived media content, such evidence may reveal a movement seeking support for a pre-determined topical agenda. This chapter advances the following research question:

> RQ3: To what extent is there a difference in the time lag for traditional media sourcing versus nontraditional media sourcing in the framing of discussion posts in the Tea Party Patriots' Facebook group?

The aforementioned three research questions were tested through a sample of discussion topics on the Tea Party Patriots' Facebook group that referenced a URL through the period October 15, 2010 to June 15, 2011.

Methodology

In order to test the research questions, it was necessary to first locate all Tea Party Patriot discussion post topics that referenced a URL source during the period designated for download of content. In order to locate these posts that mentioned a URL, all post topic data were downloaded by saving the index pages (pages with links to all discussion topic posts) during the time frame October 15, 2010, to June 15, 2011. Because all information in the Tea Party Patriots' discussion group was public, a Web script was written to visit these index pages and grab the initial post topic (the first, orienting post that began the discussion output for each topic) for each topic entry, together with other information related to these initial post topics (such as poster name, date created for post topic, total discussion posts generated from the topic, and word count). Through a web application, manual visits to each post clarified whether post topics contained URLs or not. During the period October 15, 2010, to June 15 2011, there were 1,250 total post topics with URLs. Using a table of random numbers, a skip interval of every second post was sampled for coding, resulting in a sample of 625 posts with URLs. Upon visiting these 625 posts, a further 77 posts contained unusable URLs (failure to load, 404 errors, etc.). The sample that was used to test the research questions was 548 posts with URLs, which resulted in 718 working URLs for coding.

URLs were then coded in situ (as opposed to being extracted from post topics and coded solely for their pointers) in an effort to examine the usage of the URL within the discussion post. To answer the first research question, URLs were coded for the name of the site they were pointing to (cg., CNN), the type of site that the URL was pointing to (*CNN* = Traditional Media>Cable Television), and the publish date of the source in relation to the date of the topic post. Given the prevalence of YouTube links, each YouTube link was also visited in an effort to clarify whether the content linked to repurposed traditional/other media content, or whether it was purely a citizen media creation/mashup.

Upon coding each site, more than 40 site-specific categories emerged at the micro-level of analysis. In adapting a previously published rubric (Meraz, 2009), site-specific categories were reduced into the following five main media categories of interest: *traditional media* (for example, containing sites coded as newspapers [*New York Times*], television [NBC news], radio [*NPR*], cable news [Fox, MSNBC], wire/AP news [Reuters], newsroom blogs [*New York Times* Opinionator blog], and newsroom social media sites [*New York Times* Twitter feed]); *political blogs* (for example, liberal [Huffington Post], conservative [Michelle Malkin], and moderate political blogs [Moderate Voice]); *citizen media* (for example, sites that were general citizen media sites authored by individuals in genres of technology/science [Mashable], entertainment [Gawker], social networking technologies [Facebook], video technologies [YouTube]); *political establishment sites* (for example, government [White House], political candidates

[Rick Santorum], military sites, elected officials [John Boehner], political parties [British Freedom Party.), and political movements [Taking Back America Rally]; and *online-only news sites/aggregator sites* (for example, online-only news sites [Salon], portal news [Yahoo!], and social media news aggregators [for example Digg, Reddit] etc.). All sites that did not fit into these categories were coded as *other,* with these sites primarily pertaining to business/organizational sites (profit and nonprofit), nonpolitical cause sites (religious and health), and educational sites. This chapter's main interest in media agenda setting (traditional and citizen) enabled the aforementioned five master media categories to provide the codes for answering the first and subsequent research questions.

To answer the second question, URLs were coded for whether they were used as broadcast (no comments associated with URL beyond its mention) or whether they involved commentary. These commentaries were then coded for whether they contained substantive information or affective information. Utilizing Stromer-Galley's (2007) coding schema for differentiating substantive from non-substantive deliberative information, post topics were coded as substantive if they included any of the following elaborative techniques: "a definition, a reason for holding the opinion, an example, a story, a statistic, or fact, a hypothetical example, a solution to the problem, further explanation for why the problem is a problem, a definition, an analogy, a consequence to the problem or solution, a sign that something exists or does not exist, or any further attempt to say what they mean or why they have taken the position that they have." Posts were coded as containing affective information if they presented a tone of positive or negative (for example, "I hate [negative], or "this is awesome!" [positive]). For this code, posts could contain either, both, or neither substantive and/or affective information. No attempts were made on the part of the coder to ascertain whether substantive information was accurate. Posts were coded for substantive if they portrayed any of the above criteria, regardless of whether the information was indeed accurate.

Pulling a random sample of 10% of posts (resulting in 68 URLs), Holsti's Coefficient of Reliability ranged from 0.90 to 0.98 for coding URLs according to site name, site affiliation, and create date. Intercoder ratings also fell in the aforementioned range for identification of broadcast versus commentary within the posts, and for substantive versus affective post characteristics. A single coder completed the remaining sample, coding through an online interface. Codes were automatically entered into a database and queries were made against the data for frequency counts pertaining to the research questions and hypothesis.

Results

General Statistics

Extracting user information for each post revealed a total of 273 unique users posting these 548 links. Similar to other large-scale Web networks where a few users are responsible for the majority of content (Adamic & Huberman, 2000; Barabasi, Albert, & Jeong, 2000), 20 users or 7% of the users were responsible for 208 topics or 38% of the content. Likewise, 201 users were responsible for one post on the site. The curve for user participation resembles a long tail or power law where a few users drive the majority of discussion topics posted to the site.

Traditional Media as Source Agenda Setter

The first research question sought to determine the extent of source agenda-setting influence for traditional media when compared with that of other media. Table 1.1 provides general data on the top 15 sources across all 548 discussion posts. Among the 320 unique sources cited, the video storage site *YouTube* is the most highly cited source. Further in-depth coding of *YouTube* links revealed that 92 of the 128 links, or 72%

Table 1.1 Top 15 sources across time period

Source	*Frequency*
YouTube	128
Facebook	74
Yahoo	13
The Blaze	13
Wikipedia	10
CNBC	7
New York Times	7
Wnd	6
Reason Reigns	6
British Freedom Party	6
CNN	6
Washington Post	6
Examiner	5
Stop National Debt	5
Washington Times	5

Table 1.2 Comparison of mean source links among media entities

	SS	*df*	*Mean Square*	*F*	*Sig*
Between Groups	4803.650	4	1200.91	12.59	< 0.05
Within Groups	3338.125	35	95.375		
Total	8141.775	39			

of the *YouTube* content, pointed to the creations of average citizens. *Facebook, Yahoo News,* and *Wikipedia* also topped the list of primary sources as utilized in discussion topics. In terms of traditional media, *the New York Times, CNN,* and the *Washington Post* remained as top sources. *The Blaze,* an independent news site started by Glenn Beck, similar to the liberal leaning *Huffington Post,* was also a leading source.

Aggregating relevant media type data into the aforementioned five main umbrella media entities (citizen media, traditional media, political blogs, political establishment, and online news/aggregator news), a one-way ANOVA (see Table 1.2) revealed significant differences in the total quantity of source links dedicated to each macro media entity ($f(4, 35) = 12.59$, $p < 0.05$). Tukey post-hoc tests revealed that citizen media entities were significantly more likely to be selected as a source ($M = 33.12$, $SD = 17.79$) on the Tea Party Patriots' Facebook group site when compared to traditional media ($M = 18.12$, $SD = 11.06$), political blogs ($M = 5.87$, $SD = 2.95$), political establishment sites ($M = 6.75$, $SD = 4.86$), or online news-only/aggregator news sites ($M = 4$, $SD = 2.32$). The only other significant difference was between traditional media and online news-only/aggregator news sites. Traditional media was significantly more likely to be utilized as a source when compared to online news-only/aggregator news sites. There were no significant differences between usage of traditional media and usage of political blogs as sources for political discussant topical posts.

Media's Role in Substantive Versus Affective Dialogue

In addition to traditional media's capacity to set the agenda for the Tea Party Patriots' Facebook group through source agenda setting, this chapter also questioned whether traditional media's style of news reporting would result in more substantive effects on the nature of topical posts. The question was based on the newsroom norms of traditional media institutions, which place an importance on objective, nonpartisan, and investigative reporting. These norms drive a form of reporting that is not solely based on affective, opinionated material (unless it is an opinion piece). This chapter questioned whether differences in news writing styles and news selection practices between traditional media and nontraditional media entities could manifest themselves in how Web publics use these sources to direct the agenda of discussion topics.

A series of step-by-step tests were conducted to test this research question. First, t-tests were conducted across each media macro entity for whether the source in the discussion topic was utilized purely in a broadcast way (without further commentary attached), or accompanied by commentary (either through pulling citations from the source and/or personal commentary). Results reveal that all media types, with the exception of citizen media, were significantly more likely to inspire commentary as opposed to broadcast usage. Traditional media entities were significantly more likely ($t(14) = -3.054$, $p < 0.05$) to be sourced with commentary ($M = 3.37$, $SD = 2.32$) as opposed to solely broadcast ($M = 0.75$, $SD = 0.25$). Political blogs were significantly more likely ($t(14) = -4.413$, $p < 0.05$) to be sourced with commentary ($M = 9.62$) as opposed to broadcast style ($M = 1.8$, $SD = .83$). Similarly, online news-only/aggregator news outlets ($t(14) = -3.52$, $p < 0.05$) were significantly more likely to elicit commentary ($M = 5.1$, $SD = 3.35$) as opposed to broadcast ($M = .75$, $SD = 1.03$). Likewise, political establishment sites ($t(14) = -3.61$, $p < 0.05$) were significantly more likely to elicit commentary ($M = 5.12$, $SD = 3.36$) as opposed to the broadcast style ($M = .62$, $SD = 1.1$).

Unlike all of these macro media types, citizen media was as likely to be utilized with commentary ($M = 22.1$, $SD = 16$) as with the broadcast style ($M = 10.5$, $SD = 6.48$). Deconstructing this finding among the two largest components of citizen media, social networking sites like Facebook were significantly more likely to be used for commentary ($M = 8.5$, $SD = 6.3$) as opposed to broadcast ($M = 2.87$, $SD = 1.8$), but YouTube showed no significant differences in being accompanied by commentary ($M = 8.5$, $SD = 5.87$) or being used in solely a broadcast form ($M = 7.25$, $SD = 6.78$).

Probing for differences in the usage of these sources for substantive versus affective style discursive practices yielded little difference when testing the macro media entities, suggesting that once users comment on the source, they are equally as likely to cite facts as they are to offer emotive/affective commentary. Table 1.3 provides the means and standard deviations for all media entities related to substantive, affective, and both forms of dialogue appearing in posts. Upon running a one-way ANOVA, traditional media ($f(2, 21) = 1.805$, $p > 0.05$), political establishment ($f(2, 21) = 2.544$, $p > 0.05$), political blogs ($f(2, 21) = 1.306$, $p > 0.05$), online news-only outlets ($f(2, 21) = .178$, $p > .05$), and citizen media ($f(2, 21) = 2.629$, $p > 0.05$) revealed no significant differences in the usage of sources for eliciting substantive, affective, or both forms of dialogue in discussion topics. Though these tests failed to achieve significance, close perusal of the findings reveals that traditional media is the only entity that has a higher mean in substantive dialogue when compared to affective or neutral dialogue.

Deconstructing citizen media among its largest sources yielded more interesting findings. As Tables 1.3 and 1.4 highlight, a one-way ANOVA revealed significant differences among the different discursive dialogue forms with usage of YouTube. Post hoc Tukey t-tests show that YouTube was significantly more likely to be associated with affective commentary ($M = 4.87$, $SD = 3$) in comparison to substantive commentary ($M = 0.75$, $SD = 0.71$) or both forms ($M = 2.6$, $SD = 2.97$). Unlike YouTube and similar to other media entities, social networking sites yielded no significant differences ($f(2, 21) = 1.138$, $p > 0.05$) in the appearance of substantive ($M = 1.75$, $SD = 2.05$), affective ($M = 3$, $SD = 2.32$), or both forms of dialogue ($M = 3.87$, $SD = 3.8$) in topic commentary.

Table 1.3 Means and standard deviations for source usage in substantive, affective, and both forms of dialogue among media entities

Media	*Means*	*Standard Deviations*
Traditional Media		
Substantive	5.37	6.0
Affective	2.12	1.8
Both	5.5	3.0
Political Establishment		
Substantive	1.25	1.39
Affective	1.37	1.99
Both	3.25	2.43
Political Blogs		
Substantive	1.75	2.31
Affective	1.75	1.49
Both	3.25	2.49
Online-only News		
Substantive	0.875	0.834
Affective	1	1.06
Both	1.25	1.75
Citizen Media		
Substantive	3.37	2.56
Affective	9.5	5.29
Both	7.21	6.22

Table 1.4 YouTube's usage of source links in dialogue

	SS	*Df*	*Mean Square*	*F*	*Sig*
Between Groups	68.25	2	34.13	5.5	< 0.05
Within Groups	130.25	21	6.2		
Total	198.5	23			

Time Lag, Agenda Setting, and Agenda Supporting

To what extent is traditional media's agenda-setting influence leveraged on more current news content? Agenda-setting theory on the Internet has revealed effects within a one-week time frame, and more traditional media agenda-setting effects are typically measured with a time lag of one month.

However, given easy access to the long tail of media available on the Internet, including archived content, it is possible that this Tea Party audience may reveal different motivations in searching for media on the Internet. The third research question sought to examine whether traditional media's influence in the Tea Party Patriots' group was indeed one of setting more current agendas. More importantly, evidence of significant usage of backdated media content can provide evidence for the selective usage of media in the support of preexisting, topical issue agendas.

Separating source citation dates based on different time periods (no specific date/always live, less than one month, and one month or greater) and running one-way ANOVAs provided a range of significant results, but in different directions, depending on the media entity examined. Table 1.5 provides means and standard deviations for time differences in source citations across the five macro media entities, and Table 1.6 consolidates findings of one-way ANOVAs across the different media entities on time lag. As predicted by agenda-setting theory, source citations to traditional media were significantly more likely to be within a time period of one month ($f(2, 21) = 12.452$, $p < 0.05$) when compared with no date citations or citations greater than one month. Similar to traditional media, source citations to political blogs were significantly more likely to be within a time period of one month ($f(2, 21) = 12.61$, $p < 0.05$) when compared with no date citations or citations greater than one month.

In sharp contrast, other media entities were depended on for different purposes. One-way ANOVAs revealed that the Tea Party Patriots' Facebook group was significantly more likely ($f(2, 21) = 13.19$, $p < 0.05$) to link to political establishment sources that were not dated or timestamped in comparison to references published in under a month or over a month. Similarly, links to citizen media sources were significantly more likely ($f(2, 21) = 4.64$, $p < 0.05$) to carry no set date stamp as opposed to citations under or over one month. Deconstructing citizen media, social networking sites were also more likely to be cited ($f(2, 21) = 21.330$, $p < 0.05$) in relation to undated content ($M = 10.88$, $SD = 6.47$) as opposed to timed content under ($M = 0.5$, $SD = .076$) or over ($M = 0$, $SD = 0$) one month.

No significant differences were found in citing online-only news references ($f(2, 21) = 1.995$, $p > 0.05$) in relation to no date time periods, less than one month, or over a one-month time frame. Similarly, there were no significant time differences in YouTube citation links ($f(2, 21) = 4.68$, $p > 0.05$) in relation to no date ($M = 2.38$, $SD = 2.67$), less than one month ($M = 4.88$, $SD = 3.48$), and over a one month ($M = 8.5$, $SD = 5.42$) time frames.

Table 1.5 Means and standard deviations for time differences in source citations across different media

Media	*Means*	*Standard Deviations*
Traditional Media		
No Date	2.5	2
< 1 month	12.3	7.44
> 1 month	2	2.56
Political Establishment		
No Date	4.5	2.82
< 1 month	0.87	1.36
> 1 month	0.125	0.35
Political Blogs		
No Date	2.13	1.96
< 1 month	7.37	3.74
> 1 month	1.5	1.41
Online-only News		
No Date	1.12	0.99
< 1 month	1.62	1.5
> 1 month	0.5	0.75
Citizen Media		
No Date	17.63	11.1
< 1 month	6.5	4.5
> 1 month	9	5.71

Table 1.6 Differences in source citations by time

Traditional Media					
	SS	*Df*	*Mean Square*	*F*	*Sig*
Between Groups	547.750	2	273.88	12.452	< 0.05
Within Groups	461.875	21	21.994		
Total	1009.625	23			
Political Blog					
Between Groups	166.58	2	83.29	12.61	< 0.05
Within Groups	138.75	21	6.61		
Total	305.33	23			
Political Establishment					
Between Groups	87.583	2	43.79	13.19	< 0.05
Within Groups	69.75	21	3.32		
Total	157.33	23			
Citizen Media					
Between Groups	545.08	2	272.54	4.64	< 0.05
Within Groups	1233.88	21	58.76		
Total	1778.96	23			
Social Networking Sites					
Between Groups	603.08	2	301.54	21.33	< 0.05
Within Groups	296.88	21	14.14		
Total	899.96	23			

Conclusion

This chapter sought to examine the source agenda-setting impact of different media on the Tea Party Patriots' Facebook group. Given the growing usage of Facebook for political activism among Web publics (Kushin & Kitchener, 2009) and political candidates (Williams & Gulati, 2009, 2008), this chapter analyzed how the Tea Party Patriots Movement referenced media in the creation of discussion topics on their Facebook group. Tea Party leaders have expressed disdain for traditional media (Rasmussen & Schoen, 2010), recalling the hostile media effect, an effect in which partisans interpret neutral or non-neutral media coverage as biased against their point of view. Reports credit social media as a linchpin for growing the decentralized Tea Party Movement (Preston, 2011; Smith, 2011a; Smith, 2011b), and given the growing evidence for traditional media's loss of singular power in setting the agenda in social media networked environments (Meraz, 2011a, 2011b, 2009), this chapter sought to extend analysis to a nascent political movement with a decidedly palpable effect on current U.S. politics. Given this movement's strong prearticulated agenda, this chapter also questioned whether time lags in media sourcing could indicate usage of media for agenda-setting versus agenda-supporting functions.

Similar to other "produser" environments where Web publics can select the news they wish to promote or comment on (Meraz, 2012; Meraz, 2008b), the Tea Party Patriots group showed a significant preference for citizen media, particularly in relation to video (YouTube) and social networking sites (Facebook). YouTube and Facebook outnumbered other sources in inspiring discussion topics, a finding which suggests the growing influence of citizen-created media in networked environments. Links to citizen creations outnumbered links to repurposed traditional media content on YouTube. YouTube's growing influence on U.S. politics (for example, virally spreading George Allen's politically incorrect 'macaca' moment (Hohmann, 2011), or Rick Santelli's rant about Obama's bailout program (Armey & Kibbe, 2010; Lepore, 2010; Paul, 2011; Zernike, 2010), now extends to networked environments in which individuals select media as the basis for political conversation/discussion. This selective attention to media sources provides additional support to studies that have found evidence of selective exposure in social networking sites in terms of friendship choices (boyd & Ellison, 2007), group memberships (Kushin & Kitchener, 2009; Meraz, 2012; Woolley, Limperos, & Oliver, 2010), and in motivations for interpersonal engagement with others (Kaye & Johnson, 2012).

As the results highlight, YouTube provides an avenue for more citizen voices to set the agenda through hosting their content and allowing it to go viral via the effects of citizen marketers. Like online-only news sites and news aggregators, the preponderance of links to YouTube that referenced older content suggests that this site was used to support preexisting agendas as opposed to setting current agendas. Clearly, Tea Party Web publics conducted searches on YouTube for relevant clips that further advanced their selected discussion topic. Like former studies on the blogosphere that revealed partisan linking practices due to selective attention (Adamic & Glance, 2005; Meraz, 2011a, 2011b), YouTube can also support selective political conversation and partisan political dialogue.

The uncovering of the significance of selective attention to an agenda-supporting function highlights a new theoretical advance to the agenda-setting effects theory. In this new media environment where the Internet's long tail of media comprises a never-ending supply created by citizens and traditional media entities, political Web publics can now choose to select media that support a predetermined agenda. Unlike the past, when media supply was scarce, allowing traditional media to leverage a strong agenda-setting function on attentive Web publics, the always-on, continuous supply of media on the Internet enables researchers to more fully understand what comprises a news public's agenda.

Though the Tea Party Movement has publicly scorned traditional media, Web publics on their Facebook group still actively utilize traditional media to locate discussion agendas. Unlike other sources, traditional media's content was significantly more likely to be cited within a time lag of one month or less, suggesting that these outlets play an important role in impacting current agendas of political Web publics. As former studies have found (Meraz, 2011a, 2011b), traditional media are no longer alone in this role. Similar to traditional media, political blog content was also significantly more likely to be cited within a time lag of one month or less. Unlike YouTube, both traditional media and political blog content appear to play a role in setting more current agendas as opposed to supporting preexisting agendas.

This chapter reveals two distinct public agendas: an agenda capable of being set, and more importantly, an agenda that is in need of support. For those Web publics with a hostile media perception, choosing media in a selective fashion for an agenda-supporting function bolsters a predetermined agenda of topical issues. This chapter calls for greater attention to what constitutes the public agenda, highlighting that Web publics online form their agendas through curating the Web and networking with other like-minded individuals. Source agenda-setting effects now extend beyond traditional media's time content to backdated content from alternative media sources. There is tremendous utility in viewing a select Web public's agenda as less of a singular entity and more as diverse and varied, consisting of several sub-agendas, some capable of being set and some already set but in need of support.

In focusing on an active Web public arguably hostile to traditional media's influence, this chapter also suggests that the operation of the agenda-setting process may potentially differ based on the nature of the Web public under consideration. As the agenda-setting theory has formerly outlined, a Web public's need for orientation can impact the degree of attention it pays to the daily news streams of traditional media organizations (Kim, Chapter 4, this volume; McCombs, 2004; Valenzuela, Chapter 3, this volume). In the case of more active Web publics, it may be that the agenda-setting theory functions less as a mechanism for setting the salience of news select issues and news issue attributes, and more as a process that further heightens the relevance of politics and elections. Future studies are encouraged to segment Web publics along the spectrum of engagement (passive to active) in order to decipher whether news habits and news involvement might predict both the strength of agenda-setting effects and the nature of their operation.

This chapter was limited by the examination of only one Facebook political discussion group. Future studies are encouraged to sample political communities across the political spectrum to examine whether political affiliation might impact how agenda setting operates in networked environments. This chapter did not examine the actual content of the source link. Future studies can improve on the limitations in this one by examining the topics or subject matter contained in both source links and discussion posts in order to ascertain whether linked content indeed sets, or solely supports, the groups' existing topical agenda. Though traditional media's content, when sourced, is more current, further comparisons to their daily news agendas can highlight whether active Web publics like the Tea Partiers are following the agenda as set by traditional media, or selectively choosing current media content that amplifies their pre-existing opinions.

References

ABC News. (2007). ABC news joins forces with Facebook. *ABC News*. Retrieved June 10, 2012, from http://abcnews.go.com/Technology/Politics/story?id=3899006&page = 1

Adamic, L., & Glance, N. (2005). The political blogosphere and the 2004 U.S. election: divided they blog. *HP Labs*. Retrieved June 10, 2012, from http://www.blogpulse.com/papers/2005/AdamicGlanceBlogWWW.pdf

Adamic, L. A., & Huberman, B. A. (2001). Power-law distribution of the world wide web. *Science*, 287.

Ancu, M., & Cozma, R. (2009). MySpace politics: uses and gratifications of befriending candidates. *Journal of Broadcasting & Electronic Media, 53*, 576–583.

Anderson, C. (2006). *The long tail: why the future of business is selling more of less*. New York: Hyperion.

Anderson, C. (2010). *Free: how today's small businesses profit by giving something for nothing*. New York: Hyperion.

Armey. D., & Kibbe, M. (2010). *Give us liberty: a tea party manifesto*. New York: William Morrow.

Bai, M. (2012). The tea party's not-so-civil war. *The New York Times*. Retrieved June 10, 2012, from http://www.nytimes.com/2012/01/15/magazine/tea-party-south-carolina.html?_r = 2&hp

Barabasi, A-L., & Albert, R. (1999). Emergence of scaling in random networks. *Science*, 286.

Barabasi, A-L., & Albert, R., & Jeong, J. (2000). Scale-free characteristics of random networks: the topology of the World Wide Web. *Physics A*, 281, 69–77.

Benkler, Y. (2006). *The wealth of networks: how social production transforms markets and freedom*. New Haven, CT: Yale University Press.

Berelson, B. R., Lazarsfeld, P. F., & McPhee, W. N. (1954). *Voting: a study of opinion formation in a presidential campaign*. Chicago: University of Chicago Press.

Boerma, L. (2011). Candidates vie to star in the social network 2012 edition. *The National Journal*. Retrieved June 10, 2012, from http://www.nationaljournal.com/politics/ candidates-vie-to-star-in-the-social-network-2012-edition-20110419

Boyd, d. m. & Ellison, N. B. (2007). Social network sites: definition, history, and scholarship. *Journal of Computer-Mediated Communication*, 13(1). Retrieved June 10, 2012, from http://jcmc.indiana.edu/vol13/issue1/boyd.ellison.html

Brundidge, J. (2006, June). *The contribution of the internet to the heterogeneity of political discussion networks: Does the medium matter?*" Paper presented at the Annual Convention of the International Communication Association, Dresden, Germany.

Bryant, J. & Miron, D. 2004. Theory and Research in Mass Communication. *Journal of Communication*, 54(4), 662–704.

Brynjolfsson, E., Hu, Yu, & Simester, D. (2007). From niches to riches: the anatomy of the long tail. *Sloan Management Review*, 47(4), 67–71.

Brynjolfsson, E., Hu, Yu, & Simester, D. (2011). Goodbye pareto principle, hello long tail: the effect of search costs on the concentration of product sales. *Management Science*, 57(8), 1373–1386.

Budak, A. (2010). *Facebook, twitter, and Barack Obama: new media and the 2008 presidential election*. Published thesis. Kindle Edition.

Choi, J., Park, H. S. & Chang, J. C. (2011). Hostile media perception, involvement types, and advocacy behaviors. *Journalism & Mass Communication*, 88, 23–39.

Christen, C. T., Kannaovakun, P., & Gunther, A. C. (2002). Hostile media perception: partisan assessments of press and public during the 1997 United Parcel service strike. *Political Communication*, 19, 423–436.

Cohen, R. (2010). The starfish and the tea party. *The NonProfit Quarterly*. Retrieved June 10, 2012, from http://www.nonprofitquarterly.org/index.php?option = com_ content&view = article&id = 3866:the-cohen-report—the-starfish-and-the-tea-party&catid = 149:rick-cohen&Itemid = 117

Coleman, J. S., Katz, E. & Menzel, H. (1966). Medical innovation: A diffusion study. New York: Bobbs-Merrill. Coleman, J. S., Katz, E., & Menzel, H. (1957). The diffusion of an innovation among physicians. *Sociometry*, 20, 253–270.

Coleman, R. & McCombs, M. (2007). The young and the agenda-less? Exploring age-related differences in agenda setting on the youngest generation, baby boomers and the civic generation. Journalism & Mass Communication Quarterly, 84(3), 495–508.

Dalton, R. J., Beck, P. A., & Huckfeldt, R. (1998). Partisan cues and the media: information flows in the 1992 presidential election. *American Political Science Review*, 92, 111–26.

Davis, R. (1999). *The Web of politics: The Internet's impact on the American political system*. Oxford: Oxford University Press.

Eggen, D., & Bacon, Jr., P. (2009). Tea Party conservatives gear up to affect 2010 elections with fundraising. *The Washington Post*. Retrieved June 10, 2012, from http://www.washingtonpost.com/wp-dyn/content/article/2009/12/09/AR2009120904637.html

Elberse, A., & Oberholzer-Gee, F. (2008). *Superstar and underdogs: An examination of long tail phenomenon in video sales*. Harvard Business School Working Paper No. 07–015.

Fitzgibbons, M. J. (2010). The tea party and the spider's Web. *The American Thinker*. Retrieved June 10, 2012, from http://www.americanthinker.com/2010/08/the_tea_party_ and_the_spiders.html

Fraser, M., & Dutta, S. (2008, November 19). Barack Obama and the Facebook election. *US News and World Report*, Retrieved June 10, 2012, from http://www.usnews.com/opinion/ articles/2008/11/19/barack-obama-and-the-facebook-election

Giner-Sorolla, R., & Chaiken, S. (1994). The causes of the hostile media judgments. *Journal of Experimental Social Psychology*, 30, 165–180.

Good, C. (2009). The tea party movement: who is in charge? *The Atlantic.* Retrieved June 10, 2012, from http://www.theatlantic.com/politics/archive/2009/04/the-tea-party-movement-whos-in-charge/13041/

Gunther, A. C. & Chai, C. S. (2001). Predicting pluralistic ignorance: the hostile media perception and its consequences. *Journalism & Mass Communication Quarterly,* 78, 688–701.

Gunther, A. C., & Christen, C. T. (2002). Projection or persuasive press? Contrary effects of personal opinion and perceived news coverage on estimates of public opinion. *Journal of Communication,* 52, 177.

Gunther, A. C., Christen, C. T., Liebhart, J. L., & Chai, C. S. (2001). Congenial public, contrary press, and biased estimates of the climate of opinion. *Public Opinion Quarterly,* 65, 295–320.

Gunther, A. C., Christen, C. T., Libehart, J. L. & Chia, S. (2011). Congenial public, contrary press, and biased estimates of the climate of opinion. Public Opinion Quarterly, 65, 395–320.

Gunther, A. C., & Liebhart, J. L. (2006). Broad reach or biased source? Decomposing the hostile media effect. *Journal of Communication,* 56(3), 449–466.

Gunther, A., & Schmitt, K. (2004). Mapping boundaries of the hostile media effect. *Journal of Communication,* 54, 55–70.

Hansen, G. J., & Kim, H. (2011). Is the media biased against me? A meta-analysis of the hostile media effect research. *Communication Research Reports,* 28, 169–179

Hohmann, J. (2011). George Allen apologizes for 'macaca' moment. Politico. Retrieved June 22, 2012, from http://www.politico.com/news/stories/0611/56212.html

Holt, R. (2004). *Dialogue on the Internet: Language, civic identity, and computer–mediated communication.* Westport, CT: Praeger.

Huge, M., & Glynn, C. (2010). Hostile media and the campaign trail: perceived media bias in the race for governor. *Journal of Communication,* 60, 165–181.

Jee Oh, H., Park, J., & Wanta, W. (2011). Exploring factors in the hostile media perception: partisanship, electoral engagement, and media use patterns. *Journalism & Mass Communication Quarterly,* 88, 40–54.

Katz, E., & Lazarsfeld, P. F. (1955). *Personal influence: the part played by people in the flow of mass communications.* Glencoe, IL: The Free Press.

Kaye, B. K. (2010). Between Barack and a net place: users and uses of social network sites and blogs for political information. In Z. Papacharissi (Ed.), The networked self: Identity, community and culture on social network sites (pp., 208–231). New York: Routledge.

Kaye, B. K., & Johnson, T. (2012). Net gain? selective exposure and selective avoidance of social network sites. In Francesca Comunello (Ed.), *Networked sociability and individualism: technology for personal and professional relationships.* Hershey, PA: Information Science Reference.

Kaye, K. (2009). Google grabbed most of Obama's 16 million in 2008. *TechPresident.* Retrieved June 10, 2012, from http://techpresident.com/blog-entry/google-grabbed-most-obamas-16-million-2008

Kim, Y. M., & Geidner, N. (2008, May). *Politics as friendship: The impact of online social networks on young voters political behavior.* Paper presented at the annual convention of the International Communication Association, Montreal, Canada.

Kuhn, D. P. (2010). RIP political establishment. *Real Clear Politics.* Retrieved June 10, 2012, from http://www.realclearpolitics.com/articles/2010/09/21/rip_political_establishment_ and_tea_party_netroots_107238.html

Kushin, M. J., & Kitchener, K. (2009). Getting political on social network sites: exploring online political discourse on Facebook. *First Monday, 14(11).* Retrieved June 10, 2012, from http://firstmonday.org/htbin/cgiwrap/bin/ojs/index.php/fm/article/view/2645/2350

Lang, K., & Lang, G. E. (2006). Personal influence and the new paradigm: some inadvertent consequences. *The ANNALS of the American Academy of Political and Social Science,* 608, 157–178.

Lepore, J. (2010). *The whites of their eyes: the tea party's revolution and the battle over American history.* Princeton, NJ: Princeton University Press.

Livne, A., Simmons, M. P., Adar, E., & Adamic, L. A. (2011). The party is over here: structure and content in the 2010 election. Retrieved June 10, 2012, from http://www.cond.org/partyat.pdf.

McCombs, M. (2004). *Setting the agenda: the mass media and public opinion.* Cambridge, UK: Policy Press.

McCombs, M., & Weaver, D. (1973). *Voters' need for orientation and use of mass communication.* Unpublished paper presented to the 1973 International Communication Association (Montreal, Canada).

Mehta, S. (2011). The rise of the internet electorate. *The LA Times.* Retrieved June 10, 2012, from http://articles.latimes.com/2011/apr/18/nation/la-na-social-media-20110418

Meraz, S. (2008a). The blogosphere's gender gap: differences in visibility, popularity, and authority. In Paula Poindexter, Sharon Meraz, and Amy Schmitz Weiss (Eds.), *Women, men, and news.* New York: Routledge .

Meraz, S. (2008b). The many faced 'you' of social media. In Zizi Papacharissi (Ed.), *Journalism and citizenship: New Agendas.* New York: Routledge/

Meraz, S. (2009). Is there an elite hold: traditional media to social media agenda setting influence in blog networks. *Journal of Computer Mediated Communication,* 14, 682–707.

Meraz, S. (2011a). Using time series analysis to measure intermedia agenda setting influence in traditional media and blog networks. *Journalism & Mass Communication Quarterly,* 88(1), 176–194.

Meraz, S. (2011b). The fight for how to think: traditional media, social networks, and issue interpretation. *Journalism: Theory, Practice, and Criticism,* 12(1), 107–127.

Meraz, S. (2012). The sociality of news sociology: examining user participation and news selection practices in social media news sites. In Burton Saint John III and Kirsten Johnson (Eds.), *News with a View: Essays on the Eclipse of Objectivity in Modern Journalism.* Jefferson: McFarland & Company.

Merton, R. (1949). Patterns of influence: A study of interpersonal influence and communications behavior in a local community. In P. F. Lazarsfeld and F. N. Stanton (Eds.), *Communications research.* New York: Harper and Brothers.

Monge, P., & Contractor, N. (2003). *Theories of communication networks.* Oxford, NY: Oxford University Press.

Neuman, M. E. J. (2003). The structure and function of complex networks. *SIAM,* 45(2).

Nocera, J. (2011). The tea party's war on America. *The New York Times*. Retrieved June 10, 2012, from http://www.nytimes.com/2011/08/02/opinion/the-tea-partys-war-on-america.html?_r = 1&src = ISMR_HP_LO_MST_FB

Page, B. (1996). *Who deliberates? Mass media in modern democracy*. Chicago: University of Chicago Press.

Paul, R. (2011). *The tea party goes to Washington*. Nashville: Center Street.

Perloff, R. M. (1989). Ego-involvement and the third person effect of televised news coverage. *Communication Research,* 16, 236.

Postelnicu, M., & Cozma, R. (2008, November). *Befriending the candidate: uses and gratifictions of candidate profiles on MySpace*. Paper presented at the annual meeting of the National Communication Association, San Diego, CA.

Preston, J. (2011). Republicans sharpening online tools for 2012. *The New York Times*. Retrieved June 10, 2012, from http://www.nytimes.com/2011/04/20/us/politics/20social.html?_r = 3

Purcell, K., Rainie, L., Mitchell, A., Rosenstiel, T., & Olmstead, K. (2010). Understanding the participatory news consumer. Pew Internet and American Life Project. Retrieved June 22, 2012, from http://www.pewinternet.org/Reports/2010/Online-News.aspx

Rainie, L., & Smith, A. (2008). The Internet and the 2008 election. Pew Internet and American Life Project. Retrieved June 10, 2012, from http://www.pewinternet.org/Reports/2008/The-Internet-and-the-2008-Election.aspx

Rasmussen, S., & Schoen, D. (2010). *Mad as hell: how the Tea Party movement is fundamentally remaking our Two-Party system*. New York: Harper.

Roberts, M., & McCombs, M. (1994). Agenda setting and political advertising: origins of the news media. *Political Communication,* 11, 249–262.

Saad, L. (2010). Tea partiers are fairly mainstream in their demographics. Gallup. Retrieved June 10, 2011, from http://www.gallup.com/poll/127181/tea-partiers-fairly-mainstream-demographics.aspx

Schmitt, K., Gunther, A. C., & Liebhart, J. L. (2004). Why partisans see mass media as biased. *Communication Research,* 31, 623–641.

Schudson, M. (1995). *The power of news*. Cambridge, MA: Harvard University Press.

Shirky, C. (2008). *Here comes everybody: the power of organizing without organizations*. New York: Penguin Press.

Shoemaker, P., & Reese, S. D. (1996). *Mediating the message: theories of influence on mass media content* (2nd ed.). New York: Longman.

Sigal, L. V. (1973). *Reporters and officials: the organization and politics of newsmaking*. Lexington, MA: DC Heath and Company

Siu, W. (2010). Source agenda setting: intermedia influence in the food safety scandal. *Journal of Communications Research,* 1(4), 355–381.

Smith, A. (2011a). The Internet and campaign 2010. *Pew Internet and American Life Project*. Retrieved June 10, 2012, from http://pewinternet.org/Reports/2011/The-Internet-and-Campaign-2010.aspx

Smith, A. (2011b). 22% of online Americans used social networking or Twitter for politics in the 2010 campaign. *Pew Internet and American Life Project*. Retrieved June 10, 2012, from http://pewinternet.org/Reports/2011/Politics-and-social-media.aspx

Stromer-Galley, J. (2007). Measuring deliberation's content: a coding scheme. *Journal of Public Deliberation,* 3(1) Article 12.

Stroud, N. J. (2011). *Niche news: the politics of news choice.* New York: Oxford University Press.

Sunstein, C. (2000). Deliberative trouble? Why groups go to extremes. *Yale Law Journal,* 110 (1): 71–119.

Sunstein, C. (2001). *Republic.com.* New Jersey: Princeton University Press.

Sunstein, C. (2002). The law of group polarization. *The Journal of Political Philosophy,* 10(2): 175–195.

Sunstein, C. (2009). *Going to extremes: how like minds unite and divide.* Oxford, UK: Oxford University Press, 2009.

Sweetser, K., Golan, G. J., & Wanta, W. (2008). Intermedia agenda setting in television, advertising, and blogs during the 2004 election. *Mass Communication & Society,* 11, 197–216.

Tanenhaus, S. (2011). Jefferson's tea party moment. *The New York Times.* Retrieved June 10, 2012, from http://www.nytimes.com/2011/07/31/sunday-review/us-has-lovehate-relationship-with-debt.html?ref = opinion

Tremayne, M., Zheng, N., Lee, J., & J. Jeong. (2006). Issue publics on the web: applying network theory to the war blogosphere. *Journal of Computer-Mediated Communication 12*(1). Available at http://jcmc.indiana.edu/.

Tsfati, Y. (2007). Hostile media perception, presumed media influence, and minority alienation:the case of Arabs in Israel. *Journal of Communication,* 57, 632–651.

Tsfati, Y. & Cohen, J. (2005). The influence of presumed media influence on democratic legitimacy: The case of Gaza settlers. Communication Research, 32(6), 794–821.

Vallone, R. P., Ross, L., & Lepper, M. R. (1985). The hostile media phenomenon: biased perception and the perceptions of media bias in coverage of the Beirut massacre. *Journal of Personality and Social Psychology*, 49, 577–589.

Vitak, J., Zube, P., Smock, A., Carr, C., Ellison, N. & Lampe, C. (2010). It's complicated: Facebook users' political participation in the 2008 election. Cyberpsychology, Behavior & Social Networking. Williams, C. B., & Gulati, G. J. (2008, August 28–31). *What is a social network worth? Facebook and vote share in the 2008 presidential primaries.* Presented at the 2008 Annual Meeting of the American Political Science Association, Boston, MA

Williams, C. B., & Gulati, G. J. (2009, April 2–5). *Facebook grows up: an empirical assessment of its role in the 2008 congressional elections.* Presented at the 2009 Annual Meeting of the Midwest Political Science Association, Chicago, IL.

Woolley, J. K., Limperos, A. M., & Oliver, M. B. (2010). The 2008 presidential election, 2.0: a content analysis of user-generated political Facebook groups. *Mass Communication & Society,* 13, 631–652.

Zernike, K. (2010). *Boiling mad: inside tea party America.* New York: Times Books/Henry Holt and Co.

Zernike, K., & Thee-Brenen, M. (2010). Poll finds tea party backers wealthier and more educated. *The New York Times.* Retrieved June 10, 2012, from http://www.nytimes.com/2010/04/15/us/politics/15poll.html

Zhang, W., Johnson, T., Seltzer, T., & Bichard, S. (2010). The revolution will be networked: the influence of social network sites on political attitudes and behaviors. *Social Science Computer Review,* 28(1), 75–92.

Chapter 2

Agenda Setting, Elections, and the Impact of Information Technology

Jason A. Martin

The proliferation of available news sources has raised questions about whether or not traditional news organizations still have a consistent agenda-setting effect on the public, especially on the Internet where audiences tend to be more fragmented (Takeshita, 2005). Additionally, innovations in online interactivity and information delivery have resulted in a growing interest in the roles of the public, policymakers, and other sources in the process of building and contributing to the media agenda (Denham, 2010; Meraz, Chapter 1, this volume; Ragas, Chapter 12, this volume; Roberts, Wanta, & Dzwo, 2002).

Despite increased attention to these convergent issues within the realm of agenda setting, research generally has not yet provided conclusive statements about the impact of information technology on both the media and public agendas, and their relationship in the agenda-setting process (McCombs & Reynolds, 2009).

This chapter presents findings from two studies that were designed to address this research problem by considering the impact of developments in information technology on agenda setting within the context of two different types of elections in two different years. These studies analyzed the public's interactions with a metropolitan newspaper website in the weeks preceding the elections to better understand how citizens learn about campaigns and engage with news during election periods. As the media, candidates, and the public all increasingly produce and consume more content online, it is important to understand whether and how these changes may affect long-held assumptions of agenda-setting theory.

The first study examined how the public agenda, as measured by online commentary attached to the end of news stories, was related to the media agenda during the 2008 congressional and presidential primary season. This study specifically sought to address gaps in knowledge about how agenda setting develops online as the media agendas are built and before long-term agenda-setting effects take hold.

The second study analyzed the media, candidate, and public agendas during the 2010 U. S. Senate general election to better understand whether changing patterns of campaign-related information delivery and voter media use have affected the traditional relationship between the media and public agendas. As candidates devote more resources to online messages, and as alternative sources of election-related news and commentary develop online, more empirical evidence is needed to state more conclusively whether the public is responding with greater attention to those innovations, or whether they continue to learn about local elections primarily through their local news media.

Literature Review

The analysis of the influence of developments in information technology on agenda setting during election campaigns begins with an examination of the relevant literature in two distinct but sometimes overlapping areas: agenda building and agenda setting in the Internet age.

Agenda Building

For about three decades, scholars have been interested in how the policy, media, and public agendas are built and relate to one another in reciprocal and interchangeable ways (Denham, 2010; Lang & Lang, 1991). In the context of campaigns, candidates and media have long clashed about what stories are important. Candidates decry the difficulty of getting their messages into the media, and the media gripe about the superficiality of campaign messages, the proliferation of staged events, and the lack of access to candidates (Hayes, 2010). The result is a competition for building the media agenda, which has a natural symbiotic relationship with the candidate agenda yet must have awareness of the public agenda and profitability in a commercial system, and also a struggle for the building of the public agenda, which is influenced directly by a combination of the media and candidate agendas (Denham, 2010).

Denham (2010) has termed these concepts of contention *media agenda building* and *public agenda building*. He described media agenda building as "an ongoing negotiation between media personnel and their sources of information" (p. 311) and public agenda building as "behavioral responses to mass and interpersonal communication" (p. 316). Denham compares public agenda building to traditional agenda-setting research that has focused on salience transfer, but points out that consideration for the building of the public agenda must encompass a wider scope of outcomes such as voting and measures of political and civic engagement.

Turning to how these three agendas are interrelated, Shoemaker and Reese (1996) identified various levels of influence on the media agenda, ranging from the personal values of the individual journalist to the macro-level impact of cultural values. Their fourth level, examining influences external to the media, has the most applicability to understanding the three-way relationship of candidates, media, and public during election campaigns in the studies presented in this chapter. Research looking at this level of influence on media agenda building most often looks at the impact of sources, other media, or the public in influencing news coverage (Denham, 2010). During campaigns, candidate press releases have been found to have a strong agenda building influence on news coverage (Kiousis, Mitrook, Way, & Seltzer, 2006).

Examining a campaign's indirect influence on the public agenda through the media is important because few citizens have direct experience with campaigns in the form of attending rallies, speeches, or other gatherings, and instead rely on news to decide what issues are important (Ridout & Mellen, 2007). Consideration for the specific characteristics of the election is important when examining three-way agenda building; specifically, competitiveness, intensity of news coverage, and stage of the campaign have been found to be important factors (Tedesco, 2001).

Internet-related developments have resulted in more research attention than ever before to how the media agenda is influenced by policymakers, the public, other media, advertisers, and communications professionals (Denham, 2010; Kiousis et al., 2006; Ku, Kaid, & Pfau, 2003; Lim, 2006; McCombs & Reynolds, 2009; Ragas, Chapter 12, this volume; Wallsten, 2007). Another practical reason this phenomenon has received more scholarly attention in recent years is that developments on the Internet have allowed for methodological advancements in capturing and analyzing content from sources, the public, and other entities that contribute to the building of the media agenda (McCombs, 2004).

Within an election context, campaign sources such as advertising and candidate communications have long tried to contribute to the media agenda with the hopes of influencing the public (Jamieson & Campbell, 1992). More recent research that has examined press releases, websites, and other campaign-related content has found that candidates and their campaign communications staff have had consistent success in influencing the building of both the media and public agendas (Foot & Schneider, 2006; Kiousis et al., 2006, 2009; Tedesco, 2001, 2005).

However, less research attention has been paid to how the media agenda is shaped by online audience contributions (Roberts et al., 2002), especially during election campaigns. Scholars have taken creative approaches to measuring the public agenda over the years as a means of comparison to the media agenda, including using debate questions (Benoit & Hansen, 2001). As the Internet has developed as a domain for increased public engagement with news content, scholars have become increasingly interested in different ways to archive and measure the public agenda through measures such as Internet search frequency and volume, news story comments, and electronic bulletin board posts (Jeong, Kim, & Shin, 2008; Roberts et. al 2002; Weeks & Southwell, 2010). Although these studies have found a consistent relationship between the media and public agendas, with the media mostly influencing online public discussion, more work is needed to understand in what contexts the public also may contribute to the building of the media agenda.

Agenda Setting in the Internet Age

Scholars have been publishing research about agenda setting in the Internet age in mass communication journals for about a decade (Tran, Chapter 10, this volume). Most of this research attention has examined how the media agenda is built and how the fragmentation of audience has influenced public agenda setting, especially during U. S. election campaigns (Tran, Chapter 10, this volume). In addition to aforementioned studies of online campaigning tools, research has delved into the mixture of news available online (Maier, 2010), the rise and impact of blogs (Messner & DiStaso, 2008), the development of niche news websites (Tran, Chapter 10, this volume), and the proliferation of social media and online commentary (Sayre, Bode, Shah, Wilcox, & Shah, 2010).

In general, research has found that while these Internet-related developments have not fundamentally altered the traditional understanding of agenda-setting theory, they are potentially changing the complicated relationship of how the media agenda is built (Tran, Chapter 10, this volume). Specifically in the political sphere, citizen-created blogs are helping redistribute the traditional influence of media and citizens by having variable influence on the media agenda (Meraz, 2009) and rising in prominence as both a source and topics of the news (Messner & DiStaso, 2008). The result is that citizen-created content outside the traditional media has created more of a bidirectional stream of influence with the media agenda (Wallsten, 2007), although the traditional media has retained most of the influence on the public agenda (Tran, Chapter 10, this volume).

Online agenda-setting effects have been found to have a shorter time span than for traditional media (Tran, Chapter 10, this volume; Weeks & Southwell, 2010). However, with published studies of online agenda setting representing only a fraction of the overall body of agenda-setting literature, more work is needed to better understand how continued developments on the Internet are altering multiple aspects of agenda building and agenda setting for the media, public, and campaigns.

Study 1: 2008 Primary Election

The first study investigated how news website feedback interacted with the media agenda, and the role that this form of public response played in the agenda-setting process during a two-week period prior to the

2008 primary election in Indiana. Based on the relevant agenda-building literature that has examined the public agenda's influence on the media (Denham, 2010; Jeong, Kim, & Shin, 2008; Roberts et. al 2002; Weeks & Southwell, 2010), the study's main research question investigated whether the media primarily influenced the public in the traditional agenda setting process, or whether the public agenda, measured by the online story comments, may have had an influence on which stories the newspaper chose to make salient when it formed and presented its media agenda on the following day.

Method

A content analysis was performed on the *Indianapolis Star*; its website, indystar.com; and online comments available at the end of most news stories posted online. The *Star* is Indiana's largest newspaper with a weekday circulation in excess of 200,000 copies, and indystar.com was the single most-visited Internet site of any news organization originating in Indiana in 2008, with 2.2 million unique visitors per month (Indianapolis Star Media Group, 2008).

The sample frame was a two-week period, March 10–23, 2008, during which several public affairs issues in Indianapolis were being covered. The Indiana state legislature was in session during the month and there was an emergency congressional election campaign ongoing. A new airport and a billion-dollar sports stadium were under construction with accompanying public debate about financing in the midst of an economic downturn. This sample therefore yielded rich data about how issues rise and fall in the media and public agendas over the course of two weeks, because there was more competition among issues available. With regard to the somewhat truncated time frame, while news and discussion certainly interacts quickly online, studies on the effects of online agenda setting have found one day to one week to be a sufficient time lag for measuring the media and public agendas (Roberts et. al, 2002; Tran, Chapter 10, this volume; Weeks & Southwell, 2010).

The context units were the front and metro/state sections of the *Star*'s print edition; the home page and local news page of indystar.com; and online comments attached to the relevant news stories. These user discussions were moderated by media staff to include only pertinent discussion of articles and issues, but were open to any Internet user who registered with the site. The unit of analysis was each story that covered public affairs issues, which were used to establish rank-ordered agendas for each of the three sources.

The print edition was analyzed each of the 14 days. Data collection from the two online sources required additional scrutiny due to the changing nature of online news and feedback. The online and feedback agendas were initially sampled three times per day in the morning, afternoon and evening. Data from a pilot study indicated that evening measurement was sufficient for gathering the online and feedback agendas, which were collected in that manner for the entire two-week period of the actual study. For the media agenda, salience was measured based on (a) the number of stories about the issue and (b) prominence on the newspaper section, page, or website. For the public agenda, salience was measured in terms of the number of comments accompanying each story.

The agendas of all three sources over the course of the two weeks were compared based on their daily and mean rank orders to see whether the agendas correlated and to understand the nature of their relationships. Due to the unique nature of the media and time period examined, issue categories, informed by research that has examined issues during election campaigns (Drew & Weaver, 2006; McCombs, 2004), were created by the author to reflect the consistent set of news stories.

Thus, based on similar studies that examined campaign issues and voter knowledge, the print, website, and feedback agendas had eight consistent issues in common over the two-week period: taxes; political campaigns; crime; transportation; health care; education; personal economics; and public works (Drew & Weaver, 2006).

As operationalized, tax stories included those about legislation, government options, and impact. Political campaign stories involved local, congressional, and presidential elections applicable to the Indianapolis coverage area. Crime stories involved incidents, public safety, and other police topics. Transportation stories involved roads, airport construction, and regulation issues. Health care stories involved both physical and mental health, air quality, and hospital topics. Education stories concerned schools, educational opportunities, access to education, and the arts. Personal economics stories often involved topics such as mortgages, gas prices, and employment. And public works stories involved stadium construction, parks management, and other discussions of green spaces. There were only four days in the two-week period when all eight issues were not represented on all three agendas by some form of reporting. On those rare occasions when an issue was absent, it was designated as the bottom of the rank-order.

Intra-coder reliability was tested during the March 1–5, 2008, pilot study during which the codebook and content categories were developed and refined. Inter-coder reliability was tested with a second coder who recoded three of the 14 days (or roughly 20%). This process produced an overall Scott's Pi value of 0.857 with individual values that ranged from 0.772 to 1.0.

One drawback of using reader comments as the measure of the public agenda was that the *Star* did not post most wire-generated national and foreign news on its website due to an editorial decision. Therefore, it was impossible to judge the public agenda of the wire stories since feedback about them was not available. Since those national and international stories were not consistent across the agendas, important public affairs topics such as the Iraq war, presidential election, and national economy that were quite salient in the print edition had to be discarded from analysis. However, the absence of those topics also brings into sharper focus the relationship of the media and public agendas in terms of local news during an election campaign.

Findings

Correlations between the print and online versions of the *Star* for the two-week period were nearly perfect as measured by Spearman's rho (0.996). On all but two days, the print and online versions produced identical issue agendas. One explanation for this high degree of correlation is that when the newspaper decided to devote staff resources to a story, it chose to maximize its investment. Also, the web editors took their layout salience cues from the corresponding print edition. Issue developments in print usually led to prominent placement on the home page on the next day's website. And breaking news stories given salient treatment on the website were, in turn, usually highlighted on a section front of the next print edition. This repetition of content was consistent for election updates, developments in the sports stadium construction, and legislative bargaining over alteration of statewide property tax assessments.

Given the nearly perfect correlation of the print and online agendas, the website agenda was used to represent the media agenda during comparison with the public agenda. Overall, the mean media website and mean public agendas had a Spearman's rho correlation of 0.70. Table 2.1 displays the great deal of similarity in the agendas. The mean two-week media agenda was, in order: campaigns; crime; personal economy; health care; taxes; public works; transportation; and education. By comparison, the public agenda was: campaigns; crime; taxes; public works; a tie between personal economy and transportation; health care; and education.

Based on story comments, the public had greater concern for taxes and less concern for stories about personal economy and health care than did the media. Property tax reform, which included debate over re-assessment and sharp tax increases, was under consideration by the state general assembly during the sampled time period and may have motivated people to comment about it at rates that exceeded media coverage. Personal economy and health care issues received less commentary. Finally, both agendas placed the greatest importance on political

Table 2.1 Mean rank orders of media and public agendas, March 10-24, 2008 (mean 1-8 rank-order for 14-day period, with 1 indicating most important and 8 least important)

Media	*Public*
1. Campaigns (1.4)	1. Campaigns (1.6)
2. Crime (2.2)	2. Crime (2.6)
3. Personal Economy (3.7)	3. Taxes (3.9)
4. Health Care (4.2)	4. Public Works (4.4)
5. Taxes (4.6)	T-5. Personal Economy (5.0)
6. Public Works (5.7)	T-5. Transportation (5.0)
7. Transportation (6.5)	7. Health Care (6.9)
8. Education (7.2)	8. Education (7.3)

campaigns and crime issues. They also both generally had the lowest regard for education and transportation issues of the eight topics in question, although reporting and commentary was consistently present.

On a daily basis, the correlations between the media and public agendas were even stronger than the correlations of the two-week means. There was a distinct pattern of daily agenda setting from media to public as measured through the feedback comments. Internet news consumers consistently responded to the most salient stories presented by the media with few exceptions. As shown in Table 2.2, very high correlations were found across the two-week period, ranging from 0.62 to 0.93. Of the 14 days, 11 displayed correlations of 0.79 or greater. Two of the three days that did not produce significant correlations were weekend days.

Clearly, in the short term, the media had a powerful and immediate agenda-setting effect on the public as measured through online user feedback. The Internet audience regularly responded to the issues that the *Star* made most salient in both its print and online editions. These results were remarkably consistent regardless of which of the eight persistent issues ebbed and flowed onto the front pages.

Analyzing these agendas' movement over longer periods of time proved a more difficult task. Both agendas fluctuated widely from day today even though the eight issue categories were consistently represented. While campaign stories were among the top three issues on either the media or public agenda for each of the 14 days, other issues such as taxes varied more widely: On March 20, taxes were the second-most salient issue, and a day later they were the lowest of the eight issues measured.

Table 2.2 Spearman's rho rank-order correlations for media and public agendas measured daily

Mon. March 10	0.79*
Tues. March 11	0.93**
Wed. March 12	0.86**
Thur. March 13	0.81*
Fri. March 14	0.76*
Sat. March 15	0.67
Sun. March 16	0.86**
Mon. March 17	0.86**
Tues. March 18	0.79*
Wed. March 19	0.67
Thurs. March 20	0.93**
Fri. March 21	0.83*
Sat. March 22	0.91**
Sun. March 23	0.62

*-$p < 0.05$; **- $p < 0.01$.

However, cross-lagged correlations helped to reveal some notable patterns, and also allowed for analysis of whether the public agenda led the media agenda by comparing the previous day's public agenda to the following day's media agenda. Figures 2.1 and 2.2 depict two sets of key relationships that emerged from comparisons across the 14 days.

First, Figure 2.1 indicates that the agenda emphasized by the media early in the first week had a substantial effect on the public agenda later in that week. Figure 2.1 shows the correlations with corresponding Rozelle-Campbell baselines. The Rozelle-Campbell baseline is frequently used in agenda-setting research that incorporates cross-lagged correlations to indicate the value to be expected by chance alone (McCombs, 2004). The media agenda on March 10 had a 0.643 correlation with the March 15 public agenda, which well exceeded the Rozelle-Campbell baseline of 0.343.

Figure 2.2 displays analysis of the second week, March 17–22. In this instance, early-week public agenda displayed influence on the media agenda later in the week. The public-media diagonal's correlation of 0.357 surpassed the Rozelle-Campbell baseline of 0.341. Yet the early-week media agenda (0.333) also nearly surpassed the baseline. This mixed result appears at odds with the previous week, when the media had a clear influence on the public later in the week.

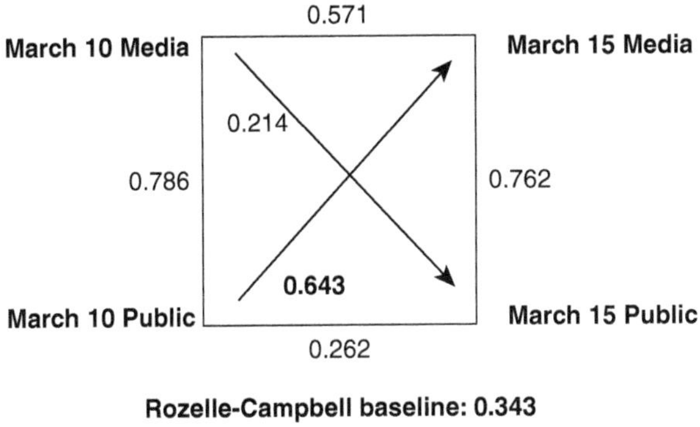

Figure 2.1 Cross-lagged correlations of Monday-Saturday agendas, March 10-15, 2008

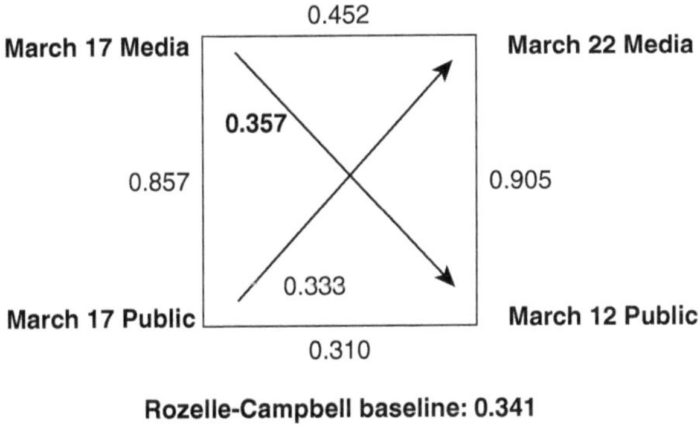

Figure 2.2 Cross-lagged correlations of Monday-Saturday agendas, March 17-22, 2008

Conclusions

This first study examined the agenda-setting relationship of a newspaper website (media agenda) and online reader commentary on news stories (public agenda) over a two-week period preceding the 2008 Indiana primary election that included presidential and congressional races on the ballot. This study sought a better understanding of how the local news media and public agendas interacted online during an election period, and whether the traditional agenda-setting relationship held, or whether the public agenda, as measured by online story comments, had any agenda-setting influence on the media in the days following those comments.

Findings indicate that the media demonstrated a powerful and immediate influence on the public on a daily basis. Consistently during the two-week period, stories presented as most salient in both the print and online editions corresponded with the stories on which website users offered the most comments. The consistent significance of daily correlations and the direction of the relationship signify that media held an agenda-setting influence on the public agenda in immediate fashion.

On the other hand, the dynamic between media and public agendas over longer periods of time yielded more complicated and more mixed results. Taken together, Figures 2.1 and 2.2 are inconclusive regarding the nature of influence among the media and public agendas beyond daily comparison. Cross-lagged correlations depict a more intricate relationship over time regarding how the online news consumer interacts with and is influenced by media salience cues. In some instances, the traditional agenda-setting relationship of media influencing the public was present. In others, there is evidence that the public might contribute to the media agenda by providing feedback that orients the media toward how its stories are being received in the short term. In the first week, the media appeared to influence the public, particularly closer to the weekend after days of coverage. However, in the second week, the public agenda actually showed influence on the media agenda measured a few days later. These results raise the possibility that there may be significant and complicated interaction between the media and public online when story commentary is immediately available.

Study 2: 2010 U. S. Senate Election

The second study sought a better understanding of whether changes in information technology use by the media, candidates, and the public may be altering agenda setting during election campaigns. This study examined media coverage, candidate press releases, and public opinion about the 2010 Indiana Senate campaign between Republican Dan Coats and Democrat Brad Ellsworth. A content analysis was conducted to understand which public affairs issues were covered by newspaper, television, and candidate press releases in the final two months of the campaign, and survey questions in the days following the election asked which issues from the campaign potential voters found most important.

The central research question of this second study investigated whether the traditional agenda-setting process remained consistent despite the increased commitment by candidates to websites and social media and the propagation of information available from other online sources such as citizen and interest group blogs. Most empirical evidence from agenda-setting studies conducted during campaigns in the Internet era have indicated the consistency of the media's agenda-setting effect on the public, but more work is needed to replicate findings and provide contextual details (Coleman & McCombs, 2007; Tran, Chapter 10, this volume). In general, the media have been found to have the greatest influence in the final few weeks and months before the election (Weaver, Graber, McCombs, & Eyal, 1981). Given these consistent findings,this study's hypothesis predicted that the media issue agenda would set the voter issue agenda during this campaign.

Method

In the tradition of McCombs and Shaw's original agenda-setting study (1972), this second study's methodology combined a representative survey of the media people consumed during the campaign and the issues they felt were most important, followed by a content analysis of the media content they identified.

First, in the survey, potential respondents were mailed a questionnaire and self-addressed stamped envelope for self-administering the survey in the days immediately following the 2010 midterm elections. The population surveyed was registered voters in Indianapolis who were exposed to the campaign information examined in the content analysis of Indianapolis market media. Multi-stage random sampling was used to assure random selection. First, a numbered list was made of the 590 voting precincts in Marion County, each containing about 1,000 registered voters for a total of about 600,000 registered voters in the county. Three precincts were chosen using a random number generator. Demographic data were consulted to determine that the selection of these precincts was representative of the population of registered voters. In those precincts, the median values for gender (48% of population and to 48% of sample were male), age (35 to 33.6), race (70% White to 68.9% White), and income ($51,613 to $43,510) all were essentially representative of the Marion County medians. In the second stage of sampling, a list of registered voters and their addresses in those precincts was acquired from the Marion County Voters Registration Office through a public records request. This list included the names of 3,173 registered voters arranged by street address, from which 500 names were randomly selected to represent the survey sample.

Respondents were asked about issue salience for comparison to the candidate and media issue agendas. The survey used an open-ended variation of the "most important problem" question: "What do you feel was the most important policy issue that was discussed during the recent Senate campaign?" The state of the economy and the lack of job creation was the top choice (31.4%), followed by taxes and spending (25.5%), health care (13.6%), lobbying and government functioning (2.7%), social order and immigration (2.3%), national defense (0.9%), and civil rights (0.5%).

Ultimately, 220 surveys were completed out of 432 valid mailing addresses, for a final response rate of 50.9%. The maximum sampling error for the survey was plus or minus 6.6 percentage points at the 95% confidence level.

The second method was a quantitative content analysis of campaign coverage of the 2010 Indiana Senate election. In the survey, open-ended survey questions prompted respondents to identify all of the sources from which they acquired Senate campaign information, and specifically which sources they used the most to learn about the campaign as a way of making a direct empirical connection between their survey responses and the content they consumed during the campaign. Respondents most frequently identified the *Indianapolis Star* (46.1%), followed by local television stations as a group (31.5%), Internet sites (10.7%), radio (7.9%), interpersonal communication (2.2%), and information direct from the campaign (1.7%).

Given their great popularity in the survey responses, campaign-related articles from the *Star* along with the four local network news channels serving Indianapolis were chosen for analysis. Additionally, press releases available on the major party candidate websites were analyzed as a means of comparing media, public, and candidate issue agendas.

Survey respondents singled out only indystar.com and the websites of the respective local television stations among local media organizations, and those stories were captured during analysis of newspaper and TV content. Many respondents only vaguely nominated "blogs," or "websites," and the *Star*'s print content captures the entirety of the indystar.com's coverage. Since most local media have been demonstrated to be similar in Senate campaign coverage qualitatively (Druckman, 2005), additional websites and radio coverage were not included for practical purposes.

The unit selected for this analysis was the individual story or press release. This selection allowed for parallel comparison across newspaper, television, and press releases. The sample frame was Sept. 3–Nov. 3, 2010, which includes the Friday prior to Labor Day through the day after the Senate election. Analyzing content in the two-month period prior to elections with some variance on the starting and ending days has become common practice for similar studies (Dalton, Beck, & Huckfeldt, 1998; Druckman, 2005; Kahn & Kenney, 2002).

During the 62 days analyzed, the *Star* published 55 articles about the Senate campaign. The four local television stations produced 68 stories during the same time period (WISH 24, WTHR 23, WTRV 11, and FOX 59 10 each) for a mean of 17. The Coats campaign posted 14 press releases on their website dated Sept. 2–Nov. 2, while the Ellsworth campaign posted 11 during that period.Overall, the content sample analyzed included 148 units of a combination of newspaper and television stories and press releases.

The list of campaign issues was adapted from relevant literature (Hayes, 2006; Petrocik, 1996) with specific attention to studies that had examined Senate campaigns (Druckman, 2004). Coders were asked to determine whether public affairs issues—including civic problems, proposed solutions, afflicted groups, and a candidate's leadership potential including related campaign promises—were a central part of each story. Coders were asked to examine the headline and first few paragraphs of print stories or press releases, or the opening statements of a broadcast story.

Seven issue categories (economy, taxes/spending, lobbying, health care, defense, societal welfare, and civil rights) were adapted from previous studies (Druckman, 2004; Hayes, 2006; Petrocik, 1996) to investigate the degree to which multiple issues were discussed in each story. If stories spent a paragraph discussing an issue, it was coded in the affirmative. Therefore, each story could have included discussion of up to seven discrete issues.

Reliability values were calculated using Krippendorff's alpha. In the pretest, conducted on 25 stories, alpha values ranged from 0.807 to 1.0 for all 24 categories. In the posttest, conducted on 20.3% of the actual sample using a random selection of 30 of the 148 stories and press releases, alpha values ranged from 0.810 to 1.0.

Findings

Figure 2.3 indicates that the media and candidate agendas presented a consistent selection of seven issues: the economy and jobs, taxes and spending, health care, lobbying, societal welfare/order, defense, and civil rights. Of those seven issues, the economy, taxes, health care, and lobbying received the most attention from all three outlets. Monetary issues received the most coverage in each, although the order differed, with newspapers covering the economy the most, and the other two

Newspaper	TV	Press Release
1. Economy/jobs (41.8%)	1. Taxes & spending (26.5%)	1. Taxes & spending (48.0%)
2. Taxes & spending (40.0%)	2. Economy/jobs (21.1%)	2. Economy/jobs (44.0%)
3. Lobbying (32.7%)	3. Lobbying (20.6%)	3. Health care (24.0%)
4. Health care (36.4%)	4. Health care (16.2%)	4. Lobbying (20.0%)
5. Defense (12.7%)	5. Societal welfare (8.8%)	5. Defense (16.0%)
6. Societal welfare (9.1%)	6. Civil rights (4.4%)	6. Societal welfare (8.0%)
7. Civil rights (1.8%)	7. Defense (0%)	7. Civil rights (0%)

Figure 2.3 Mean rank orders of media and candidate agendas, Sept. 3-Nov. 3, 2010

outlets featuring more coverage of taxes and spending. Those issues were followed by media coverage of lobbying, with slightly less coverage of health care. For the candidates, the order was reversed: their press releases discussed health care more often than the impact of lobbying. All three outlets paid far less attention to societal order (mainly immigration and gun control), defense, and civil rights.

Figure 2.3 also provides a visual display of the overall mean issue agendas produced by the media and the candidate press releases during the two-month period. They are quite similar with only a few exceptions. Using Spearman's rho rank-order correlations, the newspaper and candidate agendas correlated at 0 .929 (p <0.003); the newspaper and TV agendas correlated at 0.857 (p <0.014); and the TV and candidate agendas also correlated at 0.857 (p <0.014).

Qualitative analysis of the candidates' issue positions in media coverage supports the findings from the quantitative analysis of the press releases. Given the opportunity to identify the election's key issues during in-depth feature interviews, all of the candidates discussed taxes, spending, the overall economy, and jobs most frequently (Ruthhart, 2010). Ellsworth had a specific focus on Coats' lobbying career, and Coats spent a great deal of time making an issue of Ellsworth's vote in favor of health care reform, which accounts for the popularity of those two issues (Ruthhart, 2010). Coats tried to tie Ellsworth with plans to close federal penitentiaries in Guantanamo Bay early in the campaign, and there were some briefer exchanges about immigration policy, but those issues largely took the backseat throughout the campaign (Tully, 2010).

Turning to the public agenda, Figure 2.4 shows the description of the survey respondents. They came from diverse backgrounds based on demographic indicators such as age, gender, race, party identification, and education levels. Also, the sample data in this regard approach the statistics of the larger Marion County population in terms of gender (51.4% female in sample; 51.6% female in population); race (72.3% White in sample, 27.7% non-White in sample; 69.1% White in population, 30.9% non-White in population); and age (median age 50 in sample; 46.4 in population). The number of respondents with a bachelor's degree or advanced degree was 50.0%, which was higher than the county population rate (25.4%), but in line with education levels of respondents in similar Senate media use and knowledge surveys (see Druckman, 2005, with 57% of the sample reporting such degrees). Together with multi-stage random sampling and satisfactory response rate, these descriptive indicators provide confidence that the sample is representative of the larger Indianapolis population, and therefore provides the basis for generalizing results.

Variable	Sample Data
Voting status	Voted: 79.5% Did not vote: 20.5%
Gender	Male: 48.6% Female: 51.4%
Age	18–24: 3.2% 25–34: 18.6% 35–44: 15.5% 45–54: 22.3% 55–64: 20.5% 65–74: 12.7% 75+: 7.3%
Race	White: 72.3% Black: 16.4% Hispanic: 3.6% Native American: 0.5% Multi-racial: 4.5% Other: 2.7%
Education	High school or less: 24.5% Some college: 25.5% College degree: 32.7% Post-graduate: 17.3%
Party identification	Democrat: 33.3% Republican: 31.9% Independent: 30.5% Libertarian: 0.5% Other: 3.8%

Figure 2.4 Demographics of 2010 Senate Election survey respondents

Table 2.3 shows the descriptive patterns of media exposure in terms of respondents' self-reported days per week of usage. Newspapers and television were the media most used on an everyday basis. More than half of the sample read a newspaper, watched some form of television news, or listened to radio news at least five days per week. Respondents also were widely exposed to advertising on a daily basis. Most respondents did not access nonmainstream Web news sources, such as blogs and individual social media, as heavily, although about one-third (33.2%) went to the Internet for news at least five days per week. However, the Web-only, nonmainstream information sources also produced the largest percentage of non-users (38.6%), followed by radio and newspapers.

Table 2.3 Days per week of media exposure (% respondents), 2010 Senate Election survey

Days	*Newspaper*	*Web*	*Radio*	*Local TV*	*Cable TV*	*Advertising*
0	19.5%	38.6%	28.2%	14.1%	11.8%	6.4%
1	13.2%	8.2%	6.8%	9.5%	14.5%	4.1%
2	10.0%	7.7%	6.8%	9.1%	10.0%	7.7%
3	5.0%	5.9%	10.5%	10.5%	6.4%	10.0%
4	1.8%	6.4%	5.0%	4.5%	6.8%	6.8%
5	5.5%	12.3%	24.1%	14.5%	10.0%	20.0%
6	2.3%	3.2%	4.5%	5.0%	8.2%	7.7%
7	42.7%	17.7%	14.1%	32.7%	32.3%	37.3%

These frequencies of campaign news attention in Table 2.4 reveal some more interesting patterns. Television is the most-attended media, with 64.1% using it for "some" or "a lot" of their news, followed closely by newspapers at 59.1%. These rates of popularity support the other data from the survey in which respondents, asked to state their campaign news preferences, chose newspapers and TV first and second. Radio was third among media at 47.3%, with Web news last at 33.2%. As with exposure, nonmainstream website news also received the highest percentage of people who said they pay no attention to the format (37.7%), indicating that more than a third of respondents had no exposure and paid no attention to campaign news on the Internet.

Figure 2.5 makes a visual comparison of the rank-ordered agendas across media, campaigns, and the public. The media agenda represents the combined newspaper and television agendas. The campaign agenda reflects the frequency of issues mentioned in the press releases of the two major-party candidates. The public agenda was based on the percentage of respondents who identified the issue as the most important one in the campaign. As Figure 2.5 shows, the public agenda was nearly identical

Table 2.4 News attention by source (% respondents), 2010 Senate Election survey

	Newspaper	*Web*	*Radio*	*Television*	*Advertising*
None	16.8%	37.7%	27.3%	7.7%	25.0%
Very Little	24.1%	19.1%	25.5%	28.2%	42.7%
Some	40.5%	18.2%	31.4%	43.6%	28.6%
A lot	18.6%	15.0%	15.9%	20.5%	3.6%

Media	Campaign	Public
1. Economy/jobs (36.5%)	1. Taxes & spending (48.0%)	1. Economy/jobs (31.4%)
2. Taxes & spending (33.2%)	2. Economy/jobs (44.0%)	2. Taxes & spending (25.5%)
3. Lobbying (28.4%)	3. Health care (24.0%)	3. Health care (13.6%)
4. Health care (25.6%)	4. Lobbying (20.0%)	4. Lobbying (2.7%)
5. Societal welfare (9.0%)	5. Defense (16.0%)	5. Societal welfare (2.3%)
6. Defense (5.6%)	6. Societal welfare (8.0%)	6. Defense (0.9%)
7. Civil rights (2.2%)	7. Civil rights (0%)	7. Civil rights (0.5%)

Figure 2.5 Mean ranks orders of media, campaign, and public agendas, 2010 Senate Election

to the media agenda, with only lobbying and health care in third and fourth places alternating. By comparison, the candidates placed disproportional emphasis on taxes and defense as compared to the public.

Finally, a correlation matrix of these three agendas was compiled using Spearman's rho. The data confirm this study's hypothesis: the media agenda, which preceded the survey, produced strong and statistically significant correlations with the public agenda, and thus had an influence on setting the public agenda ($r = 0.964$, $p < 0.001$). In fact, the agendas were nearly identical, other than the public considering health care as a more important issue than the status of a candidate as a former lobbyist. The campaign and public agendas also were quite similar, although the candidate press releases did not set the public agenda as strongly as the media ($r = 0.929$, $p < 0.001$). The campaigns and the public differed on the importance of the economy and jobs versus taxes and government spending and the relative importance of defense versus societal welfare issues like education. The media and campaign also had a similar agenda ($r = 0.893$, $p < 0.001$), although it was relatively the weakest of the three agenda comparisons.

These differences, and the relative strength of the influence of the media on the public, were reinforced through partial correlations. The media and public agendas produced a correlation of $r = 0.809$, $p < 0.05$ when controlling for the campaign agenda. On the other hand, the correlation between the public and campaign agendas was not statistically significant ($r = 0.567$, $p < 0.24$).

Conclusions

The second study analyzed the issue agendas available to voters from the media and candidate websites prior to the November 2010 U. S. Senate election, and investigated whether those agendas influenced the formation of the public agenda as measured through survey response to the Most Important Problem question.

Content analysis revealed that the media and candidate agendas were quite similar; however, voters read about the election much more often from the media, and the media had a slightly stronger relationship with the public agenda than the candidates when comparing Spearman's rho rank-order correlations. This finding from the correlation analysis confirmed this study's hypothesis that had expected conventional agenda setting to be maintained for this election.

The mail survey of 220 randomly selected registered voters (50.9% response) found that the newspaper website was the most popular and most important Internet outlet for campaign news, followed by local television websites. By contrast, people chose to learn about election issues from campaign websites, nonmainstream political blogs, or other outlets such as social media much less frequently. And when voters consumed election-related information online, they most often used the websites of these news organizations. Voters less frequently used other forms of online information to learn about the election, such as archived candidate press releases, social media, and blogs.

These preferences were corroborated by statistical analysis that indicated that these media had a stronger influence on setting the agenda than the candidates. These data indicate that the traditional agenda-setting function holds for this local election context despite increased resources dedicated to conducting the campaign online. These findings align with other studies that have suggested that agenda setting at its core remains unchanged in the Internet era despite the many fundamental changes brought about by the introduction and adoption of so many forms of information technology (Coleman & McCombs, 2007; Tran, Chapter 10, this volume).

The strong similarities in the media, candidate, and public agendas indicate that the media during this campaign were quite successful in covering the candidate agendas and informing the public about the most important election-related issues. This finding is not unexpected for a Senate election that did not poll particularly closely and did not receive intense national or outside advertising interest (Kahn & Kenney, 1999). However, it is another indication that the variety of differences in media coverage and public information-seeking that shape elections require a stronger contextual understanding of mass communication theories in general and agenda setting specifically.

And while the impact of the candidate press releases on setting the media agenda was outside the scope of analysis in this study, the strong similarities in the media and candidate agendas and their similarly strong correlations with the public agenda also tend to support related findings from studies of the impact and proliferation of Internet-based campaign communications on the media, particularly in local elections (Kiousis et al., 2006, 2009).

Discussion

In recent years, mass communication scholars have grappled with how advancements in information technology, the proliferation of news sources, and the fragmentation of news audiences may be altering the agenda-setting process (Coleman & McCombs, 2007; Denham, 2010; Takeshita, 2005; Tran, Chapter 10, this volume). Findings from the two studies presented in this chapter indicate that although these developments have produced many changes in how the news is produced, transmitted, and consumed, by and large, the agenda-setting process has remained the same within the context of citizens' use of local news to inform themselves about these election campaigns. These data offer evidence that traditional news websites maintain a strong agenda-setting influence on the public despite the array of sociological and technological changes that have impacted how the news is produced, transmitted, and consumed. These studies contribute to a clearer empirical understanding of agenda setting in the context of local elections as media and candidates continue to produce more online content, and as rates of Internet political news use increase.

Results from the first study of the 2008 primaries indicate that traditional media and their websites retained a strong immediate agenda-setting influence in the period preceding primary and general elections of local and national importance. Though online comments sections have opened opportunities for the public to contribute to media agenda building (Denham, 2010), these findings indicate that the media dictated the public agenda on a daily basis in a manner consistent with results found in previous research of agenda setting and online news discussion. A similar short-term pattern of the direction of influence was revealed a decade ago when Roberts, Wanta, and Dzwo (2002) found that mainstream media set the agenda for electronic bulletin board discussions.

However, for longer time periods, cross-lagged correlations produced more mixed results, with the media and public agendas alternately influencing one another. Given the lack of research attention in this area, more work is needed to better understand whether and how the public's ability to communicate more directly, more instantaneously, and more frequently with media alters the agenda-setting process and/or influences the media agenda in the long term.

In addition to contributing to agenda-setting theory, the first study makes a methodological contribution by measuring the public agenda through online news story comments instead of survey data. Previous studies have used aggregate search history (Jeong, Kim, & Shin, 2008) and message board comments (Roberts, et. al, 2002), but this study tied online discussion directly to the news stories that provided the salient topics. However, this experimentation with public agenda measurements also may be seen as a limitation. Little was learned about the identity of these people, and it is possible that these committed news consumers were not representative of the larger public due to self-selection. However, their contributions also indicate that they are an active audience that plays some role in the media process that is as yet undefined. And given the proliferation of audience interactivity options offered by media, it is important to continue inquiries into how these feedback opportunities play a role in media operations and agendas. Certainly, future work in this area should consider ways to improve and refine various online measures of the public agenda, and perhaps triangulate this measure using multiple approaches from those that have previously proven successful, such as discussion board posts, Facebook and other social media website posts, search engine queries, and hyperlinking through blogs and social media.

In the second study of the 2010 U. S. Senate election, news and candidate websites both significantly influenced the public agenda; however, the media correlations with the public agenda were stronger, and partial correlations revealed that the media had a greater influence on the public when controlling for the campaign agenda. Respondents also indicated their consistent preferences for frequency of exposure and attention to traditional news over candidate messages and other nonmainstream Web-based information. These findings are similar to previous studies of Senate campaigns (Druckman, 2005; Kahn & Kenney, 1999, 2002), but build on that previous work to update such conclusions for Senate elections in the Web 2.0 era.

These findings also support research on agenda setting during campaigns in the Internet era that examined the increased efforts of candidates to influence the public's consumption of election information (Kiousis et al., 2006, 2009; Tedesco, 2001, 2005). In this instance, the public very rarely reported accessing candidate websites, although they also reported paying little attention to traditional candidate communication methods such as advertising and debates due to the perceived lack of closeness of the campaign (Kahn & Kenney, 1999). Respondents indicated television advertising was the candidate communication to which they were most exposed.

Likewise, in terms of accessing election-related content on the Internet, respondents selected news organizations at rates that exceeded candidate websites, social media, blogs, and other Web-based forms of news. Thus, despite the increased availability of election-related content online, when it came to this 2010 Senate election, the public chose to get its news most often from the local newspaper or its website.

These same general patterns for acquiring campaign news were found in research conducted on the 2012 presidential primaries by the Pew Research Center for the People and the Press. Respondents preferred to get news about the primaries from traditional news organizations, and few took advantage of social media or nontraditional Internet sources (Pew Research Center, 2012). Most respondents had seen or heard commercials about the election, but one in five or fewer had received printed materials or emails, had visited candidate websites, or followed candidates on social media (Pew Research Center, 2012). Pew's findings considered alongside the studies in this chapter support the notion of a general pattern of behavior of how citizens acquire campaign news in the Web 2.0 world that transcends the type of election or the office contested. Despite the propagation of media technologies and platforms, citizens still prefer to get their election news from traditional media organizations, even if the ways that they access that news are slowly diversifying and despite increased efforts of candidates to influence the public directly.

There are limitations for both studies in this chapter in terms of the case study approach to studying the role of the media in elections. Every election has several factors that tend to vary across time, location, and other campaign characteristics, especially when it comes to local elections (Kahn & Kenney, 1999; Ridout & Mellen, 2007). However, it is only through the accumulation of replicated data in the form of examining several elections that mass communication theory may be built.

In the case of the 2010 Senate election in Indiana, although it was not a campaign of exceptional newsworthiness or national appeal, the routine nature arguably makes it representative of how many other congressional elections are covered and how that coverage affects voters in many districts. In presidential elections or in the few nationally covered congressional races, the role of the local media in informing voters is more muted because of the abundance of national coverage, the opportunities for incidental learning and more instances of interpersonal discussion. In this case, and in many other similar elections, the lack of national interest allowed the content analysis to account for an almost complete range of media content within the local market and therefore the greater isolation of the effects of news on the public. The nature of this election also ensured that potential voters would not be as overwhelmed by media coverage and advertising and could better recall from what sources they learned about election issues. Certainly, the closeness of the race, the national media attention, and the amount of advertising are all important factors when considering the role of media in Senate elections (Kahn & Kenney, 1999).

The changing nature of the online media environment and the methodological advancements in studying the public and source agendas on the Internet make agenda building and Internet agenda setting in elections two areas primed for further research. Future studies should continue to consider not only ways that the proliferation of the Internet and the expanded adoption and use of information technologies may be changing agenda-setting theory, but also account for the ways that these changes may help develop new and better measures of the agenda building and agenda setting. More than ever before, data is available in many different forms that will help address previous research gaps in terms of how media agendas are built and how the public agenda may be measured. For example, research on how the media agenda is built in specific online tools such as Twitter could allow the ability to track time ordering of agenda building and agenda setting and analyze the contributions of sources, candidates, other media, and the public. Replications also are needed in terms of broadening the scope and type of election campaigns studied, and as a means of accounting for the increasing variety of information sources available.

The findings from the two studies presented in this chapter provide a small measure of greater clarification in understanding how the agenda-building and agenda-setting processes are responding as media and candidates increase and diversify their production of online content, and as rates of Internet campaign news use increase. Yet there will be more work needed in these areas as these developments accelerate and the world of election campaigns continues to move online.

References

Benoit, W. B., & Hansen, G. J. (2001). Presidential debate questions and the public agenda. *Communication Quarterly, 49*(2), 130–141.

Coleman, R., & McCombs, M. (2007). The young and agenda-less? Exploring age-related differences in agenda setting on the youngest generation, baby boomers, and the civic generation. *Journalism & Mass Communication Quarterly, 84*(3), 495–508.

Dalton, R. J., Beck, P. A., & Huckfeldt, R. (1998). Partisan cues and media: Information flows in the 1992 presidential election. *American Political Science Review, 92*, 111–126.

Denham, B. E. (2010). Toward conceptual consistency in studies of agenda-building processes: A scholarly review. *Review of Communication, 10*(4), 306–323.

Drew, D., & Weaver, D. (2006). Voter learning in the 2004 presidential election: Did the media matter? *Journalism & Mass Communication Quarterly, 83*(1), 25–42.

Druckman, J. N. (2004). Priming the vote: Campaign effects in a U. S. Senate election. *Political Psychology, 25*(4), 577–594.

Druckman, J. N. (2005). Media matter: How newspapers and television news cover campaigns and influence voters. *Political Communication*, 22, 463–481.

Foot, K., & Schneider, S. M. (2006). *Web campaigning*. Cambridge, MA: MIT Press.

Hayes, D. (2010). The dynamics of agenda convergence and the paradox of competitiveness in presidential campaigns. *Political Research Quarterly*, *63*(3), 594–611.

Hayes, D. W. III. (2006). *Beyond the horse race: The content and consequence of issue news in American elections.* (Unpublished dissertation.) University of Texas at Austin.

Indianapolis Star Media Group. (2008). Star media products. Retrieved February 17, 2008, from http://indystarmedia.com/products

Jamieson, K. H., & Campbell, K. K. (1992). *The interplay of influence: News, advertising, politics, and the mass media* (3rd ed.). Belmont, CA: Wadsworth.

Jeong, Y., Kim, K., & Shin, W. (2008). *Agenda building function of Internet searches: Measuring the unique contribution of the public agenda on the media agenda*. Paper presented to the International Communication Association, New York.

Kahn, K. F., & Kenney, P. J. (1999). *The spectacle of U. S. Senate campaigns*. Princeton, NJ: Princeton University Press.

Kahn, K. F., & Kenney, P. J. (2002). The slant of the news: How editorial endorsements influence campaign coverage and citizens' views of candidates. *American Political Science Review*, 96, 381–394.

Kiousis, S., Mitrook, M., Wu, X., & Seltzer, T. (2006). First- and second-level agenda-building and agenda-setting effects: Exploring the linkages among candidates news releases, media coverage, and public opinion during the 2002 Florida gubernatorial election. *Journal of Public Relations Research*, *18*, 265–285.

Kiousis, S., Soo-Yeon, K., McDevitt, M., & Ostrowski, A. (2009). Competing for attention: Information subsidy influence in agenda building during election campaigns. *Journalism and Mass Communication Quarterly*, *86*, 545–562.

Ku, G., Kaid, L. L., & Pfau, M. (2003). The impact of web site campaigning on traditional news media and public information processing. *Journalism & Mass Communication Quarterly*, *80*(3), 528–547.

Lang, G. E., & Lang, K. (1991). Watergate: An exploration of the agenda-building process. In D. L. Protess & M. McCombs (Eds.), *Agenda setting: Readings on media, public opinion, and policymaking* (p. 278). Hillsdale, NJ: Lawrence Erlbaum.

Lim, J. (2006). A cross-lagged analysis of agenda setting among online news media. *Journalism & Mass Communication Quarterly*, *83*(2), 298–312.

Maier, S. (2010). All the news fit to post? Comparing news content on the web to newspapers, television, and radio. *Journalism & Mass Communication Quarterly*, *87*(3/4), 548–562.

McCombs, M. (2004). *Setting the agenda: The mass media and public opinion*. Malden, MA: Blackwell.

McCombs, M., & Reynolds, A. (2009). How the news shapes our civic agenda. In J. Bryant & M. B. Oliver (Eds.), *Media Effects* (pp. 1–16). New York: Routledge.

McCombs, M., & Shaw, D. L. (1972). The agenda-setting function of mass media. *Public Opinion Quarterly*, *36*, 176–187.

Meraz, S. (2009). Is there an elite hold? Traditional media to social media agenda setting influence in blog networks. *Journal of Computer-Mediated Communication, 14*(3), 682–707.

Messner, M., & DiStaso, M. W. (2008). The source cycle: How traditional media and weblogs use each other as sources. *Journalism Studies, 9*(3), 447–463.

Petrocik, J. R. (1996). Issue ownership in presidential elections, with a 1980 case study. *American Journal of Political Science, 40*(3), 825–850.

Pew Research Center for the People and the Press. (2012). Cable leads the pack as campaign news source. Retrieved from http://www.people-press.org/2012/02/07/cable-leads-the-pack-as-campaign-news-source

Ridout, T. N., & Mellen, R. Jr. (2007). Does the media agenda reflect the candidates' agenda? *Harvard International Journal of Press/Politics, 12*(2), 44–62.

Roberts, M., Wanta, W., & Dzwo, T. (2002). Agenda setting and issue salience online. *Communication Research, 29*(4), 452–465.

Ruthhart, B. (2010, September 8). Ellsworth wants his deeds to do the talking. *Indianapolis Star*, A1.

Sayre, B., Bode, L., Shah, D., Wilcox, D., & Shah, C. (2010). Agenda setting in a digital age: Tracking attention to California Proposition 8 in social media, online news, and conventional news. *Policy & Internet, 2*(2), 1–7.

Shoemaker, P. J., & Reese, S. D. (1996). *Mediating the message: Theories of influences on mass media content.* White Plains, NY: Longman.

Takeshita, T. (2005). Current critical problems in agenda-setting research. *International Journal of Public Opinion Research, 18*(3), 275–296.

Tedesco, J. C. (2001). Issue and strategy agenda-setting in the 2000 presidential primaries. *American Behavioral Scientist, 44*(12), 2048–2067.

Tedesco, J. C. (2005). Issue and strategy agenda setting in the 2004 presidential election: Exploring the candidate-journalist relationship. *Journalism Studies, 6*(2), 187–201.

Tully, M. (2010, October 27). Ellsworth presses, but election clock favors Coats. *Indianapolis Star*, B1.

Wallsten, K. (2007). Agenda setting and the blogosphere: An analysis of the relationship between mainstream media and political blogs. *Review of Policy Research, 24*(6), 567–587.

Weaver, D. H., Graber, D. A., McCombs, M. E., & Eyal, C. H. (1981). *Media agenda-setting in a presidential election: Issues, images, and interest.* New York: Praeger.

Weeks, B., & Southwell, B. (2010). The symbiosis of news coverage and aggregate online search behavior: Obama, rumors, and presidential politics. *Mass Communication & Society, 13*(4), 341–360.

Chapter 3

Value Resonance and the Origins of Issue Salience

Sebastián Valenzuela

It is almost a truism that not everyone is equally susceptible to media effects. Within the agenda-setting research tradition, need for orientation, issue obtrusiveness, level of political involvement, media trust, and contextual differences have all been related to variations in the transfer of salience from the media to the public agenda (Soroka, 2002; Tsfati, 2003; Weaver, 1980; Zucker, 1978). Despite the richness of this literature, few scholars have paid attention to higher-order psychological variables as regulating factors of agenda-setting effects. Chief among these long-lasting orientations are human values—people's worldviews and goals they hold for their society and community. As it turns out, a variety of cognitive, attitudinal, and behavioral decisions can be explained by the values individuals hold. Allport (1961) even argued that values were the "dominating force in life" (p. 543) because they directed all of an individual's behavior toward their realization. In the political communication field, empirical work has repeatedly demonstrated that values can be powerful moderators of media effects, by either amplifying or dampening them. For instance, postmaterialist values (e.g., protecting the environment) have been positively related to news use and political discussion, whereas materialist values (e.g., fighting crime) have been linked to entertainment media and less frequent political discussions (McLeod & Sotirovic, 1998). Framing effects on decision making have also been related to how strongly people hold certain values (Shah & Domke, 1996).

In this chapter, I examine the role of human values on agenda-setting effects by tackling three different aspects. First, I elaborate on the theoretical and empirical implications of the values-issues consistency hypothesis, which posits that materialist and postmaterialist values moderate the strength of agenda-setting effects (Valenzuela, 2011). Second, I explore the relationship between values and need for orientation, the key psychological process that researchers have used to explain variations in the size of agenda-setting effects. Last I examine the consequences of values in the agenda-setting process for the development of attitudes and behaviors among citizens, also known as media priming effects. The ultimate purpose is to open new venues for research on the psychology of agenda setting.

The Values-Issues Consistency Hypothesis

Any discussion of the impact of individuals' values on issue salience needs to start by defining what values are in the first place. According to Schwartz (1999), values are "conceptions of the desirable that guide the way social actors (e.g., organizational leaders, policy-makers, individual persons) select actions, evaluate people and events, and explain their actions and evaluations" (p. 24). In other words, values provide criteria for evaluating human actions and external stimuli based on perceptions of desirable (and undesirable) end-states of human life. Some authors group values with attitudes under the label of predispositions and/or orientations. Although both operate at the psychological level, they are conceptually and operationally different. Attitudes tend to be object-specific (e.g., liking or disliking the Republican party); values, in contrast, can be applied across different contexts and thus form a higher-order concept (e.g., holding egalitarian versus libertarian values). Due to their stability over time, values are usually conceived as determinants of attitudes, although they may well influence each other in the long run—particularly if attitudes are conceived as manifestations of one's values. The "transsituational" (Schwartz, 1994, p. 21) nature of values is what makes them a key determinant of individuals' decisionmaking and behavior.

Among existing theories of values, Inglehart's (1977, 1990, 1997) work regarding materialist and postmaterialist values seems most consistent with agenda setting and the notion of issue salience. Materialist values refer to security and survival goals that can be attained by material means, such as economic growth and maintaining public order. Postmaterialist values, in turn, emphasize autonomy and quality-of-life aspects, prioritizing goals such as self-expression and achieving gender equality. They are postmaterialist in the sense that they do not refer to material or economic conditions *per se*. In his seminal *The Silent Revolution* (1977), Inglehart developed and tested two propositions regarding the development of these values:

1. *The scarcity hypothesis*: People's goals are shaped by their socioeconomic environment so that the greatest value is placed on aspects of life that are in relatively short supply.
2. *The socialization hypothesis*: The effect of the socioeconomic environment on goal priorities is not immediate but takes place over the long run.

The scarcity hypothesis implies that wealth promotes postmaterialist values, whereas economic decline promotes materialist values. This relationship, though extensively documented with data from the

Eurobarometer and World Values Surveys (Inglehart & Welzel, 2005), is not immediate, as if individuals are materialist when poor and post-materialist when rich. On the contrary, Inglehart (1977) argued that people's values reflect their pre-adult life experiences and are shaped by the cultural and social forces prevailing at the time they were raised. Therefore, the scarcity hypothesis is qualified with the socialization hypothesis: Most values develop before reaching adulthood and change relatively little thereafter.

Applying Inglehart's theory of value change to agenda setting, we should expect that the transfer of salience from the media agenda to the public agenda is smoother when there is consistency between individuals' values and the type of issues covered in the media. In hypothesis form, the expectation is that *The power of the media to set the public's issue priorities is larger for issues that better reflect the values of individuals. The more the news media cover materialist issues, the stronger the agenda-setting effects on materialist individuals (rather than on postmaterialist individuals). Conversely, the more the news media cover postmaterialist issues, the stronger the agenda-setting effects on postmaterialist individuals (rather than on materialist individuals).*

One can think of three alternative explanations for this values-issues consistency (VIC) hypothesis. Work on selective perception and selective retention suggests that individuals tend to automatically process more attentively and recall messages that are congruent with their existing values and attitudes (Atkin, 1985). Political psychologists also speak of motivated reasoning, the tendency to discount information with which one disagrees, and confirmation bias, the notion that people use their prior beliefs when processing information and arguments (Taber, 2003). Therefore, people with materialist values should be biased in their information processing toward materialist issues, whereas people with postmaterialist values should be biased toward postmaterialist issues. A third possible explanation lies in public opinion formation theories. For instance, Zaller's (1992) Receive-Accept-Sample model posits that the greater a person's level of cognitive engagement with an issue, the more likely she or he is exposed to, and comprehends, media messages about that issue. At the same time, people tend to resist arguments that are inconsistent with their values and beliefs. Extrapolating these propositions to the notion of VIC, the issue agendas of materialist and postmaterialist individuals may be different because their comprehension and acceptance of the mix of materialist and postmaterialist issues in the media agenda may be quite different. A similar mechanism for value resonance has been proposed for framing effects by other researchers (Schemer, Wirth, & Matthes, 2011), which validates the proposition that congruence between people's value predispositions and exposure to value-laden content can make a difference in agenda setting.

The empirical evidence supporting the VIC hypothesis comes from a panel survey matched with a content analysis of press coverage during the 2006 Canadian elections, and a laboratory experiment with college students manipulating issue salience. The methodological details of these analyses are available elsewhere (see Valenzuela, 2011), so I concentrate here on the main findings. First, the content analysis indicated higher salience in the Canadian press of materialist issues such as the economy compared to postmaterialist issues such as the environment. Subsequently, using the World Values Survey 4-item index to measure values and the Most Important Problem (MIP) question, the agenda of materialist and postmaterialist respondents was compared to the media agenda, defined as a ranking of issues according to the aggregated number of stories about each topic covered by different newspapers over the course of the campaign. As was expected, the rank-order correlation between the media agenda and the materialists' agenda was higher (+0.37) than that between the media and postmaterialists' agendas (+0.18). Of course, determining which issues are materialist and postmaterialist is not always clear-cut, so the comparison was repeated, this time considering only the issues that were clearly materialist- or postmaterialist-oriented (e.g., crime and minority rights). Again, in line with the notions of VIC, the difference in correlations between the media agenda and materialists' and postmaterialists' agendas grew significantly: 0.48 versus 0.22, respectively.

To further probe the robustness of these findings, an individual-level analysis of the same survey data was performed, in which the dependent variable was an agenda-setting susceptibility score (see Wanta, 1997) and the independent variables were the 4-point values index and an extensive battery of control variables. The results of this stringent analysis showed that the index of postmaterialist values was significantly associated with agenda-setting susceptibility, further validating the initial findings.

As a final test of the VIC, an experiment was conducted among Canadian college students in which the salience of clearly materialist and postmaterialist issues was manipulated. Here, the expectation was that students with a materialist orientation would rank as more salient materialist issues regardless of the frequency of exposure to materialist issues, and, conversely, students with a postmaterialist orientation would rank as more salient postmaterialist issues regardless of the frequency of exposure to postmaterialist issues. The results showed that in three out of four possible instances, values had a statistically significant effect on the strength of agenda-setting effects. Specifically, when materialist and postmaterialist participants read more stories about the postmaterialist issue, the postmaterialist participants rated the issue as far more important than the materialist participants. Likewise, when participants read more stories about the materialist issue, materialist subjects rated the issue as more salient than their postmaterialist peers. Thus, the results of the experiment also supported VIC; agenda-setting effects are stronger when the topics in the news agenda match individuals' values. At the same time, they seem consistent with the assumption that values precede attitude formation.

Though the evidence collected so far tackles individual-level differences in agenda setting as related to human values, one could also apply the notion of VIC at the aggregate-level—the traditional context of agenda-setting research. In this case, the VIC hypothesis could be adapted to the following: *The power of the media to set the public's issue priorities is larger for issues that better reflect the values of society-at-large. In materialist societies, the agenda-setting power of the media is larger for materialist issues and smaller for postmaterialist issues, whereas in post-materialist societies the agenda-setting power of the media is larger for postmaterialist issues than for materialist issues.* Testing this proposition would require cross-national data and exhaustive controls for differences in media systems and institutional arrangements that are known to affect agenda setting at the cross-national level (Walgrave & Van Aelst, 2006). A good starting point is South America, where most countries have fairly similar media systems (i.e., market-driven) and political regimes (i.e., developing democracies), yet vary widely in the proportion of materialist and postmaterialist segments of the population. Certainly, keeping constant extraneous factors is an important hurdle, but the effort could pay off by opening a new way of conducting comparative research on agenda-setting effects.

Values and Need for Orientation

So far, values have been considered as exerting a unique effect on agenda setting, operating in isolation from other processes that influence the transfer of issue salience. Although this is justified when one is making the case for the importance of bringing in a new variable to the study of agenda-setting effects, it is entirely possible that values are related to agenda setting in conjunction with other relevant psychological variables. Here, I make the case for need for orientation (NFO), which remains the most studied individual-level factor for attributing differences in the strength of agenda-setting effects (see Camaj, Chapter 5, this volume; Kim, Chapter 4, this volume; McCombs, 2004). More specifically, in this section I advance the proposition that values can be an antecedent of NFO.

Ever since Weaver (1980) highlighted relevance and uncertainty as the key components of NFO, researchers have agreed that relevance is the most substantive aspect. In other words, variations in the relevance people attach to issues, public figures and other elements of the media agenda are directly related to variations in the strength of agenda-setting effects. If a topic, say, the environment, is not relevant to an individual, the amount of coverage devoted to urban pollution in the media will not exert a noticeable influence on the salience of the environment for that individual. Evatt and Ghanem (2001) identified two substantive aspects of relevance, social relevance and personal relevance, and an affective aspect, emotional relevance. People can recognize that an issue may be important for society even if it is not important for their personal life. Emotions, in turn, can increase the relevance of an issue even if it is not personally or socially relevant.

Values could be related to all three of these sources of relevance. Holding postmaterialist values should increase people's awareness of social issues related to quality of life and self-expression, such as the protection of the environment and giving people more say in political decisions. The same should be true of materialist values and the relevance of crime, inflation, and other collective issues related to physical and economic security. Connecting these societal issues to how they affect people's personal lives could also be made easier if there's consistency between people's values and the issues on the media and public agendas. This is because individuals' values can provide frameworks of reference to process this type of information more effectively. In other words, value resonance can help people determine which issues they find personally relevant. Values can also indicate domains of emotional relevance. The clearest example is hot button issues. For people who hold strong conservative or liberal values, media coverage of abortion, same-sex marriage, and similar issues will more likely elicit strong emotional reactions compared to other topics. And because they elicit these responses, it is more likely that these individuals will pay more attention to these issues and name them when a survey asks what is the most important problem facing the country.

There are several ways of empirically testing the proposition that values are an antecedent of NFO. Using survey data and specific measures for values and the different types of relevance, for instance, researchers could estimate the path model depicted in Figure 3.1 below. Rather than values directly influencing issue salience judgments as cued by media exposure, they are assumed to exert an indirect effect by bolstering all three modes of issue relevance (social, personal, and emotional). These dimensions of NFO, in turn, are conceptualized as more immediate regulators of issue salience. The reason for this expectation is related to the fact that when people find something relevant, they are more likely to seek information about it on the media, increasing the likelihood of being influenced by it.

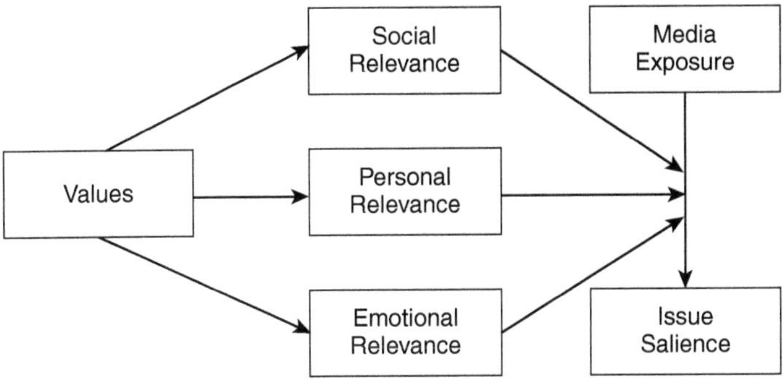

Figure 3.1 A proposed path model connecting values, NFO, and agenda-setting effects

The model in Figure 3.1 also suggests another aspect of the relationship between values and relevance, the key component of NFO. Values are regarded as primary sources of relevance rather than the other way around because they tend to be quite stable over time, operating almost like personality traits. In the Canadian panel survey mentioned earlier, for instance, the test-retest correlation coefficient of the index measuring postmaterialist values over a 2-year period was a strong +0.55. Excluding individuals in the mixed category and comparing pure materialists and postmaterialists only, the reliability coefficient increased to an impressive +0.85. The same was not true for the agenda-setting susceptibility index; the reliability coefficient over the same period for this variable was +0.25 only (see Valenzuela, 2011, for further details). Put another way, it is more likely that values operate as a moderator of issue salience judgments (again, as cued by the media) than the other way around. Of course, future experimental work could confirm this causality quandary.

Values and the Consequences of Agenda Setting

Lastly, I examine the consequences of the moderating role of values in the agenda-setting process for the development of attitudes and behaviors among citizens, also known as media priming effects. Previous research on this area has demonstrated that the media can have a decisive impact on the outcome of elections by altering the criteria used by citizens to evaluate candidates and political parties (Iyengar & Kinder, 1987; Krosnick & Kinder, 1990; Miller & Krosnick, 1996). A popular example is the 1992 U. S. presidential election campaign. According to polling data, the public perceived the domestic economy as poor early in the race (by late 1990) but only when the news media shifted attention from foreign policy to the economy in 1992 did President Bush's approval ratings plummet (Hetherington, 1996). From the perspective of the values-issues consistency hypothesis, it could be expected that voters with stronger materialist values should have been more primed by the coverage of the economy than individuals with stronger postmaterialist values. This is because the transfer of salience from the media agenda to the public agenda of the country's economic situation is augmented for materialist individuals and, thus, it is easier for them to use the economy as a standard on which to evaluate the president's job. Figure 3.2 depicts such a model of priming effects in terms of values moderating the paths of influence between media coverage and issue salience, and between issue salience and individuals' evaluations (in addition, of course, to the main effect of values on salience judgments and attitudes and opinions). Thus, values are conceived as affecting the priming process at two stages. In the first stage, as explained earlier, values influence the magnitude of agenda-setting effects. In the second stage, values help explain differences in the impact of issue salience on political attitudes and opinions.

One way of assessing the empirical validity of these propositions would be to replicate previous media priming studies where the influence of the media in shaping citizens' judgments of presidents and government has been documented, with the novel addition of a path between values and the general process of media effects. Unfortunately, most existing priming research has been conducted using survey data from the American National Election Studies, which does not usually include proper measures of materialism and postmaterialism. This has the regrettable consequence of preventing a direct estimation of the model depicted in Figure 3.2. Thus, media priming studies using other data sets, such as the Canadian Election Studies (e.g., Mendelsohn, 1996), could be explored in future research.

Another route for connecting values with priming effects is through attribute agenda setting and the notion of compelling arguments. As McCombs (2004) and others have argued, issues in the news have

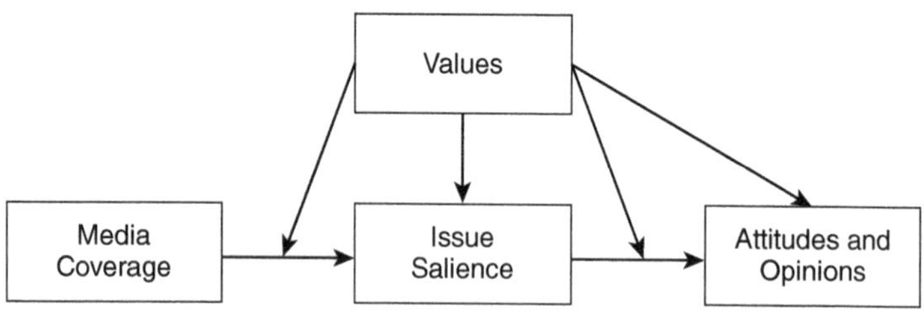

Figure 3.2 A model for estimating the intervening role of values on media priming and agenda setting

both substantive and affective attributes. Substantive attributes are the cognitive elements of messages that describe the denotative characteristics of an object. The affective dimension is the positive, neutral, or negative tone in the descriptions of an object's substantive attributes. Some attributes are emphasized, others receive less attention, and many receive no attention at all. Just as objects vary in salience, so do the attributes of each object. Thus, for each object there also is an agenda of attributes.

In many instances, the issues emphasized in the news have moral attributes, either substantive—as when a politician is portrayed in terms of his or her personal honesty—or affective—as when public figures are shown as corrupt. The moral attributes of these stories, be they substantive or affective, can operate as compelling arguments (Ghanem, 1997), resonating with the public's values so strongly that the salience of these attributes affects not only the salience of the issue but also the direction of opinion toward that issue. More germane to the values-issues consistency hypothesis, it could be argued that there are materialist and postmaterialist attributes in news stories that may function as compelling arguments. For instance, at the beginning of the Iraq War, news coverage about the existence of "weapons of mass destruction" could be thought of as emphasizing a materialistic attribute (i.e., an aspect related to safety and national security), whereas coverage about the promotion of democracy in Iraq could be regarded as a postmaterialist attribute (i.e., values of personal freedom and expression). A testable proposition would thus involve comparing the degree to which either of these attributes resonated with the American public.

Previous experimental work by Domke and colleagues (1998) lends some credence to the links between news attributes, values, and political attitudes. In their study, they argued that the process of candidate choice was influenced by an interaction between individuals' values and the moral framing of news stories related to the candidates. To test this, they had participants read stories that presented the same information about the candidates with varying emphasis on ethical dimensions. The analysis suggested that voters with an ethical interpretation of issues (e.g., discussing it in terms of religious morals or human rights) used values as the main yardstick on which to evaluate the candidates; that is, value resonance was the main driver of vote choice. In contrast, participants with a materialist interpretation (e.g., discussing the issue in economic terms) weighted values and other considerations about equally when it came to evaluating candidates and deciding for whom to vote.

Certainly, a more specific test of these propositions is warranted. The important point made here is that values can alter the process of attribute agenda setting and priming, which are compatible processes of attitude formation and opinions.

Conclusion

To date, most agenda-setting research has traced variance in the transfer of salience from the media agenda to the public agenda back to the concept of need for orientation—the intuitive idea that people's judgments about the relevance and uncertainty about issues is the main regulator of their acceptance or rejection of the media agenda. However, there are other psychological processes at play when people decide which issues are most important to them. I have suggested that human values may prove to be a valuable direction for future research in the agenda-setting tradition.

This is an important research topic because it sheds light on the origins of issue salience, which previous research has found to influence vote choice and policymaking (Valenzuela, 2009; Walgrave, Soroka, & Nuytemans, 2008). The initial evidence collected in Valenzuela (2011) suggests that values moderate the process by which people determine the most important issues of the day, that is, issues that deserve to be addressed by government and policymakers. From a normative perspective, this is good news. Rather than a passive electorate that blindly follows media cues, the proposition here is that people actively filter media cues by relying, in part, on their own values. From a more practical perspective, this line of research is relevant, too. For instance, the values-issues consistency hypothesis suggests that media and campaign agendas that resonate with individuals' values have a higher likelihood of being accepted by the public.

There are several benefits of bringing in values to agenda-setting theory. First, in the spirit of parsimony, we can build a bridge between agenda setting and value change theory, well-traversed paradigms in political communication and political science, respectively. Second, studying contingent conditions of media effects is usually a rewarding experience, as it allows us to have a more fine-grained answer to the question of how media affects us. Last the evidence collected so far suggests that the relationship between values and agenda setting operates not only at the aggregate level—the traditional emphasis of agenda-setting research—but also at the individual level, the proper level of analysis for psychological moderating variables, and both in the real-world context of surveys and the artificial context of the laboratory. Thus, the study of values has followed a two-pronged approach that future research should emulate in order to bolster the robustness of findings related to moderators of agenda setting.

References

Allport, G. W. (1961). *Pattern and growth in personality.* New York: Holt, Rinehart & Winston.

Atkin, C. K. (1985). Informational utility and selective exposure. In D. Zillman & J. Bryant (Eds.), *Selective exposure to communication* (pp. 63–87). Hillsdale, NJ: Lawrence Erlbaum Associates.

Domke, D., Shah, D. V., & Wackman, D. B. (1998). "Moral referendums": Values, news media, and the process of candidate choice. *Political Communication, 15*, 301–321.

Evatt, D., & Ghanem, S. I. (2001, September). *Building a scale to measure salience.* Paper presented at the annual conference of the World Association of Public Opinion Research, Rome, Italy.

Ghanem, S. (1997). Filling in the tapestry: The second level of agenda setting. In M. McCombs, D. L. Shaw, & D. Weaver (Eds.), *Communication and democracy: Exploring the intellectual frontiers in agenda-setting theory* (pp. 3–14). Mahwah, NJ: Lawrence Erlbaum Associates.

Hetherington, M. J. (1996). The media's role in forming voters' national economic evaluations in 1992. *American Journal of Political Science, 40*, 372–395.

Inglehart, R. (1977). *The silent revolution: Changing values and political styles in advanced industrial society.* Princeton, NJ: Princeton University Press.

Inglehart, R. (1990). *Culture shift in advanced industrial society.* Princeton, NJ: Princeton University Press.

Inglehart, R. (1997). *Modernization and postmodernization: Cultural, economic and political change in 43 societies.* Princeton, NJ: Princeton University Press.

Inglehart, R., & Welzel, C. (2005). *Modernization, cultural change, and democracy: The human development sequence.* New York: Cambridge University Press.

Iyengar, S., & Kinder, D. R. (1987). *News that matters: Television and American opinion.* Chicago, IL: The University of Chicago Press.

Krosnick, J. A., & Kinder, D. R. (1990). Altering the foundations of support for the president through priming. *American Political Science Review, 84*, 497–512.

McCombs, M. (2004). *Setting the agenda: The mass media and public opinion.* Cambridge, UK: Polity Press.

McLeod, J. M., & Sotirovic, M. (1998). Values as sociotropic judgments influencing communication patterns. *Communication Research, 25*, 453–485.

Mendelsohn, M. (1996). The media and interpersonal communications: The priming of issues, leaders, and party identification. *Journal of Politics, 58*, 112–125.

Miller, J. M., & Krosnick, J. A. (1996). News media impact on the ingredients of presidential evaluations: A program of research on the priming hypothesis. In D. C. Mutz, P. M. Sniderman & R. A. Brody (Eds.), *Political persuasion and attitude change* (pp. 79–100). Ann Arbor, MI: Michigan University Press.

Schemer, C., Wirth, W., & Matthes, J. (2011). Value resonance and value framing effects on voting intentions in direct-democratic campaigns. *American Behavioral Scientist, 56*, 334–352.

Schwartz, S. (1994). Are there universal aspects in the structure and contents of human values? *Journal of Social Issues, 50*, 19–45.

Schwartz, S. (1999). A theory of cultural values and some implications for work. *Applied Psychology: An International Review, 48*, 23–47.

Shah, D. V., & Domke, D. (1996). "To thine own self be true": Values, framing, and voter decision-making strategies. *Communication Research, 23*, 509–560.

Soroka, S. N. (2002). Issue attributes and agenda-setting by media, the public, and policymakers in Canada. *International Journal of Public Opinion Research, 14*, 264–285.

Taber, C. S. (2003). Information processing and public opinion. In D. O. Sears, L. Huddy, & R. Jervis (Eds.), *Oxford handbook of political psychology* (pp. 433–476). New York: Oxford University Press.

Tsfati, Y. (2003). Does audience skepticism of the media matter in agenda setting? *Journal of Broadcasting & Electronic Media, 47*, 157–177.

Valenzuela, S. (2009). Variations in media priming: The moderating role of knowledge, interest, news attention, and discussion. *Journalism and Mass Communication Quarterly, 86*, 756–774.

Valenzuela, S. (2011). Materialism, postmaterialism and agenda-setting effects: The values-issues consistency hypothesis. *International Journal of Public Opinion Research, 23*, 437–463.

Walgrave, S., Soroka, S., & Nuytemans, M. (2008). The mass media's political agenda-setting power: A longitudinal analysis of media, parliament, and government in Belgium (1993 to 2000). *Comparative Political Studies, 41*, 814–836.

Walgrave, S., & Van Aelst, P. (2006). The contingency of the mass media's political agenda setting power: Toward a preliminary theory. *Journal of Communication, 56*, 88–109.

Wanta, W. (1997). *The public and the national agenda: How people learn about important issues*. Mahwah, NJ: Lawrence Erlbaum Associates.

Weaver, D. H. (1980). Audience need for orientation and media effects. *Communication Research, 3*, 361–376.

Zaller, J. R. (1992). *The nature and origins of mass opinion*. New York: Cambridge University Press.

Zucker, H. G. (1978). The variable nature of news media influence. *Communication Yearbook, 2*, 225–240.

Chapter 4

Contingent Factors of Agenda-Setting Effects

How Need for Orientation, Issue Obtrusiveness, and Message Tone Influence Issue Salience and Attitude Strength

Yonghwan Kim

Agenda-setting effects have been supported across numerous topics and media types, although agenda setting does not work in the same manner across individuals. The extent to which people engage in information seeking may vary; and some people are more susceptible to agenda-setting effects than others. The concept of need for orientation (NFO) provides a psychological explanation for these differences and is "the most prominent of the contingent conditions for agenda-setting effects, those factors that enhance or constrain the strength of these effects" (McCombs, 2004, p. 67).

This chapter aims to extend the literature on the relationships between NFO and agenda-setting effects in several ways. First, beyond the basic relationship between NFO and agenda-setting effects, this study examines how NFO differently influences agenda-setting effects depending on issue obtrusiveness. Second, the present study explores the role of NFO for individuals' attitude strength, which has gone unstudied in the literature. Another purpose of this study is to explore the moderating influence of message tone (i.e., negative vs. positive). Studies have shown that negative messages are more powerful than positive ones in terms of forming opinions or attitudes because negative emotion activated by exposure to the negative tone of messages tends to facilitate individuals' information processing (e.g., Marcus, Neuman, & MacKuen, 2000). Given that few, if any, studies have examined the moderating role of message tone, the present study aims to see whether tone of message influences the association of NFO and attitude strength.

Need for Orientation and Agenda-setting Effects

The concept of need for orientation, which is grounded in the idea that individuals have an innate curiosity about the world around them, refers to the tendency of individuals to seek information about public issues from the media (McCombs & Weaver, 1973; Weaver, 1977, 1980). The NFO is a psychological explanation for how and why the media can influence the public's perceived issue salience. Agenda-setting research demonstrates that some people are more susceptible to agenda-setting effects than others because people do not have the same need for orientation. The greater people's desire for information, or NFO, the more attention they pay to the media agenda, which leads to the greater degree to which they accept the media agenda.

NFO is defined in terms of two lower-order concepts: relevance of the information to the individual and the individual's degree of uncertainty concerning the message (Weaver, 1977, 1980). The levels of NFO are determined by an individual's relevance and uncertainty. Basically, the higher the relevance and uncertainty, the higher the NFO, which in turn leads to greater agenda-setting effects. More specifically, Figure 4.1 outlines the levels of NFO according to each level of relevance and uncertainty. Low relevance results in a low NFO regardless of the degree of uncertainty; high relevance and low uncertainty results in a moderate NFO. NFO is highest under the condition of high relevance and high uncertainty (McCombs, 2004).

Theoretical explanations of NFO and agenda-setting effects suggest that individuals with a high need for orientation are more susceptible to media agenda-setting effects than those with a moderate or low NFO.

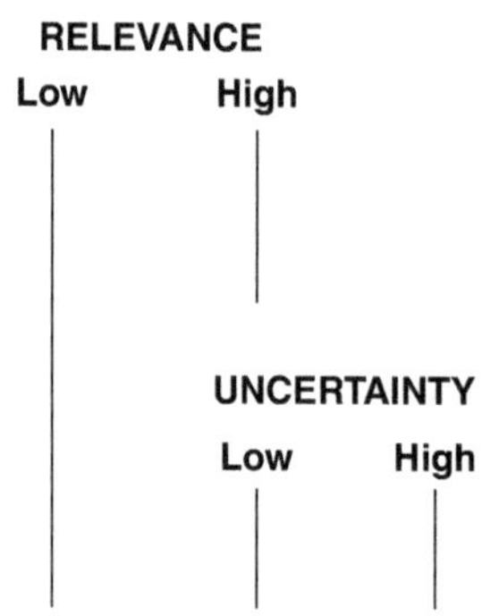

Need for Orientation:		LOW	MODERATE	HIGH
Charlotte				
Agenda-	June	+.15	+.35	+.41
Setting	October	+.29	+.59	+.68
Effects	Japan	+.39	+.62	+.86

Figure 4.1 Need for orientation and agenda-setting effects

These theoretical distinctions explain the pronounced agenda-setting effects in the Chapel Hill study (McCombs & Shaw, 1972), because the sample of the study was undecided voters with high uncertainty. Thus their NFO was high. Subsequent studies also demonstrated a positive relationship between individuals' NFO and agenda-setting effects. For example, as summarized in Figure 4.1, a study of the 1972 U.S. presidential election (Weaver, 1977) and a study in Japan (Takeshita, 1993) provided evidence that agenda-setting effects increased as NFO increased. Recent studies also support the positive relationship between NFO and agenda-setting effects. Matthes argued that NFO should be conceptualized in terms of three components, summed to create a single NFO scale: individuals' motivation or need for orientation toward the issue itself, specific facts or aspects of an issue, and journalistic evaluations (Matthes, 2006).[1] He found that this new NFO positively influenced the first level of agenda-setting effects (Matthes, 2008). A more recent study also found that the traditional NFO scale (based on relevance and uncertainty) and the new NFO subscale toward issues predicted basic first-level agenda-setting effects. Based on this literature, as previous research has demonstrated, this study expects a positive relationship between NFO and agenda-setting effects:

> **Hypothesis 1:** NFO will have an effect on the participants' perceived salience of an issue, which is the basic first level of agenda-setting effects.

Issue Obtrusiveness and NFO

Issue obtrusiveness, which refers to the extent to which issues obtrude into people's daily lives and are directly experienced, is an important factor in whether agenda-setting effects take place (Zucker, 1978). Mass media's agenda-setting effects tend to be larger for *unobtrusive* issues, with which people may not have direct experience and for which media are the only source of information, than for *obtrusive* issues, with which people have direct experience (Zucker, 1978). Subsequent studies focusing on issue obtrusiveness have supported this notion (e.g., Winter, Eyal, & Rogers, 1982). For example, unobtrusive issues, including government credibility, foreign affairs, the environment and energy, and race relations showed more substantial agenda-setting effects than obtrusive issues such as unemployment and taxes (Weaver, Graber, McCombs, & Eyal, 1981).

As scholars have advocated that the concept of issue obtrusiveness should be added to the agenda-setting theory (Severin & Tankard, 2001), the concept of obtrusive and unobtrusive issue can be linked to NFO, a key concept of agenda-setting theory. The concept of NFO provides explanations for differences in agenda-setting effects between obtrusive and unobtrusive issues.

> Because obtrusive issues are defined as issues obtruding into people's everyday lives, personal experience in many instances will sufficiently orient individuals to the situation at hand. The result is a low need for any additional orientation, a circumstance that predicts low correlations between the media agenda and public agenda. . . . On the other hand, personal experience is not a sufficient source of orientation for unobtrusive issues. For these, the theoretical assumption is that the media agenda is commonly the primary source of orientation, the source to which people turn to reduce their uncertainty. (McCombs, 2004, p. 62)

Whereas previous studies and recent research did focus on the relationships between NFO and agenda-setting effects, little is known about how NFO is differently associated with agenda-setting effects depending on issue obtrusiveness. What can be expected based on the above explanation of how NFO (relevance and uncertainty) works differently for obtrusive and unobtrusive issues (McCombs, 2004), then, is that relevance and uncertainty—two lower-order concepts of NFO—might work differently for obtrusive and unobtrusive issues. Swanson (1988) noted long ago that the NFO construct, on the one hand, "might predict that uncertainty would be highest for unobtrusive topics, since people have no direct experience with such topics, so unobtrusive topics should produce more agenda-setting. On the other hand, personal relevance should be highest for topics with which people have direct experience, so coverage of obtrusive topics should produce more agenda-setting" (pp. 607–608). Based on these notions, this study proposes that relevance and uncertainty differently predict agenda-setting effects for obtrusive issues and unobtrusive issues, and it asks about the interaction relationships of relevance and uncertainty on participants' perceived issue salience for both issues:

Hypothesis 2a: For an obtrusive issue (i.e., job), relevance will be a stronger predictor of agenda-setting effects than uncertainty.

Hypothesis 2b: For an unobtrusive issue (i.e., real estate), uncertainty will be a stronger predictor of agenda-setting effects than relevance.

Research Question 1: Do relevance and uncertainty interact in influencing participants' issue salience? If so, how do they work differently depending on issue obtrusiveness?

NFO, Attitude Strength, and Negativity

The literature is primarily limited to exploring the role of NFO in influencing first-level agenda-setting effects (e.g., Chernov, Valenzuela, & McCombs, 2011). Little attention has been devoted to the relationship between NFO and second-level (attribute) agenda-setting effects and attitudinal consequences of agenda setting. A notable recent exception is Matthes (2006), who created a new version of NFO composed of three dimensions: the NFO (1) toward issues, (2) toward facts or specific aspects of an issue, and (3) toward journalistic evaluations. With this new scale, he examined the role of NFO for both the first level and the second level of agenda-setting effects; NFO toward facts or specific aspects of an issue is the second level, as NFO toward issues is the first (see Matthes, 2008, for more about a rationale of why NFO might influence the second level of agenda setting). The study, however, found that NFO had no influence on the perceived affective attribute salience, although it predicted first-level agenda-setting effects (Matthes, 2008).

Matthes' (2008) findings indicate that NFO fails to influence which specific attributes (i.e., tone of information or affective attribute in the study) will be chosen among the audience, though it does predict that people will turn to media to gain information. These findings suggest that NFO influences how much media people consume, but not their perceived affective attributes.

> Put more bluntly, if you read more, that does not necessarily mean that you read more positive information. In contrast, you read more positive and negative information. That means, when an individual's NFO is high, the perceived salience of all possible issue attributes increases. Therefore, no assumptions can be derived which specific issue attributes will be the most salient ones for those news stories and articles a high NFO individual consumes. (pp. 450–451)

These explanations are important for extending the role of NFO to individuals' attitude strength. If NFO results in media attention and issue salience and all possible attribute saliences, this further suggests that people will begin to hold stronger or non-neutral attitudes as attention to media messages and both perceived issue salience and attribute salience (whether it is positive and negative) increase. Studies examining the attitudinal consequences of agenda setting have provided evidence that media attention and public issue salience are linked to stronger opinions or non-neutral attitudes (e.g., Weaver, 1991). Kiousis and McCombs (2004) found that increased media attention to political figures is correlated with higher public salience and attitude strength. Guided by Matthes' arguments that an individual's high NFO will lead to more attention to media, issue salience, and attribute salience regardless of direction of attributes, and the literature from attitudinal consequences of agenda setting, this study proposes:

Hypothesis 3: NFO will have an effect on the audience's attitude strength about a political candidate appearing in ads.

In addition to audience characteristics such as NFO, McCombs (2004) noted that media (e.g., newspapers and television) characteristics can be other contingent conditions of agenda-setting effects. Likewise, media content—such as tone of message—can influence the agenda-setting process. Marcus, Neuman, and MacKuen (2000) argued that political judgments are influenced by emotion. The basic proposition of their theory of affective intelligence is that individuals begin the process of responding to new stimuli (news media, political ads, other social events, etc.) with their emotional reactions. Their study revealed that subjects exposed to negative advertising were likely to display negative emotions, while those exposed to positive advertising showed no significant emotional changes. Despite evidence of the influence of tone of information, previous research has ignored the role of tone of messages as a potential moderator of influence of NFO. Thus, the current study asks:

> **Research Question 2:** Does tone of ads moderate the association of NFO with attitude strength?

Method

Design and Stimuli

To test the four hypotheses and the two research question, a 2×2 experiment was conducted. The first factor was issue obtrusiveness, with two issues ("job" for the obtrusive issue; "real estate" for the unobtrusive issue), and the second factor was the tone of ads (negative and positive). A convenience sample of 107 Korean young adults participated in this study. One may think that the use of a convenience sample of young adults is not ideal for generalizing the results to the population. However, one of the key purposes of the study is to explore how individuals' need for orientation works differently depending on issue obtrusiveness. Because young adults have similar direct and indirect experience concerning public issues—jobs and real estate, in this study—and thus they have similar levels of obtrusiveness toward each issue, this sample made it possible to explore the effects of issue obtrusiveness in the associations of the concept of NFO and agenda-setting effects. Of the participants, 43% were male and 57% were female; their ages ranged from 19 to 29 ($M = 24.5$), all of legal voting age.

The advertising stimuli were composed of text and photos containing either the job issue or the real estate issue with a negative or positive tone. The job issue and real estate issue were selected as the obtrusive and unobtrusive issue, respectively, because in general Korean young adults directly experience the process of getting a job—the job issue obtrudes more into their lives—while they only indirectly encounter the real estate issue, as they typically have not yet purchased a home. The emerging challenger in the 2007 presidential election was the sponsor of the advertisements (Candidate A), while the well-known leading candidate was the target of the negative advertisements (Candidate B). Ads with negative tone commented unfavorably on or negatively described a target candidate's personal traits and issue positions—e. g., "Candidate B claimed that he can overcome the economic crisis by providing a million job positions . . . but B is a person who dismissed lots of employees . . . the record showed that B did speculation in real estate." Conversely, the positive ads favorably stated a sponsoring candidate's characteristics and policy positions—for example, the positive ads stated "Candidate A will make strong regulations against speculation in real estate, which caused a serious economic crisis" and described a policy for making job openings and biographic profiles. A sponsoring candidate was presented in all four treatment ads, both attacking and self-promoting, making it possible to ask all participants their attitudes about a sponsoring candidate.

The 107 subjects were randomly assigned to one of the four treatment groups; and they were exposed to different advertisements in equivalent formats: (1) obtrusive issue (job) with negative tone; (2) obtrusive issue with positive tone; (3) unobtrusive issue (real estate) with negative tone; and (4) unobtrusive issue with positive tone.

Procedure

The experiment was conducted via email. First, participants were instructed to follow the procedure of the study. They were informed that this study is for understanding young adults' political consciousness and were asked to fill out pretest questionnaires asking key variables of the study, such as need for orientation (i.e., relevance and uncertainty of each issue) and obtrusiveness of each issue; and control variables such as demographics and pre-attitude about a candidate. Next, the advertising stimuli were sent to each group. After participants were exposed to the stimuli, they were asked to complete posttest questionnaires having items for issue salience (e.g., most important problem a country is facing), and attitude with regard to a sponsoring candidate.

Measurements

Issue salience. The key dependent variable of this study is issue salience. To explore the agenda-setting effects of political advertising, participants' beliefs about the importance of the two issues were measured in the posttest questionnaire. Participants were asked to rate their perceived importance of seven campaign issues—national economy, unity and national security, political reform, environment, education, unemployment, and real estate. Two target issues were unemployment (obtrusive) and real estate (unobtrusive); five issues served as distracters. Two items based on previous agenda-setting studies were used to measure participants' perceived issue salience. Participants were asked to respond on a 7-point scale "How important are the following issues facing this country?" In addition to the traditional MIP question, they also rated on a 7-point scale "To what extent do you think the following issues need the next government's additional action?" The scores of both items were added and divided by two to create an index of public issue salience (unemployment issue: $r = 0.635$, $p < 0.001$; real estate issue: $r = 0.655$, $p < 0.001$).

Need for orientation (NFO). Before being exposed to political advertising, participants were asked questions measuring the relevance and uncertainty toward two issues—unemployment and real estate. Relevance was measured by two items. Participants were asked how relevant the issues were to them on a 7-point Likert-type scale, where 1 was *not at all relevant* and 7 was *extremely relevant*. They also indicated the extent to which they were concerned about the issues on a 7-point Likert-type scale, where 1 was *no concern to me* and 7 was *a lot of concern to me*. The scores were combined and divided by two to create an index of issue relevance. The internal consistency of these two items was satisfactory both for unemployment ($r = 0.543$, $p < 0.001$) and real estate ($r = 0.661$, $p < 0.001$). Uncertainty toward the issues was gauged by asking participants the extent to which they know about the target issues on a 7-point scale anchored by 1 (*nothing*) and 7 (*a lot*). The items were reverse coded, so that a higher number indicates a greater level of uncertainty toward the issue. To create an index of NFO, the notion that relevance is the initial defining condition was considered—that is, NFO is low if levels of relevance are low regardless of the level of uncertainty, while, at high levels of relevance, NFO is moderate and high if uncertainty is low and high respectively. Following this, low NFO was coded 1; moderate NFO was coded 2; and high NFO was coded 3. This ordinal level of NFO has been the most commonly used approach in the agenda-setting literature (McCombs, 2004; see Figure 4.1 for the ordinal level of NFO by levels of relevance and uncertainty).

Attitude strength. Participants were asked to rate the extent to which they feel favorable toward a sponsoring candidate appearing in political ads on a 7-point scale in the posttest questionnaire where 1 = *very unfavorable* and 7 = *very favorable*. This item was folded into a 4-point scale, ranging from weak attitude (= 1) to strong attitude (= 4)—that is, 1 & 7 were recoded to 4; 2 & 6 were recoded to 3; 3 & 5 were recoded to 2; and 4 was recoded to 1.

Control variables. The participant's gender (57% females), age ($M = 24.5$, $SD = 2.0$), partisanship (36% partisans)—whether they identified themselves as partisan or independent—and pre-attitudes toward a sponsoring candidate with regard to his or her personal traits (e.g., leadership, sincerity, honesty, prudence, competence, knowledge, and intelligence) were included in the regression analysis predicting participants' attitude strength to control for potential confounds.

Data Analysis

In order to test Hypothesis 1, two sets of zero-order correlations were employed for each issue. To examine how relevance and uncertainty influence the participants' perceived issue salience both for obtrusive issue and unobtrusive issue (Hypothesis 2 and Research Question 1), I conducted two sets of hierarchical ordinary least squares (OLS) regressions, one for each issue salience. The main effects of relevance and uncertainty were entered in the first block, followed by interaction terms (relevance × uncertainty for each issue) in the second block. In addition, OLS regression analyses with a dependent variable of attitude strength were employed to test the effects of NFO on participants' attitude strength about a sponsoring candidate in the ads. To examine the interaction relationship between NFO and tone of ads on attitude strength, interaction terms (NFO for job issue × tone of ads, and NFO for real estate issue × tone of ads) were created and entered in regression models.

Results

The manipulation check, using *t*-tests, showed that participants rated negative advertisements more negatively than positive ads ($t = 7.352$, $p < 0.001$) and perceived the job issue as more obtrusive (had more direct experience or involvement with it) than real estate ($t = 12.039$, $p < 0.001$).

Hypothesis 1 predicted that NFO will affect the participants' perceived salience of an issue—participants with higher levels of need for orientation would show more pronounced agenda-setting effects. To test this hypothesis, two sets of zero-order correlations were performed for each issue. This showed that NFO has positive and significant influences on issue salience for both issues (job issue: $r = 0.46$, $p < 0.001$; real estate issue: $r = 0.40$, $p < 0.01$). As NFO toward job and real estate increased, participants' perceived issue salience of each issue increased after they were exposed to political ads describing those issues. Hypothesis 1 was supported.

Hypothesis 2a, which expected that relevance will be a stronger predictor of agenda-setting effects than uncertainty for an obtrusive issue, was supported. Multiple regression analysis in which relevance and uncertainty toward the job issue were entered simultaneously in the model (first block of the model 1) showed that relevance was a significant predictor of agenda-setting effects ($\beta = 0.452$, $p < 0.01$),whereas uncertainty was not significant ($\beta = -0.078$, $p =$ n.s.) (see Table 4.1). However, Hypothesis 2b, predicting that for an unobtrusive issue, uncertainty will be a stronger predictor of agenda-setting effects than relevance, was not supported. Regression analysis revealed that uncertainty toward real estate was not significant in predicting agenda-setting effects ($\beta = 0.152$, $p =$ n.s.), whereas relevance had a significant influence on agenda-setting effects ($\beta = 0.530$, $p < 0.001$) (see the first block of the model 2).

Table 4.1 OLS regressions predicting participants' job issue salience and real estate issue salience

	Job issue salience (Model 1)	*Real estate issue salience (Model 2)*
Block 1: Relevance & Uncertainty		
Relevance	.452 (.052)**	.530 (.040)***
Uncertainty	-.078 (.055)	.152 (.078)
Incremental R^2 (%)	48.1	22.6
Block 2: Interaction term		
Relevance × Uncertainty	.298 (.083)*	-.267 (.111)*
Incremental R^2 (%)	4.5	6.6
Total R^2 (%)	52.6***	29.2***
N	55	52

Note: Cell entries are final-entry OLS standardized coefficients; all entries in parentheses are standard errors. Job issue salience and real estate issue salience (dependent variables) were measured with an index of two questions: "How important are the following issues facing this country?" and "To what extent do you think the following issues need the next government's additional action?"

$^{*}p < 0.05$ $^{**}p < 0.01$ $^{***}p < 0.001$

In response to RQ 1 about the interaction between relevance and uncertainty in influencing agenda-setting effects, interaction terms (relevance × uncertainty for each issue) were created and entered in separate blocks of the regression analyses. As shown in Table 4.1, both interaction terms were significant. For a better understanding of the interactions between relevance and uncertainty, both significant interaction relationships were plotted. Figure 4.2 presents the interaction relationship of relevance and uncertainty toward the job issue on participants' perceived job issue salience. Those with high relevance and high uncertainty showed the greatest agenda-setting effects.

A similar analysis was employed for the significant interaction between relevance and uncertainty for the real estate issue. As presented in Figure 4.3, for those with low relevance, greater uncertainty showed greater agenda-setting effects than lower uncertainty. For those with high relevance, however, lower uncertainty produced greater agenda-setting effects than greater uncertainty. That is, agenda-setting effects were highest among participants with higher relevance and lower uncertainty.

Hypothesis 3, stating that NFO will affect the participants' attitude strength, was not supported. NFO for both the job issue and the real estate issue was not significant in influencing participants' attitude strength about a political candidate in the ads (NFO for job issue $\beta = -0.096$, p = n.s.; NFO for real estate issue $\beta = 0.109$, p = n.s.).

RQ 2, about the interaction relationship between NFO and tone of ads on participants' attitude strength, was examined by creating interaction terms (NFO for job issue × tone of ads, and NFO for real estate issue × tone of ads) and entering them in separate blocks of regression models. Interaction between NFO toward job issue and tone of ads was

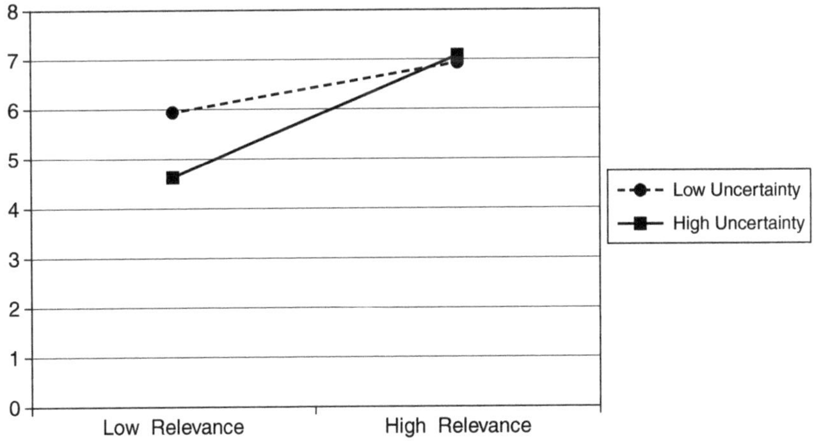

Figure 4.2 Interactions of relevance and uncertainty on job issue salience

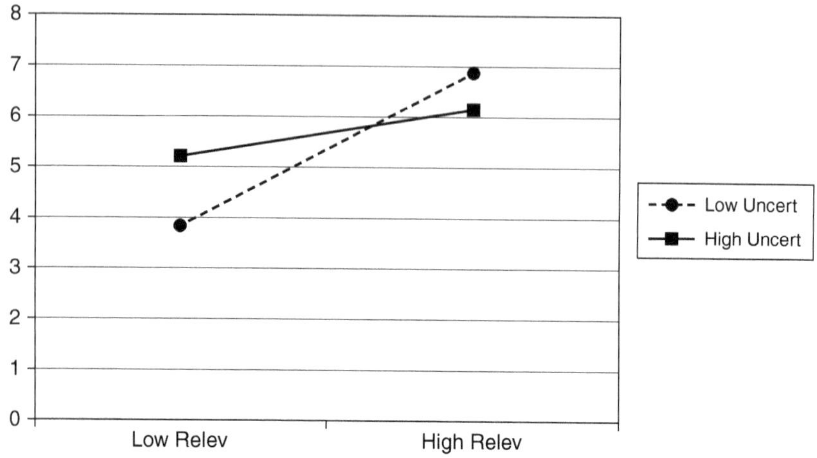

Figure 4.3 Interactions of relevance and uncertainty on real estate issue salience

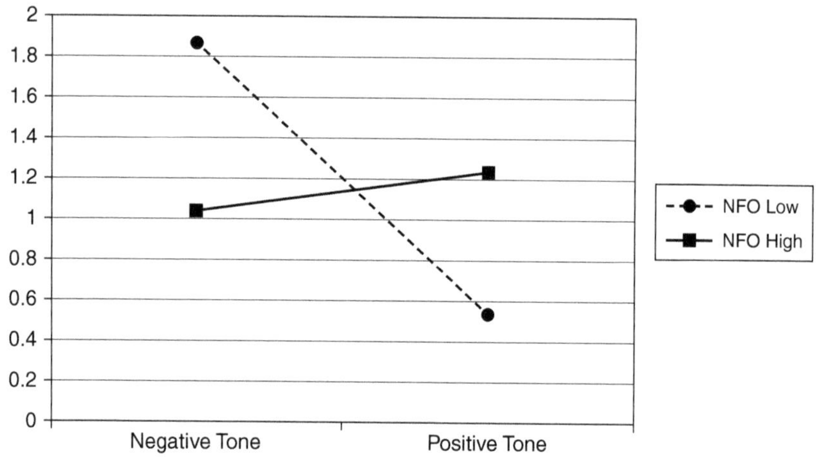

Figure 4.4 Interactions of NFO and tone of ads on attitude strength

found to be significant, although there was no significant interaction between NFO toward real estate issue and ad tone. Specifically, tone of ads (negative and positive) significantly moderated the relationship between NFO toward the job issue and participants' attitude strength about the sponsoring candidate of the ads ($\beta = 1.315$, $p < 0.05$). As shown in Figure 4.4, those with high NFO were less likely to be affected by the tone of ads even though, among them, those who were exposed to a positive ad showed a slightly stronger attitude than a group exposed to a negative ad. However, among those with low NFO, attitude strength was significantly higher for individuals exposed to a negative ad than for those exposed to a positive ad.

Discussion

This chapter attempts to expand the literature on need for orientation in the tradition of agenda-setting studies. First, following the traditional and most recent literature, this study expected that individuals' NFO would have a positive effect on their issue salience—that is, the first level of agenda-setting effect. Results support the idea that NFO positively and significantly affects individuals' perceived issue salience, which is consistent with previous studies. In particular, this significant role of NFO in predicting issue salience was found for both an obtrusive issue and an unobtrusive issue, indicating that NFO predicts basic first-level agenda-setting effects across issues.

Second, this study examined how relevance and uncertainty independently influence issue salience for both an obtrusive issue and an unobtrusive issue, and how relevance and uncertainty interact in influencing agenda-setting effects. Results showed that, regardless of issue obtrusiveness—that is, for both the obtrusive and the unobtrusive issue—only relevance was found to be a strong predictor of agenda-setting effects, while uncertainty failed to influence individuals' issue salience. These findings indicate that "relevance is the initial defining condition of need for orientation" (McCombs, 2004, p. 54) when it comes to predicting agenda-setting effects.

Furthermore, this study examined how relevance and uncertainty interact in influencing agenda-setting effects depending on issue obtrusiveness by employing separate regression analysis for the obtrusive issue (job) and the unobtrusive issue (real estate). Interestingly, results showed a different pattern of interaction effects between relevance and uncertainty on salience for the two issues. For the job issue (obtrusive), the greatest agenda-setting effects were found among those with higher relevance and higher uncertainty. These results support the idea that the greater individuals' need for orientation, the more likely they will show agenda-setting effects (McCombs, 2004), which is a linear relationship between individuals' need for orientation and agenda-setting effects.

While the results demonstrated a linear relationship between NFO and agenda-setting effect for obtrusive issue, the greatest agenda-setting effects were found at moderate levels of NFO for the real estate issue (unobtrusive issue)—that is, those who were in higher relevance and lower uncertainty produced the greatest level of issue salience. A possible explanation may be found in the characteristics of issue obtrusiveness. People would be more likely to desire more information during an election campaign even though they already know about public issues. In particular, if an unobtrusive issue appeared in campaign media such as news and political ads, people with knowledge about that issue would pay more attention because they already have some information resources to process; however, people with less knowledge about an unobtrusive issue may not pay much attention because they don't have a certain amount of information to process further and they don't have direct experience with that issue.

These results indicate that relevance and uncertainty work differently in influencing agenda-setting effects across issues, and that the association of individuals' NFO and agenda-setting effects would be either linear or nonlinear. In this study, linear relationships between NFO and agenda setting were found in an obtrusive issue while nonlinear associations were found in an unobtrusive issue. Further studies need to confirm these linear and nonlinear relationships between NFO and agenda-setting effects according to issue obtrusiveness, and further investigate what factors (e.g., content and context such as campaign period versus non-campaign period) would produce linear or nonlinear relationships between individuals' need for orientation and agenda-setting effects.

This study, as an exploratory examination, found no significant role of NFO in influencing individuals' attitude strength. However, this may be because of the moderating role of tone of ads in influencing the association of NFO and attitude strength. While there was no significant direct influence of NFO on participants' attitude strength, results showed significant interaction effects between NFO toward the job issue and the tone of ads on attitude strength. That is, participants with lower NFO tended to show significantly greater attitude strength when they were exposed to negative advertising than when they were exposed to positive advertising; however, subjects with higher NFO were less likely to be affected by the tone of ads. These results indicate that people are more likely to be affected by negative message tone than positive tone while they have low NFO toward an issue, forming a greater level of attitude strength. This is consistent with previous studies showing that negative information has more influence in forming individuals' perceptions and impressions (Skowronski & Charlston, 1989; Wanta, Golan, & Lee, 2004). And it supports the negativity bias in the theory of affective intelligence, which posits that individuals' negative emotions are likely to be activated when they are, in particular, exposed to negative messages, thus leading to greater effect on political attitudes and behaviors (Marcus, Neuman, & McKuen, 2000); but these effects of negativity were found only for those with low NFO. In sum, this study's findings suggest that individuals' need for orientation may interact with message characteristics (tone of ads, in this study) in influencing attitude formation or attitude strength; thus, it further suggests that future research should take into consideration potential moderating factors such as message tone and visual messages (e.g., Coleman & Wu, 2010) in the study of NFO, attribute agenda-setting effects, and consequences of agenda-setting effects.

This study has some limitations. A major one is that this study had relatively weak control of the experimental procedure, compared with a laboratory experiment, because it was conducted via email, which may lack external validity. Even though this may not be a fatal concern since all subjects voluntarily participated in this study, future studies would overcome this limitation by employing more controlled experimental designs such as a lab experiment. Another limitation may be the use of convenience samples of young adults, making a generalization of the study difficult. However, as mentioned earlier, young adults were recruited for this study examining issue obtrusiveness because young adults should have similar experience concerning public affairs, making possible a distinction between an obtrusive issue (job) and an unobtrusive issue (real estate). No significant direct relationship between participants' NFO and attitude strength was found in this study. A key reason for these results could be that this study measured attitude strength toward a sponsoring candidate appearing in ads rather than toward issues. Since NFO was conceptualized with participants' relevance and uncertainty toward job and real estate issues, it might be the case that NFO toward the issue would have led to individuals' increased attitude strength about each issue, rather than attitude strength about the candidate. Thus, future research should address this concern by measuring attitude strength about issues (i.e., opinion toward certain issues) for a better understanding of the relationship between NFO and the public's formations of attitude strength.

Another direction for future studies is to take other relevant psychological factors into consideration in this line of research. Though NFO remains the most studied psychological factor conditioning the strength of agenda-setting effects, other psychological variables can be related to agenda-setting effects, including issue salience and attitude strength. For instance, Valenzuela (Chapter 3, this volume) contends that human values (i.e., materialist versus postmaterialist) may moderate the agenda-setting process by demonstrating evidence that people with a materialist orientation, compared to postmaterialist people, showed greater agenda-setting effects for materialist issues (e.g., economy) whereas postmaterialist issues (e.g., environment) were more influential to postmaterialist individuals than materialist ones. He further suggests that individuals' values can be an antecedent of NFO and values may also moderate the influence between issue salience and individuals' attitudes and opinions (see Chapter 3 for details). Based on the results of the current study and these suggestions, future research could examine how individuals' NFO and values interact in influencing the formation of attitudes such as citizens' attitude strength. Testing the potential moderating relationship between other psychological variables and NFO in the agenda-setting process, therefore, should provide a new venue for future research.

Note

1 Examples of items for NFO toward issues, facts, and journalistic evaluations are "It is important for me to observe this issue constantly," "I want to know many different sides about that topic," and "It is interesting to see how several journalists comment on that issue" respectively.

References

Chernov, G., Valenzuela, S., & McCombs, M. E. (2011). An experimental comparison of two perspectives on the concept of need for orientation in agenda-setting theory. *Journalism & Mass Communication Quarterly, 88*, 142–155.

Coleman, R., & Wu, H. D. (2010). Proposing emotion as a dimension of affective agenda setting: Separating affect into two components and comparing their second-level effects. *Journalism & Mass Communication Quarterly, 87*, 315–327.

Kiousis, S., & McCombs, M. E. (2004). Agenda-setting effects and attitude strength: Political figures during the 1996 presidential election. *Communication Research, 31*(1), 36–57.

Marcus, G. E., Neuman, W. R., & MacKuen, M. (2000). *Affective intelligence and political judgment*. Chicago, IL: University of Chicago Press.

Matthes, J. (2006). The need for orientation towards news media: Revising and validating a classic concept. *International Journal of Public Opinion Research, 18*, 422–444.

Matthes, J. (2008). Need for orientation as a predictor of agenda-setting effects: Causal evidence from a two-wave panel study. *International Journal of Public Opinion Research, 20*, 442–453.

McCombs, M. E. (2004). *Setting the agenda: The mass media and public opinion*. Malden, MA: Blackwell

McCombs, M. E., & Shaw, D. L. (1972). The agenda-setting function of mass media. *Public Opinion Quarterly, 36*, 176–187.

McCombs, M. E., & Weaver, D. (1973). *Voters' need for orientation and use of mass communication*. Paper presented at the International Communication Association.

Severin, W. J., & Tankard, J. W. (2001). *Communication theories: Origins, methods, and uses in mass media* (5th ed.). New York: Addison Wesley Publishing.

Skowronski, J. J., & Charlston, D. E. (1989). Negativity and extremity biases in impression formation: A review of explanations. *Psychological Bulletin, 105*, 131–142.

Swanson, D. L. (1988). Feeling the elephant: Some observations on agenda-setting research. *Communication Yearbook, 11*, 603–619.

Takeshita, T. (1993). Agenda-setting effects of the press in a Japanese local election. *Studies of Broadcasting, 29*, 193–216.

Wanta, W., Golan, G., & Lee, C. (2004). Agenda setting and international news: Media influence on public perceptions of foreign nations. *Journalism & Mass Communication Quarterly, 81*, 364–377.

Weaver, D. (1977). Political issues and voter need for orientation. In D. Shaw & M. E. McCombs (Eds.), *The Emergence of American Political Issues: The Agenda-Setting Function of the Press* (pp. 107–109). St. Paul. MN: West.
Weaver, D. (1980). Audience need for orientation and media effects. *Communication Research, 7*, 361–376.
Weaver, D. (1991). Issue salience and public opinion: Are there consequences of agenda-setting? *International Journal of Public Opinion Research, 3*(1), 53–68.
Weaver, D., Graber, D. A., McCombs, M. E., & Eyal, C. H. (1981). *Media agenda-setting in a presidential election: Issues, images, and interest.* New York: Praeger Publishers.
Winter, J., Eyal, C., & Rogers, A. (1982). Issue-specific agenda setting: the whole as less than the sum of the parts. *Canadian Journal of Communication, 8*(2), 1–10.
Zucker, H. (1978). The variable nature of news media influence. In B. Ruben (Ed.), *Communication Yearbook 2* (pp. 225–240). New Brunswick, NJ: Transaction Books.

Chapter 5

Setting the Political Culture Agenda

The Impact of Media Use on Political Trust and Participation in Kosovo

Lindita Camaj

Agenda-setting research undisputedly gives mass media the credit for citizens' political learning (McCombs, 2004). Moreover, it emphasizes the consequences of agenda-setting effects for the formation of political attitudes (Iyengar & Kinder, 1987) and attitude strength (Kiousis & McCombs, 2004; Weaver, 1991). Yet, as Kiousis and McDevitt (2008) point out, "only a small body of research has examined agenda-setting behavioral outcomes" despite the importance of the research that expands "the boundaries of the theory to connect it with broader processes of public opinion, civic participation, and political development" (p. 482).

A few studies in this direction point to the potential of the agenda-setting theory to explain the link between media use, political attitudes, and political behavior. Still, they mainly conceptualize media effects in terms of change. First- and second-level agenda-setting studies explore citizens' knowledge gain through the transfer of salience of issues, objects and their attributes (McCombs, 2004), whereas priming and attribute priming research explores how citizens form and change their opinions as a consequence of these effects (Krosnick & Kinder, 1990). Kiousis and McCombs (2004) recognize that "The greatest shortcoming of media-effects scholarship on attitudes over the years has been the primary emphasis on attitude change," even though as they claim, "This is not the only facet of attitudes that researchers should examine when exploring the impact of mass media, especially in the context of agenda setting" (p. 38).

The diversification of the media outlets combined with the establishment of the Internet as an accessible source of information has substantially changed and arguably made the environment in which people seek out information more perplexing. Thus, as Mutz and Martin (2001) note, "As the number of potential news sources multiplies, consumers must choose among them, and that exercise of choice may lead to less diversity of political exposure" (p. 111). In this sense, the notion of selective exposure and reinforcement effects developed at the dawn of the communication research (Blumler, 1979; Klapper, 1960) is more relevant to the current environment than ever before.

The aim of this study is twofold. Relying on a theoretical framework that merges attribute agenda setting and affective priming, this study explores the consequences of media use for political trust and participation—two aspects of political culture critical for the development and stability of democracy (Diamond, 1993). Second, this study explores if the interaction between attribute agenda setting, affective priming effects, and political culture represents attitudinal and behavioral change or reinforcement. These interactions are explored through Structural Equation Modeling (SEM) techniques. Two SEM models are developed to probe the multiple influences of media source, need for orientation, media use, attribute agenda setting, and affective priming on political trust and participation as the outcome variables. The first SEM model measures these relationships for people who use independent (objective) media as their main source of political information, while in the second SEM model the same relationships are measured for people who primarily rely on partisan media for information gain.

This study represents the first attempt to link agenda-setting and priming effects to the development of political culture in a society undergoing political transition. The consequences of media use and agenda-setting effects for political culture can have important implications for established political systems around the world, but maybe most so for countries caught in the struggle of democratization. In countries undergoing democratic transition, citizens can be confused as to the rules of the game under the new political system, their expectations and attitudes toward the system, and their role in it.[1]

Agenda Setting and Political Culture

As originally defined by Almond and Verba (Almond, 1963), political culture refers to the psychologically internalized political orientations of a society toward social objects and processes. It includes "cognitive orientation" (knowledge of and belief about political system and its outputs); "affective orientation" (feelings about the political system and its performance); and "evaluational orientations" (judgments and opinions about political subjects) (Almond & Verba, 1963 p. 15). As such, political culture represents a cognitive map or a set of expressive symbols that orient people toward evaluating the operations of a political system (Samuel Beer, 1974; cited in Almond, 1989, p. 27) providing "structure and meaning to the political sphere" (Pye, 1965, p. 8). An important dimension of political culture is disposition toward authority and political trust in particular, which itself "drive(s) to the very heart of what democracy is about" (Diamond, 1993, p. 12). Gamson (1968) defines political trust as "the probability, P_b, that the political system (or some part of it) will produce preferred outcomes even if left untended. In other words, it is the probability of getting preferred outcomes without the group doing anything to bring them about" (p. 54).

Two theoretical traditions compete to explain the origins of political trust: cultural theories, a bottom-up approach that assumes political trust originates in beliefs rooted in cultural norms developed through early-life socialization (Almond & Verba, 1963; Easton, 1965; Inglehart, 1997), and institutional theories, a top-down approach that regards political trust endogenous and thus shaped by the performance of political institutions (Coleman, 1990; Dasgupta, 1988; Hetherington, 1998). These theories are not mutually exclusive (Norris, 1999), however, as they "share a fundamental assumption that trust is learned and linked at some level to experience" (Mishler & Rose, 2001, pp. 37–38). According to "Lifetime learning models" (Mishler & Rose, 1997, 2001; Rose & McAllister 1990), the initial predispositions to trust, developed by pre-political experiences, may be reinforced or revised depending on the degree to which they are challenged by later experience with political (Diamond, 1999) and economic (Przeworski, Alvarez, Cheibub, & Limongi, 1996) performance of the institutions.

This study is interested particularly in the effect of performance perceptions on political trust. In this regard, media scholarship offers compelling evidence suggesting that media use for political information provides an important criterion for the formation of citizens' perception, attitude, and opinion regarding polity performance. The vigorous debate on the role of mass media in promoting and sustaining political culture and civic orientation (Almond & Verba, 1963) has produced contradicting results, though. On one hand, the "media malaise" thesis predicts that mass media cause political alienation by fueling citizens' cynicism (Patterson, 2003; Robinson, 1976). The "mobilization" approach, on the other hand, asserts that media contribute to citizens' political awareness, interest, learning, efficacy, and participation (Bennett, Rhine, Flickinger, & Bennett, 1999; O'Keefe, 1980; Moy & Pfau, 2000, Newton, 1999; Norris, 2000).

Recent literature on media effects on political mobilization suggest that people are at the same time simultaneously educated and confused, motivated and alienated, empowered and politically weakened by their exposure to the media and its interaction with their more deeply held knowledge, values, and beliefs (Aarts & Semetko, 2003; Delli Caprini, 2004). The relationship between media use and political attitudes is highly dependent on the type of content citizens consume (Aarts & Semetko, 2003; Bennet et al., 1999; Johnson & Kaye, 1998; Moy & Pfau, 2000) and on audience characteristics (Bennet, 1998; Moy & Pfau, 2000; Norris, 2000; Pinkleton & Austin, 2001, Pinkleton,

Austin, & Fortman, 1998). These findings accumulated through the years support the "dual effects hypothesis"(Aarts & Semetko, 2003), according to which "viewing behavior separates the more knowledgeable, the efficacious, and the politically involved from those who are not, revealing what might be described as a 'virtuous circle' for some and a 'spiral of cynicism' for others" (p. 778).

Agenda-setting theory can help refine and integrate existing models by providing a framework that explicates media impact on political cognitive and affective orientations, and its consequences for citizens' participation in political processes. Through the process of agenda-setting effects, mass media provide their audiences with "food for thought," engaging public thinking and discussion (McCombs, 2004). By emphasizing certain attributes or characteristics while describing objects or issues, mass media draw their audiences' attention to those properties so that when people think or talk about those issues (or objects) the same attributes are salient in their minds (Weaver, McCombs, & Shaw, 2004, p. 259). Thus, the issue/object agenda setting focuses on the early step in opinion formation—gaining the public's attention—while the attribute agenda setting elaborates on the comprehension step of this process (McCombs, 2004; Takeshita, 1997). Hundreds of empirical studies have demonstrated that mass media play the "civic teacher role" in a democratic society by setting citizens' agenda on issues and subjects and their respective attributes—the first step in public opinion formation and participation (McCombs, 2004).

Consequences of Agenda Setting for Political Culture

Media Effects on Opinion Formation

Adopted from cognitive psychology, priming theory provides a psychological link between agenda-setting effects and the formation of political judgment by offering a comprehensive explanation on how citizens formulate their political attitudes as a consequence of media content usage (Kim & McCombs, 2007; Kim, Scheufele & Shanahan, 2002; Krosnick & Kinder, 1990). By calling attention to some issues, problems, or personal attributes and ignoring others, mass media provide their audiences with specific political knowledge they tend to use when forming political judgments. In this manner, mass media constantly create new foundations for judgment, thus "changing the standards that people use to make political evaluations" (Iyengar & Kinder, 1987, p. 63).

Increase in the salience of an issue and the valence associated with it are two separate aspects of media priming mechanism (Pan & Kosicki, 1997, p. 25). Thus, a comparative parallel exists between first- and second-level agenda setting with priming and attribute (or affective) priming (Kim et al., 2002; Kim & McCombs, 2007; Sheafer, 2007). Kim et al. (2002) define attribute priming as the process during which "certain issue attributes emphasized in the media will become significant dimensions of issue evaluation among the public" (p. 12). During the attribute agenda-setting process, media make various aspects of issues and traits of institutions more or less accessible, priming the pieces of information people will rely on when forming opinions about political institutions.

Mass media have the potential to transfer to the audience cognitive and affective attributes of political issues and objects they cover in their reports (Kiousis, 2005; Kiousis, Bantimaroudis, & Ban, 1999; McCombs, Lopez-Excobar, & Lamas, 2000, McCombs, Llamas, Lopez-Escobar, & Rey, 1997). Kiousis et al. (1999) define cognitive attributes as "pertaining to those characteristics of news that help us cognitively structure news and discern among various topics," whereas affective attributes are referred to as "those facets of news coverage that elicit emotional reactions from its audience members" (pp. 416–417). Media can transfer the qualification and personality traits of political objects (Kiousis, 2005; Kiousis et al., 1999) as well as the positive and negative valence of those traits.

Empirical evidence suggests that media use can induce positive and negative evaluations of institutional performance based on the valence of issues emphasized in media reports. Pan and Kosicki (1997) found that the affective valence associated with issues emphasized in the media is an important factor for the evaluation of the U. S. president. Kim (2005) proposed the "attitude structure modeling" as an alternative to assessing priming effects, which tested the directionality or the tone of the primed concepts when judging the president's performance during the Gulf War. He found that when making a judgment, people attend to all aspects of the issues, not only the issues salience per se. Sheafer (2007) argues for the affective compelling arguments according to which "an increased salience of a certain issue in the media signals voters about the direction of the evaluation of the incumbent's performance" (p. 35). Thus, the attribute priming effects occur through a combination of message strength and direction. Kim and McCombs (2007) applied the concept of attribute priming to the evaluation of the political candidates during the 2002 Texas gubernatorial and U. S. senatorial elections. They found that candidate attributes that received extensive media attention were more likely to affect attitudinal judgments about candidates, and that the positive and negative sides of the attributes were related to the corresponding opinions.

The present study tests the relationship between media use and the transfer of the positive institutional attributes. Second, it relates media use and attribute agenda setting with the positive priming on institutional issue performance. Based on theoretical assumptions and empirical evidence presented above, this study is guided by the following hypothesis:

> H1a: There will be a significant correlation between media use and agenda-setting effects on positive institutional attributes.
> H1b: There will be a significant correlation between media use and priming of positive issue performance by institutions.

Moreover, the present study assumes that by emphasizing different attributes of political institutions—their performance and integrity in dealing with issues—mass media prime citizens to rely on that information when forming their attitude of trust toward institutions. When the primed institutional and issue attributes in the media are positive, then trust in institutions will be higher.

> H2a: Agenda-setting effects on institutional positive attributes will be positively associated with higher trust in political institutions.
> H2b: Positive evaluation of institutional issue performance will be positively associated with higher trust in political institutions.

Agenda Setting and Political Participation

The concept of the two-step model that linked agenda-setting theory with behavioral effects during marketing campaigns (Ghorpade, 1986; Sutherland & Galloway, 1981) has been successfully implemented and supported in political communication research as well (Ghorpade, 1986; Roberts, 1992; Sheafer & Weimann, 2005). Ghorpade's (1986) study of the 1984 North Carolina senatorial race found that voters reasons in choosing a candidate were positively correlated with political advertising, implying that the transfer of salience can lead to voting outcomes. In the 1990 Texas gubernatorial election study, Roberts (1992) found that issue concerns emphasized in the media were associated with citizen's votes for candidates, concluding "that the mass media may not only tell us what to think about, but they influence what actions we take regarding those thoughts" (p. 878).

Other studies imply that the consequences of agenda setting are not confined to the vote choice of the already participating voters, but can also mobilize citizens into different types of participation. Kiousis and McDevitt (2008) claim that agenda setting provides a "theoretical model that reaches even farther out into the realm of political behavior, with agenda setting leading to electoral engagement among first-time voters" (p. 498). Their study suggested that more attention to the news media led to knowledge and emphasis of the Iraqi issue, stronger feelings about governmental handling of the Iraqi issue, and voting turnout. Similarly, Weaver (1991) found that during 1988, media salience of the federal budget deficit was correlated with public knowledge and opinion about this issue, and with "likelihood of participating in political behavior such as signing petitions, voting, attending meetings, and writing letters regarding the deficit issue" (p. 66).

Similarly, relying on a sophisticated analytical model that engaged individual-level data, Moon (Chapter 8, this volume) found a positive relationship not only between agenda-setting effects and political engagement, but also for civic participation. However, while agenda-setting effects showed a significant relationship with civic participation, just sheer media attention did not, implying that "only those audiences who, themselves, recognize the importance of issues highlighted in the news are willing to join community meetings and other organizations" (p. 22).

Sheafer and Weimann (2005) added another step to the hierarchical effects that connect agenda setting with behavior through priming. Their data from Israeli elections suggested that "By emphasizing certain issues and not others, the media may thus influence electoral results, because it appears that people tend to vote for parties that own the issues primed by the media" (p. 360).

Through agenda-setting effects, mass media transfer the salience of certain (problematic) issues that need to be addressed through concrete actions. Thus, agenda-setting effects fulfill the first criterion for participation—the citizens' feeling that institutions need to be influenced to address these problems. Moreover, through attribute agenda-setting and affective priming, media empower citizens with the necessary knowledge and strong attitudes that facilitate and fuel participation.

H3a: Attribute agenda-setting effects will be positively correlated with political participation.

H3b: Attribute priming effects will be positively correlated with political participation.

Besides the specific level of political trust toward government or political leadership, scholars emphasize the importance of also distinguishing the diffuse levels of political trust, those at a more abstract level (Easton, 1965). These are deeply rooted perceptions about the rules of the game in a political system as a whole, thus they are resilient to the perceptions of daily governmental performance. High levels of trust in the core principles of democracy at the diffuse level are needed for a greater participatory tendency (Norris, 1999). Yet, the relationship between specific type of trust (trust toward political authorities and leadership) and political participation is not so clear cut. As Gamson

(1968) claims, excessive trust in authorities can diminish citizens' necessity to take action. High levels of trust can lead to apathy and the loss of citizen control of government, both of which undermine democracy (Gamson 1968, pp. 46–8). According to him, "a combination of high sense of political efficacy and low political trust is the optimum combination for mobilization" (Gamson, 1968, p. 48). Empirical studies imply that low assessment of institutional performance and low levels of trust in government can fuel citizens' feeling that action is necessary (Catterberg, 2003).

H4: People with low trust in political institutions will be more likely to participate in political processes than people with high trust in institutions.

Agenda Setting, Priming, and the Reinforcement Hypothesis

Multiple contingent factors have been found to mediate agenda-setting effects of mass media, among which media exposure and need for orientation are the "only two contingent factors. . .for which there is totally unambiguous evidence" (Winter, 1981, p. 239). Need for orientation (NFO), a combination of relevance and uncertainty (McCombs & Weaver, 1973), explains individual differences in agenda-setting effects by clarifying the reasons that guide people's media use in the first place. According to the NFO concept, people who are very interested in politics but are uncertain about it, have a high tendency to turn to mass media in order to understand the political environment in which they live (McCombs & Weaver, 1973; Weaver, 1977; Weaver, 1980).

Research shows that need for orientation explains variations in media effects on political discussion and political knowledge (Weaver, 1980), and it explains variations in voters' adoptions of the media issue agenda during an electoral campaign (McCombs & Weaver, 1973; Weaver, 1977) and in a non-electoral context (Matthes, 2008). A recent study (Kim, Chapter 4, this volume) published in this volume examined the interaction effects between NFO and issue obtrusiveness and the role of the NFO for individuals' attitude strength. The results suggested that NFO was significantly related to both obtrusive and unobtrusive issues, however, this relationship was significant only for the relevance part of the NFO.

The concept of NFO is primarily based on the cognitive utilitarian concepts that explain people's media information-seeking behavior based on utility. Blumler (1979) defines this disposition as "cognitive orientation." In the case when people's information needs are at the cognitive level, the people generally turn to the mass media for environment surveillance and "reality exploration." In the context of politics, they might turn to mass media to learn more about governmental policies, party platforms, or institutional performance. In this case, as Blumler claims, "*cognitive motivation will facilitate information gain*" with "the person who is more *strongly* and more *exclusively* moved to consume media materials for their informational content is more likely to acquire knowledge from them" (1979, p. 18). Thus, this line of research explores the use of the overall media content for information gains of the people who are highly interested in politics but do not necessarily have much knowledge or attitudes about it.

Yet, Blumler also emphasizes the "personal identity" function of mass media. Often people have the tendency to use media content to highlight some important facets of audience's interests or life situations. In the context of political environment when the audience has well established attitudinal predispositions, people will be more likely to rely on them when they "determine the communications to which the individual is exposed" (Klapper, 1963, p. 67). Empirical research points to political ideology as a strong predictor of media use, suggesting that people with established political ideologies expose themselves primarily to information consonant with their political predispositions (Mutz, 2006; Mutz & Martin, 2001). As Mutz and Martin (2001) claim, "selective exposure clearly occurs under the right real world conditions; when people have a choice, they tend to use it to reduce their exposure to cross-cutting political views" (p. 109).

However, according to Mutz (2006), the consequences of selective exposure for public attitudes have not yet been well studied" (p. 230). Earlier scholars theorized that due to the audience selective exposure, mass media can act as a "more likely agent of reinforcement than change" (Klapper, 1963, pp. 68–69). Blumler (1979) elaborates that "*involvement in media materials for personal identity reasons is likely to promote reinforcement effects*" (p. 19, italics by the author). Citing McGuire (1974), Blumler claims that "when people throw their identities into mass communication offerings, more often than not they will probably seek (and therefore presumably find) a reinforcement of what they personally appreciate, stand for, and value" (p. 20). Slater (2007) proposes the *reinforcing spirals* perspective, which conceptualizes the relationship between media selection and media effects as a dynamic, mutually reinforcing process. According to this model,"the use of a particular type of media content is likely to reinforce the needs and preferences that led to its use in the first place, helping sustain those particular needs and desired gratifications" (Slater, 2007, p. 299).

Selectivity of media use has been mainly connected with the polarization of citizens' political attitudes. Partisans are more likely to consume consonant partisan news media and to express polarized attitudes toward the political objects (Mutz, 2006; Stroud, 2007, 2010). Yet, research also suggests that partisanship and media credibility mitigate media effects at both the cognitive and attitudinal levels as well. Kiousis and McCombs (2004) found that during the 1996 presidential election, the increased media salience of the candidates was highly correlated with the emergence of non-neutral attitudes toward candidates, and attitude polarization—the emergence of strong positive and negative attitudes at the far end side of each other. Iyengar and Kinder (1987) found that partisanship was one of the important contingent factors for agenda setting effects, with independents undergoing the highest agenda-setting effects when exposed to the network news in an experiment. Yet, Iyengar and Kinder (1987) fell short of examining the mediating role of partisan media for agenda-setting effects. Partisans are more likely to trust consonant partisan media sources, while media credibility is found to be an important contingent factor for agenda-setting susceptibility (Wanta, 1997).

Similar results have been found in media priming studies. An experiment participant who liked or disliked the candidate experienced different priming effects when exposed to messages that were explicitly for or against the candidate, suggesting a stronger effect of the resistance to dissonant messages (Bizer & Petty, 2005). Political leaning was found to be specifically important for the negative primes (Carpentier, Roskos-Ewoldsen, & Rosko-Ewoldsen, 2008). When asked to evaluate the presidency of Ronald Reagan, subjects who would agree with the messages exhibited priming effects whereas people who would disagree resisted these effects. Research also suggests that priming is most likely to occur among people who trusted the media the most (McGraw & King, 2003; Miller & Krosnick, 2000).

This literature provides solid theoretical ground to assume that people who consume partisan media have more established political attitudes, which guide media use in the first place, making those sources of information more reliable. Consequently, we can expect high agenda-setting and priming effects among people consuming partisan media, which serve to reinforce previously held attitudes and opinions about political objects. However, this does not imply that only people who use partisan media will be susceptible to agenda-setting and priming effects. We can expect similar results from people who rely on mainstream (objective) media for political information, especially the ones with higher NFO. Yet, we can expect the effects to be stronger for the first group.

<u>H5a:</u> People using partisan media will be more susceptible to attribute agenda-setting effects than people using mainstream media.

<u>H5b:</u> People using partisan media will be more susceptible to attribute priming effects than people using mainstream media.

Methods

Content Analysis

The goal of this study was to investigate news coverage of political institutions in Kosovo. This study was particularly interested in variations in media reports between independent (non-partisan) media and pro-government (partisan) media outlets. The following media organizations were included in the media content sample: three national television channels—*Kosovo Public Television* (RTK), *Koha Television* (KTV), and *Television 21* (TV21); two national daily newspapers—*Koha Ditore* and *Express*; and two national radio stations—*Radio Kosova* and *Radio Dukagjini*. *RTK* and *Radio Kosova* are both publicly owned broadcasting media, which have been under the control of the ruling party during the last five years in Kosovo. In the same line, the daily newspaper *Express* is aligned with the Democratic Party of Kosovo (PDK), the political party of prime minister Hashim Thaci, which has the majority in the Kosovo Parliament. It is expected that these three media outlets to have a more positive coverage of Kosovo institutions, given their allegiance with the ruling parties. On the other hand, *KTV* and *TV21* are both privately owned national television channels, as is *Radio Dukagjini* and the *Koha Ditore* daily newspaper. All four of these news media are considered to be the most objective and trusted media in Kosovo, independent of governmental and party pressure. A more objective/even critical reporting from these outlets is expected.

The political news content of the selected media was collected and analyzed from September 15 to October 24, 2010, six weeks prior to the administration of the survey (this is considered be the optimal span for agenda-setting effects to occur) (McCombs, 2004). The main evening news of all three national television channels was collected, but only political stories in each newscast were analyzed (excluding news stories on sports, culture, international news, and weather). Television news stories that mentioned one of the political institutions included in this study were manually selected for further analysis and coding. The same practice was implemented with radio news content. For the newspaper content, only the front-page stories that mentioned one of the political institutions were analyzed. About 710 television news stories (389 for RTK, 320 for KTV, and 289 for TV21), 399 radio stories (133 for RadioKosova, and 226 for Radio Dukagjini), and 347 newspaper articles (167 from *Koha Ditore*, and 180 from *Express*) were analyzed for this study.

Adopting the approach of McCombs et al. (2000), the unit of analysis for this study was each paragraph in a story that mentioned one of the institutions. The Kosovo institutions analyzed for this study included NATO military forces in Kosovo (KFOR), European Union Rule of Law

Mission in Kosovo (EULEX), local police, local courts, International Civilian Office (ICO), Kosovo Parliament, and Kosovo Government. The coding of the news content included a list of positive and negative institutional attributes consisting of corrupt/fair, honest/dishonest, caring/selfish, efficient/inefficient, and knowledgeable/unwise. Similar to previous agenda setting studies (Kim & McCombs, 20007; Kiousis, 2005; McCombs et al., 1997), the list of attributes captured two lines of institutional assessment: the job-related attributes or the degree to which institutions performed in the expected manner (efficient, knowledgeable) and personality traits or the degree to which the job performance was done within ethical bounds (fair, honest, caring). A long list of attribute synonyms and antonyms from conventional dictionaries was initially generated, which was expanded after the pretesting to include additional keywords that were relevant to the Albanian language linguistics (see Kiousis, 2005, for a similar approach).

Additionally, this study was interested in the issues that dominated the media agenda. The list of 18 issues coded for this study included political dialog with Serbia[2]; rule of law, corruption, foreign policy/independence recognitions[3], inter-ethnic relations[4], crime and violence, environment, health system, education system, economy/infrastructure, poverty/living standards, employment, energy policy, social security, agriculture, privatization of the publicly owned enterprises[5], administration issues, and elections. Each paragraph was coded for the presence of an issue and the valence of the issue (positive, neutral, negative).

A graduate student in sociology, an Albanian native speaker, was hired and trained to code 10% of the news content for the inter-coder reliability test. Scott's Pi inter-coder reliability coefficients were within the acceptable standards of 0.70 and higher (Riffe, Lacy, & Fico, 2005). The coefficients for institutional attributes ranged from 0.74 to 0.88; the coefficients for issues ranged from 0.80 to 0.97; coefficients for issue valence ranged from 0.72 to 0.77.

Survey Data

Social Research Kosova, a public opinion research company based in Pristina, was hired to implement the public opinion survey used in this study. The survey data were collected from October 22 to November 1, 2010, with a representative sample of the Albanian population aged 18 + living in the region of Pristina, the capital of Kosovo.

The total sample size for this study was 717, of which 550 respondents completed the survey. A stratified three-stage probability sampling strategy was implemented to collect the data. The first stratification was done according to type of settlement, with 65% of respondents from urban and 35% from rural settlements. Next, the households were selected by random route technique starting from polling stations (as designated in the last Kosovo elections prior to the survey). The household member interviewed for the study was selected using the Kish scheme developed for each household, a method that allows the data collection team to randomly select participants from households, avoiding sampling bias (Frey, 1989). The survey was conducted face-to-face at the respondent's home. Fieldwork control was done with 30% of total interviewed in all strata proportional to sample size. Sampling error calculated for this survey is +/- 3.7 at 95% confidence level.

Media use. Following McLeod and McDonald (1985), media use was measured with three different measures: media exposure, media attention, and public affairs content use. Media exposure was measured by asking "During a typical week, how many days do you read news in a printed NEWSPAPER/watch news on national TELEVISION/listen to news on national RADIO/read news on INTERNET, not including sports?" The answers ranged from 0 days to 7 days per week. Further, the respondents were asked: How much attention do you pay to news about national politics in a printed NEWSPAPER/national TELEVISION/ national RADIO/on INTERNET? The respondents could choose from the following answers: none at all, a little, a moderate amount, a lot, and a great deal. Finally, in an open-ended question, respondents were asked to identify the primary source of information for all media types: Which daily NEWSPAPER/national TELEVISION/national RADIO/ INTERNET page is your primary source of information for news? Two separate media use indexes (independent (nonpartisan) media and partisan media index) were created for television, radio, and newspaper use, combining the media exposure, attention, and source questions.

Need for orientation (NFO). Following previous theoretical specification that NFO is a construct of relevance and uncertainty (McCombs & Weaver, 1973; Weaver, 1980), two separate variables were created for these concepts, which were subsequently combined into a single measurement. Relevance was measured by a question that assessed respondents' political interest: How interested are you in information about what's going on in government and politics? If respondents were not interested at all, slightly interested, or moderately interested in politics news, they were classified in the "low relevance" category. If the respondents said they were very interested, or extremely interested in politics, they were put into the "high relevance" category.

The uncertainty measure was composed from two questions about the strength of respondents' political party identification. Respondents were asked if they identified themselves with one of the main political parties operating in Kosovo (Democratic League of Kosovo-LDK, Democratic Party of Kosovo-PDK, Alliance for the Future of Kosovo-AAK, Democratic League of Dardania-LDD, New Kosovo Alliance-ARK, "Vetvendosja," Other Party, No preference). Second, respondents were asked to what degree they identified themselves with the party. If respondents identified themselves as "very strong" or "strong" partisans of a particular political party, they were considered to have low uncertainty; if they claimed to be "somewhat strong," or "not very strong" partisans or did not have party preferences, they were categorized as highly uncertain. Finally, the two sub-measures were combined to form the NFO measure, which contained four values, with the lowest NFO value including respondents with low levels of political interest and low uncertainty and the highest NFO value including respondents with high uncertainty and high political interest.

Attribute agenda setting and priming indexes. This study is interested in agenda-setting and priming effects at the individual level, so it uses two alternative measures of media effects adopted by similar studies that measured agenda-setting effects at the individual level (Moon, 2009; Wanta, 1997). The two scales developed in this study are different from previous measures of attribute agenda setting and priming in that they do not differentiate between "cognitive" and "affective" attribute measures, but rather they represent a combination of cognitive/affective attributes of issues and objects. Similar to Moon (2009), the "*Attribute agenda setting index*" measures the distance between the salience of institutional attributes in the media and the emphasis on these attributes by the audience, capturing the basic concept of second-level agenda setting. More precisely, this index represents a scale that measures how closely associated the media and the public agendas are when it comes to the emphasis of the positive attributes of various institutions in Kosovo. Higher scores symbolize closer ties between media and public agendas, thus stronger attribute agenda-setting effects. Likewise, the "*Attribute priming index*" measures the distance between the emphasis of the positive valence of issues covered by the media and the likelihood that citizens will have a positive evaluation of institutional performance. When calculating this index, only the percentage of the positive coverage of the issues in the media is considered, whereas public opinion is measured by considering only respondents' positive evaluation of how the institutions are dealing with those issues. Higher scores in the "*Attribute priming index*" represent the public's tendency to rely on the positive tone of issues covered in the media when judging institutional performance, thus stronger attribute priming effects.

Attribute priming index. The first step in creating this index was to identify the most important issues in Kosovo and explore how citizens and the media evaluated institutional performance in relation to these issues. Survey respondents were asked in an open-ended question "What are the three most important problems in Kosovo?" This question was followed by "How do the institutions deal with those problems?" (not well at all, not that well, quite well, and very well). Four major issues on the public agenda were the economy, crime, corruption, and foreign affairs. The same four issues mentioned by the public were high on the news media agenda as well. Moreover, these four issues had the most expressed valence in the news reports, with the issues of "rule of law/crime" and "foreign affairs" containing the most positive coverage, and the issues of "economy" and "corruption" having the most negative news coverage of all issues.

These four issues were coded into four variables and each person who mentioned any or all of the issues was given the performance question score for each of the issues mentioned. For issues not mentioned by a respondent, they received a score of zero (0). For example, if a respondent mentioned corruption as an issue, and thought that institutions were handling this issue 'not that well' (answer 2), that respondent was given the score "2" for the issue of corruption. But, if the same person did not mention the issue of foreign affairs, he received the score "0" for this issue.

The second step in the process was to create media points for all four issues. Only the percentage of the positive coverage of each issue was taken in consideration. This score was then divided by 25. This represented 1/4 of the total (100%) possible chances on the performance question, because the question that measured citizens' opinion on institutional performance had four possible answers (1. not well at all, 2. not that well, 3. quite well, 4. very well). For example, if the issue of 'corruption' received 35% of positive coverage, then the media point was 35/25 = 1.4.

The third step involved creating the *distance value* between media points and opinion poll answers for each issue. This was calculated by subtracting the media point from the survey answer point. For example, if a respondent said that institutions are handling the problem of corruption "not that well" (answer 2 on the performance question), then 2–1.4 = 0.6 was the distance value. A distance value was created for each of the four possible answers to the "performance question" in the survey for all four issues included in the study.

The final step in the process was to weight the issue distance scores by the percentage of stories the issues received in the media during the six weeks prior to the survey. Issues and attributes that are more salient in the media should have a higher weight than issues emphasized less

(Wanta & Hu, 1994) and thus have a higher probability to influence citizens' judgment about institutional performance on those issues. For example, if the issue of corruption received 30% of the total media coverage of all issues, the weighted score for corruption was 0.30. Consequently, the distance values for the issue of corruption were multiplied by the media weight (0.60 x 0.30 = 0.18). The weighted distance values for all four issues were summed. However, not all respondents received a distance value for all issue, because for the issues not mentioned the respondents received the score zero (0). Thus, the summed distance value for all issues was divided by 4 (the number of total issues).

The final score was an estimate of the magnitude of attribute priming effects for each individual in our survey, with higher scores representing stronger attribute priming effects.

Attribute agenda-setting index. The procedures to create this index were similar to the ones used to build the *attribute priming index*. *Public opinion points* were extracted by a set of questions that examined citizens' evaluation of a range of five attributes for seven political institutions in Kosovo. Respondents were asked to what degree the following attributes describe Kosovo political institutions: corrupt/fair; dishonest/honest; inefficient/efficient; and selfish/caring. People emphasizing the negative valence of an attribute received the value "1" for that attribute, while the positive valence of an attribute received the value "2."

To create *media points*, only the percentage of the positive media coverage of each attribute was considered and divided by 50, because the questions that measured the salience of institutional attributes in the public opinion pool had two possible answers (1. negative valence and 2. positive valence). For example, if the government attribute 'corrupt' received 35% of positive coverage, then the media point was 35/50 = 0.7.

The *distance* value, calculated by subtracting the media point from the public opinion point, was weighted by the total media coverage of a particular attribute. For example, if a respondent said the government was "corrupt," the absolute distance (1-0.7 = 0.3) was weighted by 0.30 (30% of the total coverage of the government attribute 'corrupt'), producing the score 0.09 (0.30 x 0.30 = 0.09).

First, an *attribute agenda-setting index* was created for each institution by summing up the scores from each of their attributes. Second, an overall *attribute agenda-setting* was created by summing the individual institutions' indexes. Higher scores represented stronger agenda-setting effects.

Institutional trust. The operationalization of political trust in this study is similar to measures used in comparative literature that focuses on political institutions as a whole rather than on a particular institution leader (Catterberg, 2003; Mishler & Rose, 1997, 2001). The trust questions in this study were adopted from the New Democratic Barometer (developed by the Paul Lazarsfeld Society in Vienna) asking respondents: *"I am going to name a number of political institutions. For each one, could you tell me to what extent you trust each of these political institutions. Please, indicate on a scale from 1, for the complete absence of trust, to 7, for great trust."* The institutions listed were government, parliament, courts, police, ICO, EULEX, KFOR. For the purpose of the analysis on this study, a political trust index was created by summing trust scores in each of the institutions measured on a 7-point scale (Cronbach's Alpha: 0.804). The constructed variable is an index that goes from 7 to 49, with higher values expressing higher levels of trust in institutions.

Political participation. This study adopts the Verba and Nie (1987) definition of democratic political participation as citizens' acts that aim at influencing the government through institutionalized forms of participation. The measurements of institutional political participation for this study are similar to Verba & Nie's (1987) measurements that have successfully been used and validated by ANES studies in the United States. Thus, institutional participation was measured with the following four items: "How often do you attend political meetings or rallies," How often do you spend time working for a political party or candidate?," How often do you vote in national or local elections?" and "How often do you sign a petition?" The response choices were "often," "sometimes," "seldom," and "never."

Moreover, this study expands beyond Verba and Nie's (1987) limitation on political activities "within the system" to include political acts outside of the system that aim to challenge elites. The elite-challenging or uninstitutional forms of participation were assessed by measures developed by Barnes and Kaase (1979) and successfully used by some of the most influential studies on contentious politics (Verba, Schlozman, & Brady 1995). Each respondent was asked if they "have actually done, would do, might do, or would never do the following actions: join in boycotts, attend lawful demonstrations, join unofficial strikes, and occupy buildings or factories." An overall participation index was constructed after all participation measures showed acceptable internal consistency (Cronbach's Alpha: 0.793)

Control variables. Three demographic variables were selected as control variables in this study's analysis: age, education, and gender. Communication scholarship has demonstrated that people with different characteristics tend to have different media use habits, with age, education, and gender making a difference for news media use as a whole (Eveland & Scheufele, 2000) and news source/platform selection (Norris, 2000) . Further, political science literature points to the relevance of these three variables for political attitudes (Inglehart, Basanez, & Moreno, 1998) and participation (Norris, 1999; Putnam, 2000).

Age was measured in years whereas gender represented a binary variable (0 for females, 1 for males). Education was assessed with a four-point scale measure of "elementary school," "high school," "college," and "graduate school." Additional controls for partisanship and political interest were added in the analysis.

Results

Model specification and modification. This study employed Structural Equation Modeling (SEM) techniques to test the hypothesized assumptions, which provides the ability to test models with multiple dependent variables while testing coefficients across multiple between-subjects groups. Further, it provides the means to model mediating variables rather than be restrictive to an additive model and to include in the model the error terms. Finally, SEM statistics allow for the overall model testing rather than just testing coefficients individually, and SEM strategy provides comparisons of alternative models to assess relative model fit, which makes it a more robust statistical procedure than the mere regression analysis (see Kline, 2011). Two different SEM models were examined using AMOS software in order to investigate the relationship between media use, agenda setting, and priming effects with political trust and participation. The relative performance of tested models was determined across several "goodness-of-fit" statistic tests, which compare the observed covariance matrix to the one estimated on the assumption that the model being tested is true.

I started by fitting a saturated model with all structural paths freed up to be estimated for both models in our study, the "Independent Media Model" and the "Partisan Media Model." That is, all news media use variables were set to have a direct influence on agenda-setting and priming indexes, political trust, and participation. Then, each model was trimmed by removing all non-significant paths within both models, which we present in this study as our final models. As indicated in Table 5.1, these models fit the data best and explain the same amount of variance in the criterion variable (political participation) as the saturated and theorized models, while providing the most parsimonious solutions. Examining the performance of these models, we notice a decrease in the values of χ^2/df, RMSEA, and AIC in the trimmed models, indicating a better fit for our data. Given these results and the principle of parsimony, these models were accepted as final models to be discussed further.

Table 5.1 Fit statistics of the Structural Equation Models

	AIC	*CFI*	*RMSEA*	$\chi^2(df, p)$	χ^2/df	$R^2(\%)$
Independent Media Model						
Model 1: Saturated[a]	88.00	1.00	---	---	---	
Model 2: Theorized[b]	88.02	0.991	0.043	10.02 (df=5, p=0.075)	2.00	13
Model 3: Trimmed[c]	74.75	0.993	0.020	20.75 (df=17, p=0.238)	1.221	13
Partisan Media Model						
Model 1: Saturated[a]	88.00	1.00	---	----	---	
Model 2: Theorized[b]	98.549	0.973	0.075	20.549 (df=5, p=0.001)	4.110	16
Model 3: Trimmed[c]	97.928	0.957	0.055	39.93 (df=15, p=0.000)	2.662	17

Note: AIC is the Akaiki information criterion. RMSEA is the root mean squared error of approximation. R^2 is the squared multiple correlation coefficients for structural equations.
[a]This is the fully saturated model with all structural paths freed to be estimated.
[b]This is the model with only the theorized paths freed to be estimated.
[c]This is the trimmed model (final model) with all non-significant paths removed.

Results. Figure 5.1 presents the AMOS estimates of the structural relationships between NFO, independent media use, agenda-setting and priming indexes, political trust, and participation. Overall, this model fits the data very well, yielding a χ^2 = 0.238 (χ^2/df = 1.221), RMSEA =0 .020, and CFI = 0.993.

As predicted, the relationship observed in this model suggests that people with high NFO are more likely to use independent media for political information; the media use contributes to citizens learning about institutional traits and issue performance; which in turn affects political trust and participation. However, this relationship is not consistent across all media types.

The path for independent media showed a direct relationship between NFO and newspaper use only. The significant path coefficient between NFO and newspaper use (γ = 0.189, p = 0.000) suggested that people who were interested in politics but did not have strong partisanship ties were most likely to use independent newspaper for political information. However, the same results were not found for television and radio news. Contrary to previous research (Matthes, 2008), this analysis also found a significant direct relationship between NFO and *attribute agenda setting* (γ = 0.092, p = 0.03). This result suggests that people with higher needs for information were more susceptible to attribute agenda-setting effects, adopting the positive tone of institutional attributes reflected in the media coverage.

The use of newspapers had a positive relationship with *attribute priming index* (γ = 0.076, p = 0.070), while television news was negatively associated with *attribute agenda-setting index* (γ = -0.078, p = 0.060),. These data suggest that the people who relied on independent newspaper news were more likely to adopt the positive tone of issue coverage when judging institutional performance. Yet, the same people who watched independent television channels were less likely to adopt the positive tone of institutional attributes. Indeed, our content analysis data suggest media reporting on institutional attributes was more negative (58%) than their reports on political issues, where the majority of news items ranged from neutral (36%) to positive (32%).

Further, both *attribute agenda setting* (β = 0.639, p = 0.000) and *attribute priming* (β = 0.088, p = 0.005) measures yield significant positive association with political trust. That is, people who adopt the media agenda on institutional positive traits and issue performance were more trusting of political institutions in general. These two variables account for about 53% of variance in political trust. Finally, political participation had a positive relationship with *attribute agenda setting* measure (β = 0.393, p = 0.000), while it had a negative relationship with trust (β = -.115, p = .040). A significant positive direct relationship was also observed between participation and newspaper (β = 0.140, p = 0.000) and radio use (β = 0.074, p = 0.060). These variables account for 13% of variance in participation.

These results indicate that independent newspaper use had a dual relationship with trust and participation. It had a positive direct and indirect effect operating through the priming measure. Moreover, television news use for political information yielded only an indirect relationship with trust and participation that operated through *attribute agenda setting* (see Figure 5.1). Worth noting is that independent radio use was found to have only a slightly direct relationship with political participation (γ = 0.074, $p < 0.060$) when all other paths were considered simultaneously.

Figure 5.2 represents the results for the partisan media model. This model also yields a good fit to the data, with χ^2 = 39.93, 15 degrees of freedom, for a ratio below 3 (χ^2/df = 2.66); RMSEA = 0.055, and CFI = 0.957.

As predicted, this model suggested a significant positive relationship between NFO and partisan television use (γ = 0.082, p = 0.05) and partisan radio use (γ = 0.089, p = 0.03), even after controlling for media effects at other levels. These data suggest only an indirect relationship between NFO and attribute agenda-setting and priming effects, mediated via electronic media use.

The results presented here show a different relationship between partisan media use and other variables of interest, providing support to the thesis that people who use partisan media might be more susceptible to attribute agenda-setting and priming effects. Both partisan television use (γ = .0103, p = 0.001) and radio use (γ = 0.137, p = 0.01)

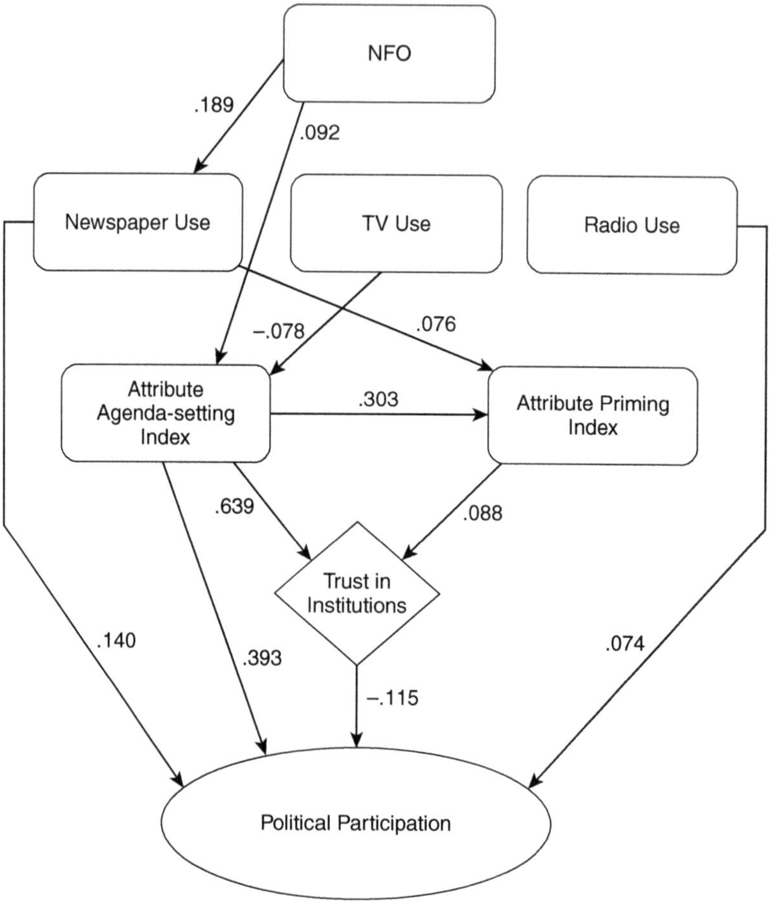

Figure 5.1 Structural Equation Model for independent media

had a significant positive relationship with the *attribute agenda setting* measure. Given that the *attribute agenda-setting index* measured the distance between media emphasis on institutional positive attributes and citizens' positive evaluations of institutional traits, our data suggest that the more people relied on partisan television and radio for their political information, the more likely they were to adopt the positive valence of institutional traits from the media agenda. Further, partisan newspaper use ($\gamma = 0.072$, $p = 0.08$), television use ($\gamma = 0.079$, $p = 0.05$), and radio use ($\gamma = 0.072$, $p = 0.05$), had a slight positive relationship with *attribute priming index*. This suggests that people who used partisan media for political information were more likely to rely on the positive valence of the issues emphasized in the news when judging institutional performance on those issues.

However, only the *attribute agenda setting* measure had a significant relationship with political trust (β = 0.719, p = 0.000) and participation (β = 0.375, p = 0.000). Similar to the "Independent Media model," partisan media use for political information had no direct effects on trust, while newspaper use (β = 0.219, p = 0.000) and radio use (β = 0.076, p = 0.050) had a significant positive direct relationship with participation. Overall, partisan television and radio use had an indirect effect on political trust operating through attribute agenda setting. Moreover, newspaper and radio use had a direct relationship with participation, as well as an indirect one, operating through agenda-setting and priming effects and trust. These variables account for about 52% of political trust variance and more than 16% of variance in participation.

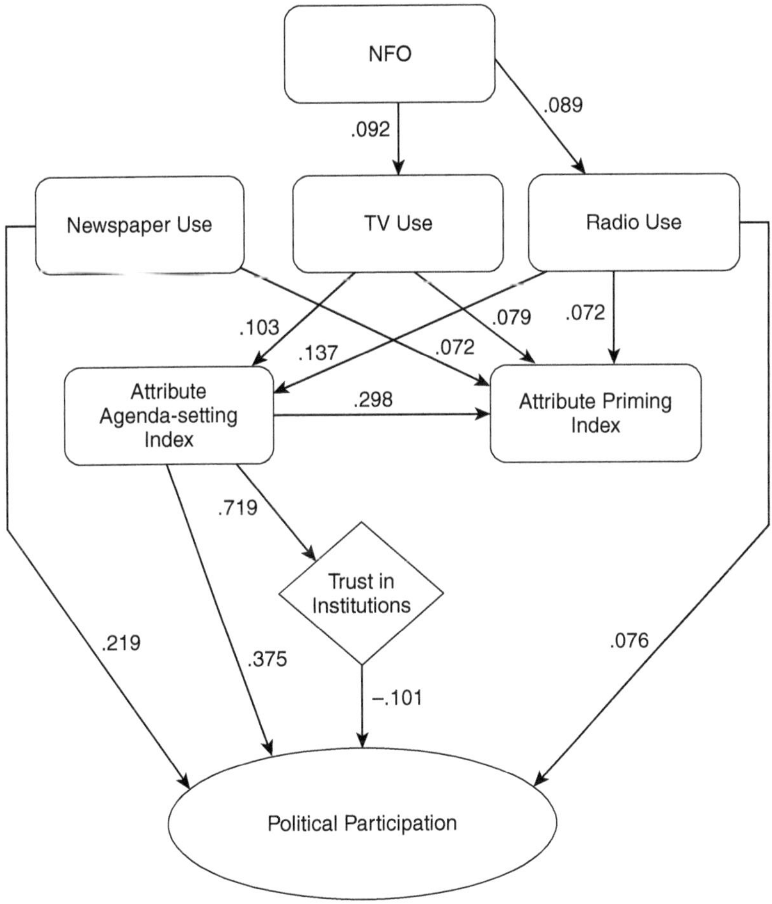

Figure 5.2 Structural Equation Model for partisan media

Discussion and Conclusion

The present study confirms previous agenda-setting research while going one step beyond—presenting empirical data that agenda-setting effects of mass media have important consequences for citizens' political culture. Hypotheses H1a and H1b, which predict a significant correlation between media use and agenda-setting and affective-priming effects, are both supported. The significant correlations between media use measures and the *attribute agenda setting* and *attribute priming* indexes suggest that citizens of Kosovo learn from the news media about the efficacy and integrity of political institutions in their country, achieving an in-depth understanding of their general performance on the most important issues. Similar to the second-level agenda-setting results found in Western countries, our study suggests that media content is related "to people's perceived salience of public issues (and objects), but also to their understanding of the substance of those issues (and subjects)" (Takeshita, 1997, p. 27).

Moreover, confirming previous research on attribute priming (Kim et al., 2002), the results of this study suggest that the valence of issues and institutional attributes emphasized in the media content become important dimensions when people judge political institutions. Hypothesis H2a, which predicted agenda-setting effects will be associated with political trust, is supported in both models. The significant coefficient between *attribute agenda-setting index* and the trust measure explains how political knowledge is used to form perceptions that affect political trust. However, H2b, which predicted a significant relationship between attribute priming of issue performance and political trust, was only partially supported in the SEM models for independent media use.

Further, supporting previous research that emphasizes political knowledge as an intervening variable between media use and political participation (McLeod et al., 1996; Moy, Torres, Tanaka, & McCluskey, 2005; Norris, 2000), this study found that agenda-setting effects were positively and significantly related with political participation (Weaver, 1991). Hypothesis H3a is fully supported in both models. However, H3b, which predicted a significant correlation between priming and political participation, was not supported. Further, confirming Gamson's (1968) thesis, this study suggests that political trust has a significant negative relationship with participation, implying that participation is higher for those who are dissatisfied with institutional performance.

These data corroborate the relevance of the NFO concept to explain agenda-setting effects, suggesting that "as an individual strives to map the political issues through the use of the mass media he is susceptible . . .to the agenda-setting effects of the mass media" (McCombs & Weaver, 1973, p. 3). Yet, these findings are contrary to Matthes' (2008) study finding that NFO had no influence on the perceived salience of the affective attributes of the issue of unemployment. One possible explanation for these divergences might have to do with the measurement of NFO. Matthes' (2006, 2008) measurement of the NFO is mainly based on the concept of "relevance." It can be argued that at the second-level agenda setting, when people are asked to evaluate issues or objects at the attitudinal level, uncertainty (as measured by partisanship strength) might prove to be of higher importance than when people are asked only about issues at the first-level agenda setting. In a partially polarized media environment (Baum & Groeling, 2008), the level of political interest or relevance might be more likely to determine the insensitivity of media exposure and attention, while the level of political uncertainty might determine the media source. Consequently, as this study suggests, differences in the attribute agenda-setting effects occur when differentiating between exposure to independent and partisan-oriented media.

Thus, most importantly, this study makes a significant contribution in understanding how the source of political information mediates attribute agenda-setting and attribute-priming effects. When comparing the relationship between media use and attribute agenda-setting and priming effects in the two models, the coefficients for partisan media are significantly stronger across all media platforms, thus supporting the H5a and H5b hypotheses. The data suggest that people exposed to partisan media with a pro-government bias were more likely to adopt the media agenda on positive institutional attributes and were more likely to use the positive issue valence when judging institutional performance.

These results imply that the salience of issues, objects, and their attributes on the media agenda is more likely to strengthen preexisting attitudes (Kiousis & McCombs, 2004) or to "contribute(ing) to the crystallization of political predispositions, which lead to electoral participation" (Kiousis & McDevitt, 2008, p. 495). This study gives credit to Slater's (2007) supposition that the "Use of media content consistent with a given social identity will result in greater (a) salience or accessibility of that social identity and (b) influence of values and attitudes associated with that identity in assessments and decisions, at least briefly, after such media exposure" (p. 291). These findings go hand in hand with the concept of priming that emphasizes the activation of previous concepts and attitudes. Media affiliated with the government will probably articulate frames that interpret issues consistent with the values of the governing political party, which in turn will provide interpretive filters for the audiences that are already government supporters. This cycle generates citizens with substantive trust in their country's institutions. Research shows that in countries undergoing political transition, political parties tend to be very volatile with floating constituencies, consequently weakening democratic representation and hindering consensus building (Bielasiak, 2005). Thus, in these countries, building partisan loyalties and stabilizing the volatile electorate might be one of the most important contributions of a free media. However, such a role for the partisan press can be beneficial only in societies with a pluralistic media system that offers citizens access to a variety of opinions.

Inherently, this study has its own limitations that call for a cautious interpretation of the results. Despite employing Structural Equation Modeling, a somewhat more rigorous statistical tool than previous studies that relied on sheer correlations, the cross-sectional data used in this study might lack the desired rigid controls. Thus, further studies should seek to employ panel or longitudinal datasets. Further, the theoretical models of this study offer grounds for more elaboration of the interactions between variables in search of better models. Of particular interest would be differentiating between elite-directed or institutional forms of participation and elite-challenging or uninstitutional participation, and their interaction with the media use and agenda-setting effects measures. As Gamson's hypothesis presupposes, the interaction between political trust and participation is dependent on citizens' sense of self-efficacy. These theoretical grounds imply the necessity to include this valuable variable in the equation that measures the relationship between media use, agenda-setting effects, political attitudes, and participation.

Notes

1 The backdrop for this study is Kosovo, a supervised parliamentary democracy in Southeastern Europe. Kosovo represents a unique case in which the democratic transition and the engineering of the democratic institutions occurred under the influence and supervision of the international community. The United Nations and the Western European institutions took over the governance of the country following the 1999 NATO intervention in the former Yugoslavia, which halted the ethnic conflict in this region. During the last decade of the twentieth century, civil conflict shattered ex-Yugoslavia into multiple independent nations in the Balkan Peninsula. The beginning of the twenty-first century has found this post-communist region struggling to build and maintain a democratic system modeled on Western values.

2 During the period this study was implemented, under the supervision of the international community, the representatives of Kosovo and Serbia were negotiating a solution regarding the status of Kosovo and other administrative issues.

3 Kosovo declared its independence in 2008, but not everybody in the international community had recognized its new political status. Thus, there was a proactive effort and pressure on the Kosovo foreign ministry to seek further recognitions.

4 Despite the fact that the war in Kosovo ended in 1999, the inter-ethnic tensions between the Albanian majority and Serb minority (that lives in the north of the country) have persisted for the last decade.

5 As the country was transitioning from a centrally organized economy into a private market, most of the publicly owned enterprises went through the process of privatization–which was one of the dominant economic issues of this time.

References

Aarts, K., & Semetko, H. A. (2003). The divided electorate: Effects of media use on political involvement. *Journal of Politics* 65, 759–784.

Almond, G. A. (1989). The intellectual history of the civic culture concept. In G. A. Almond & S. Verba (Eds.), *The civic culture revisited* (pp. 1–36). Newbury Park, CA; London, New Delhi: Sage.

Almond, G., & Verba, S. (1963). *The civil culture*. Princeton: Princeton University Press.

Barnes, S. H., & Kaase, M. (1979). *Political action: Mass participation in five Western democracies*. Beverly Hills: Sage Publications.

Baum, M. A., & Groeling, T. (2008). New media and polarization of American political discourse. *Political Communication*, 25, 345–365.

Bennett, W. L.(1998). The media and democratic development: The social basis of political

communication. In P. H. O'Neil (ed.), *Communicating democracy: The media and political transitions* (pp. 195-207). Boulder: Lynne Rienner Publishers.

Bennett, S. E., Rhine, S. L., Flickinger, R. S., & Bennett, L. M. (1999). Video malaise" revisited. Public trust in the media and government. *Harvard International Journal of Press/Politics*, 4, 8–23.

Bielasiak, J. (2005). Party competition in emerging democracies: Representation and effectiveness in post-communist and beyond. *Democratization*, 12, 331–356.

Bizer, G. Y., & Petty, R. E. (2005). How we conceptualize our attitudes matters: The effects of valence framing on the resistance of political attitudes. *Political Psychology*, 26, 553–568.

Blumler, J. G. (1979). The role of theory in uses and gratifications studies. *Communication Research*, 6, 9–36.

Carpentier, F. R., Roskos-Ewoldsen, D. R., & Roskos-Ewoldesen, B. B. (2008). A test of the network models of political priming. *Media Psychology*, 11, 186–206.

Catterberg, G. (2003). Evaluations, referents of support, and political action in new democracies. *International Journal of Comparative Sociology*, 44, 173–198.

Coleman, J. (1990). *Foundations of social theory*. Belknap Press, Cambridge.

Dasgupta, P. (1988). Trust as a commodity. In D. Gambetta (Ed.), *Trust: Making and breaking cooperative relationships* (pp. 49–72). New York: Basic Blackwell.

Delli Carpini, M. X. (2004). Mediating democratic engagement: The impact of communications on citizens' involvement in political and civic life. In L. Lee Kaid (ed.), *Handbook of political communication research* (pp. 395–434). London: LEA.

Diamond, L. (1993). Introduction. In L. Diamond (Ed.), *Political culture and democracy in developing countries* (pp. 1–36). Boulder, CO: Lynne Rienner.

Easton, D. (1965). *A systems analysis of political life*. New York, London, Sydney: John Wiley & Sons.

Eveland, W. P., & Scheufele, D. A. (2000). Connecting news media use with gaps in knowledge and participation. *Political Communication*, 17, 215–237.

Frey, J. H. (1989). *Survey research by telephone* (2nd ed.). Beverly Hills, CA: Sage Publications.

Gamson, W. A. (1968). *Power and disconnect*. Homewood, IL: Dorsey.

Ghorpade, S. (1986). Agenda setting: A test of advertising's neglected function. *Journal of Advertising Research*, 26, 23–27.

Hetherington, M. J. (1998). The political relevance of political trust. *American Political Science Review*, 92, 791–808.

Inglehart, R. (1997). *Modernization and post-modernization: Cultural, economic, and political change in 43* Societies. Princeton, NJ: Princeton University Press.

Inglehart, R., Basanez, M., & Moreno, A. (1998). *Human values and beliefs: A cross cultural sourcebook*. Ann Arbor, MI: University of Michigan Press.

Iyengar, S., & Kinder, D. (1987). *News that matters*. Chicago: University of Chicago Press.

Johnson, T. J., & Kaye, B. K. (1998). Cruising is believing: Comparing internet and traditional sources on media credibility measures. *Journalism & Mass Communication Quarterly*, *75*, 325–340.

Kim, Y. M. (2005). Use and disuse of contextual primes in dynamic news environments. *Journal of Communication*, *55*, 737–55.

Kim, K., & McCombs, M. (2007). News story descriptions and the public's opinions of political candidates. *Journalism and Mass Communication Quarterly*, 84, 299–314.

Kim, S. H., Scheufele, D. A., & Shanahan, J. (2002). Think about it this way: Attribute agenda-setting function of the press and the public's evaluation of a local issue. *Journalism & Mass Communication Quarterly*, *79*, 7–25.

Kiousis, S. (2005). Compelling arguments and attitude strength—Exploring the impact of second-level agenda setting on public opinion of presidential candidate images. *Harvard International Journal of Press-Politics*, *10*, 3–27.

Kiousis, S., Bantimaroudis, P., & Ban, H. (1999). Candidate image attributes: Experiments on the substantive dimension of second level agenda setting. *Communication Research*, 26, 414–428.

Kiousis, S., & McCombs, M. E. (2004). Agenda-setting effects and attitude strength: Political figures during the 1996 presidential elections. *Communication Research*, *31*, 36–57.

Kiousis, S., & McDevitt, M. (2008). Agenda setting in civic development: Effects of curricula and issue importance on youth voter turnout. *Communication Research*, 35(4), 481–502.

Klapper, J. T. (1960). *The effects of mass communications*. Glencoe, IL: Free Press.

Klapper, J. T. (1963). The social effects of mass communication. In W. Schramm (Ed.), *The Science of Human Communication*. New York: Basic Books.

Kline, R. B. (2011). *Principles and practice of structural equation modeling* (3rd ed.). New York, London: The Guilford Press.

Krosnick, J. A., & Kinder, D. R. (1990). Altering the foundations of s support for the President through priming. *American Political Science Review*, 84, 497–512.

Matthes, J. (2006). The need for orientation towards news media: Revising and validating a classic concept. *International Journal of Public Opinion Research*, *18*, 422–444.

Matthes, J. (2008). Need for orientation as a predictor of agenda-setting effects: Causal evidence from a two-wave panel study. *International Journal of Public Opinion Research*, *20*, 440–453.

McCombs, M. E. (2004). *Setting the agenda: The mass media and public opinion*. Malden, MA: Blackwell.

McCombs, M., Lopez-Escobar, E., & Llamas, J. P. (2000). Setting the agenda of attributes in the 1996 Spanish general election. *Journal of Communication*, *50*, 77–92.

McCombs, M., Llamas, J. P., Lopez-Escobar, E., & Rey, F. (1997). Candidate images in Spanish elections: Second-level agenda-setting effects . *Journalism & Mass Communication Quarterly*, *74*, 703–717.

McCombs, M. & Weaver, D. H. (1973). Voters' need for orientation and use of mass communication. Presented at the annual conference of the International Communication Association, Montreal, Canada.

McGraw, K. M., & Ling, C. (2003). Media priming of presidential and group evaluations. *Political Communication*, 20, 23–40.

McGuire, W. J. (1974). Psychological motives and communication gratification. In J. G. Blumler and E. Katz (eds.) The uses of mass communications: Current perspectives on gratifications research. Beverly Hills, CA and London: Sage.

McLeod, J. M., Daily, K., Guo, Z., Eveland, W. P., Jr., Bayer, J., Yang, S., & Wang, H. (1996). Community integration, local media use and democratic processes. *Communication Research*, *23*, 179–209.

McLeod, J., & McDonald, D. (1985). Beyond simple exposure: Media orientations and their impact on political processes. *Communication Research*, *12*, 3–33.

Miller, J. M., & Krosnick, J. A. (2000). News media impact on the ingredients of Presidential evaluations: Politically knowledgeable citizens are guided by a trusted source. *American Journal of Political Science*, 44, 295–309.

Moon, S. J. (2009). *Attention, attitude and behavior: Second-level agenda-setting effects as a mediator of media use and political participation*. Presented at the Association for Journalism and Mass Communication Research annual convention, Boston, MA.

Mishler, W., & Rose, R. (1997). Trust, distrust and skepticism: Popular evaluations of civil and political institutions in post-communist societies. *Journal of Politics*, 59, 418–451.

Mishler, W., & Rose, R. (2001). What are the origins of political trust? Testing institutional and cultural theories in post-communist societies. *Comparative Political Studies*, *34*, 30–62.

Moy, P., & Pfau, M. (2000). *With malice toward all? The media and public confidence in democratic institutions*. Westport, CT Praeger.

Moy, P., Torres, M., Tanaka, K., & McCluskey, M. R. (2005). Knowledge or trust? Investigating linkages between media reliance and participation. *Communication Research*, *32*, 59–86.

Mutz, D. C. (2006). How the mass media divide us. In P. S. Nivola & D. W. Brady (Eds.), *Red and blue nation? Characteristics and causes of America's polarized politics* (Vol. 1) (pp. 223–242). Washington: Brookings Institution Press.

Mutz, D. C., & Martin, P. S. (2001). Facilitating communication across lines of political differences. *American Political Science Review*, 95, 97–114.

Newton, K. (1999). Social and political trust. In P. Norris (Ed.), *Critical citizens: Global support for democratic government* (pp. 169–187). Oxford: Oxford University Press.

Norris, P. (Ed.) (1999). *Critical citizens: Global support for democratic government*. Oxford: Oxford University Press.

Norris, P. (2000). *A virtuous circle: Political communication in postindustrial societies*. Cambridge: Cambridge University Press.

O'Keefe, G. J. (1980). Political malaise and reliance on media. *Journalism Quarterly*, 57, 122–128.

Pan, Z., & Kosicki, G. M. (1997). Priming and media impact on the evaluations of the President's performance. *Communication Research*, 24, 3–30.

Patterson, T. E. (2003). *The vanishing voter: Public involvement in an age of uncertainty*. New York: Vintage Books.

Pinkleton, B. E., & Austin, E. W. (2001). Individual motivation, perceived media importance, and political disaffection. *Political Communication*, *18*, 321–334.

Pinkleton, B. E., Austin, E. W., & Fortman, K. J. (1998). Relationships of media use and political disaffection to political efficacy and voting behavior. *Journal of Broadcasting and Electronic Media*, 42, 34–49.

Przeworski, A., Alvarez, M. E., Cheibub, J. A., & Limongi, F. (1996). Classifying political regimes. *Studies in Comparative International Development*, *31*, 1–37.

Putnam, R. (2000). *Bowling alone: The collapse and revival of American community*. New York: Simon and Schuster.

Pye, L. W. (1965). Introduction: Political culture and political development. In L. W. Pye and S. Verba (Eds.), *Political culture and political development* (pp. 3–26). Princeton: Princeton University Press.

Riffe, D., Lacy, S., & Fico, F. G. (2005). *Analyzing media messages: Using quantitative content analysis in research* (2nd ed). Mahwah, NJ: Lawrence Erlbaum Associates.

Roberts, M. S. (1992). Predicting voting behavior via the agenda-setting tradition. *Journalism Quarterly*, 69, 878–892.

Robinson, M. (1976). Public affairs television and the growth of public malaise. *American Political Science Review*, 70, 409–432.

Rose, R., & McAllister, I. (1990). *The loyalties of voters*. London. Sage Publications Ltd.

Sheafer, T. (2007). How to evaluate it: The role of story-evaluative tone in agenda setting and priming. *Journal of Communication*, 57, 21–39.

Sheafer, T., & Weimann, G. (2005). Agenda building, agenda setting, priming, individual voting intentions, and the aggregate results: An analysis of four Israeli elections. *Journal of Communication*, 55, 347–365.

Slater, D. M. (2007). Reinforcing spirals: The mutual influence of media selectivity and media effects and their impact on individual behavior. *Communication Theory*, 17, 281–303.

Stroud, N. J. (2010). Polarization and partisan selective exposure. *Journal of Communication*, 60, 556–576.

Stroud, N. J. (2007). Media effects, selective exposure, and Fahrenheit 9/11. *Political Communication*, 24, 415–432.

Sutherland, M., & Galloway, J. (1981). Role of advertising: Persuasion or agenda setting? *Journal of Advertising Research*, 21, 25–29.

Takeshita, T. (1997). Exploring the media's roles in defining reality: From issue-agenda setting to attribute-agenda setting. In M. McCombs, D. L. Shaw, & D. Weaver (Eds.), *Communication and democracy: Exploring intellectual frontier in agenda-setting theory* (pp. 15–27). Mahwah, NJ: Erlbaum.

Verba, S., & Nie, J. (1987). *Participation in America. Political democracy and social equality.* Chicago: The University of Chicago Press.

Verba, S., Schlozman, K. L., & Brady, H. E. (1995). *Voice and equality: Civic voluntarism in American politics.* Cambridge, MA: Harvard University Press.

Wanta, W. (1997). *The public and the national agenda: How people learn about important issues.* Mahwah, NJ: Lawrence Erlbaum.

Wanta, W., & Hu, Y. (1994). The effects of credibility, reliance, and exposure on media agenda-setting: A path analysis model. *Journalism Quarterly*, 71, 1, 90-98.

Weaver, D. H. (1977). Political issues and voter need for orientation. In D .L. Shaw & M. E. McCombs (eds.), *The emergence of American political issues: The agenda-setting function of the press* (pp. 107–119). St. Paul, MN: West.

Weaver, D. H. (1980). Audience need for orientation and media effects. *Communication Research*, 3, 361–376.

Weaver, D. (1991). Issue salience and public opinion: Are there consequences of agenda-setting? *International Journal of Public Opinion Research*, 3, 53–68.

Weaver, D. H., McCombs, M., & Shaw, D. L. (2004). Agenda-setting research: Issues, attributes, and influences. In L. L. Kaid (Ed.), *Handbook of political communication research* (pp. 257–282). New Jersey: Lawrence Erlbaum Associates.

Winter, J. P. (1981). Contingent conditions in the agenda-setting process. In G. C. Wilhoit & H. deBock (eds.), *Mass communication review yearbook* 2 (pp. 235–243). Beverly Hills, CA: Praeger.

Chapter 6

Toward the Third Level of Agenda-Setting Theory

A Network Agenda Setting Model[12]

Lei Guo

The world is networked. So is the human mind. Consider the following two cases. During the 2008 presidential election, some people made an association between Barack Obama's racial identity and his intention to care about people of color, whereas his opponents questioned the connection (e.g., Harris, 2011). A *New York Times* poll in 2010 showed that more than two-thirds of respondents said illegal immigrants weakened the economy, whereas others argued that a struggling economy needs immigrants (Campo-Flores, 2010). As these two cases illustrate, in our brains are numerous associative networks composed of interconnected objects, such as the economy and immigration, and attributes, such as political candidates' background and their willingness to care about people. The question is how these networks are built in our minds.

As early as 1922, Walter Lippmann suggested that the news media construct a pseudo-environment for the public bridging "the world outside and *pictures* in our heads" (Lippmann, 1922, p. 3). Attributing agenda setting's intellectual origin to Lippmann's thesis, scholar Maxwell McCombs (2004) paraphrased Lippmann, stating that "the news media, our windows to the vast world beyond direct experience, determine our *cognitive maps* of that world (p. 3)." As the metaphors "pictures" and "cognitive maps" imply, agenda-setting theory suggests that news media are arguably one of the factors that can influence the audience's network-like mental structure.

Subsequent to the pioneering 1968 Chapel Hill study (McCombs & Shaw, 1972), hundreds of empirical studies have explored these "pictures" or "cognitive maps" both in various types of media and in people's heads around the world. As the theoretical focus of this research expanded, a second level of agenda setting was proposed to shift the attention to attribute agenda setting and to scrutinize the actual details of these pictures (McCombs, Lopez-Escobar, & Llamas, 2000; Shaw & McCombs, 1977; Weaver, Graber, McCombs, & Eyal, 1981). However, the elements of the "pictures" investigated to date are in fact discrete objects and attributes, which are disconnected elements of the whole. The news media's potential to transfer the salience of a more integrated, network-shape image has not yet been investigated.

The chapter presents a Network Agenda Setting Model, which primarily proposes that the salience of the network relationships among objects and attributes can be transferred from the news media to the public. This model suggests that news media are not only successful at telling us what to think about and how to think, but also are capable of telling us what and how to associate—a step forward to construct the third level of media effects. Although the specific empirical version of the Network Agenda Setting Model presented here is limited to attribute agenda setting, this theoretical model has the potential to encompass both basic agenda-setting effects (object salience) and attribute agenda-setting effects (attribute salience).

Theoretical Framework

The core theoretical proposition of the agenda-setting theory is the transfer of salience from the media agenda to the public agenda (McCombs, 2004). At the first level of agenda setting, researchers usually compare the rank-order of object salience in the news coverage with the salience of those objects among the public to determine the degree to which the two agendas are correlated. Regarding the second level of agenda-setting theory—the hierarchies of attributes describing the object—researchers perform a similar comparison of the salience of attributes on the media and public agendas (McCombs, Lopez-Escobar, & Llamas, 2000).

For both levels of the agenda-setting theory, this chapter considers that two assumptions underlie the traditional approach to agenda-setting research. First, it assumes that a human's mental representation operates primarily in a logical and linear model. As for the second assumption, the existing approach implies that the transfer of salience of different agendas occurs discretely. Moving beyond these assumptions, a Network Agenda Setting Model is proposed here to theorize the third level of media effects, in which the associations between elements on the media agenda (objects and/or attributes) are transferred to the public agenda.

Associative Network Model of Memory

Central to the Network Agenda Setting Model is an associative network model of memory. In fact, scholars in various disciplines including cognitive psychology, philosophy, geography, and communication have theorized this associative memory model in similar ways yet under different terms. Examples are "associative network model"(Anderson, 1983; Anderson & Bower, 1973), "cognitive mapping"(Kaplan, 1973), "cognitive network model"(Santanen, Briggs, & de Vreede, 2000), "connectionist model"(Monroe & Read, 2008), and "spreading activation model"(Collins & Loftus, 1975).

Rather than conceptualizing a human's mental representations as a hierarchical or linear structure as implied in the traditional understanding of agenda-setting theory (i.e., the first assumption), this associative network model indicates that the representation operates as pictures, diagrams, or cartography (Armstrong, 1973; Barsalou, 1998; Braddon-Mitchell & Jackson, 2007; Cummins, 1996). In a network model, individuals' cognitive representation of constructs is presented as a network-like structure where any particular node will, in general, be connected to numerous other nodes (Kaplan, 1973). Here, construct or node in the network can refer to any unit of information, ranging from social objects and their attributes; goals, values, and motivation; an affective or emotional state; and even as schemas or frames (Lindsay & Norman, 1977; Price & Tewksbury, 1997; Rumelhart & Norman, 1978). Take political communication as an example. When an individual considers a political candidate and uses some attributes to describe the candidate, it is not necessary for the individual to articulate a hierarchy of attributes ranked by their importance. Instead, in that individual's mind could be an assortment of attributes such as leadership, experience, credibility interconnected to each other,which constitute a network-shaped picture describing the candidate.

In addition, any network is composed of a mixture of sub-networks. Schemas, for example, can be a sub-network comprising objects, attributes, and other constructs. By the same token, linear and hierarchical thinking structures could also be subparts of the network, but they are not sufficient in themselves (Kaplan, 1973).

News Media and Cognitive Networks

By nature, a human's cognitive networks are dynamic. They are developed for different events as well as influenced by diverse external factors (Kitchin, 1994; Siegel & Cousins, 1985). News media arguably are a crucial factor impacting some of our cognitive networks, especially in regard to public affairs. In agenda-setting research, McCombs (2004) suggested that news media have the capability to influence the audience's network-like mental structure, a restatement of the basic proposition of the agenda-setting theory. As he paraphrased Lippmann's (1922) thesis, "the news media, our windows to the vast world beyond direct experience, determine our *cognitive maps* of that world" (p. 3).

Along this line, one less-noticed agenda-setting hypothesis—"compelling argument"—also implicitly suggests that news media may have the potential to affect the audience's cognitive map by transferring the *relationships*, or the *connections*, among various elements to the public's mind. Specifically, the "compelling argument" hypothesis suggests that the news media emphasis on certain attributes of an object provides people with cues to modify their perceived salience of the object that possesses these attributes (Ghanem, 1997; McCombs, 2004; McCombs & Ghanem, 2001; Severin & Tankard, 2001; Yioutas & Segvic, 2003). According to this hypothesis, news media may not only shape the perceived importance of attributes and objects separately, but also can actually bundle an object with an attribute and make them salient in the public's mind simultaneously. As such, the audience may not only treat a certain attribute as a part of the object, but also regard the two as connected, integrated elements in their "pictures." This hypothesis is in fact a challenge to the second assumption of traditional agenda-setting theory as stated earlier (i.e., the transfer of agenda salience occurs discretely), yet few scholars have pointed this out before. In addition, although there is some empirical support for the "compelling argument" hypothesis (e.g., Ghanem, 1997; Schoenbach & Semetko, 1992; Williams, Shapiro, & Cutbirth, 1983), the "compelling argument" effect remains greatly understudied in the field (Kiousis, 2005). It still remains largely unknown how the news media affect a human's cognitive networks.

Information-Processing Mechanism

In the field of media psychology, Lang's (2000) Limited Capacity Model provides a valuable framework to examine the ways in which news media interact with a human's cognitive networks. In this model, three sub-processes are described to investigate how people process mediated messages: (1) encoding, (2) storage, and (3) retrieval. In the first step, people receive the message from the news media and into their mind. These messages form a temporary network in the audience's short-term memory, or working memory.

The second step—storage—is the process wherein newly encoded information becomes linked to individuals' associative memory network. As mentioned earlier, each person has an existing associative memory network, or long-term memory. Here, new information is stored based on its relationship with the audience's underlying schemas, or previously coded information (Cortese, 2007). It is worth mentioning that this storage process can be either automatic or controlled. According to Lang (2000), individuals might unconsciously link two random messages; on the other hand, they might also actively elaborate the information, thus purposefully connecting two messages they think are meaningfully related (Eveland, 2001).

Interestingly, the "applicability effect" proposed in agenda-setting and framing research can be used to explain how news media impact this "storage" process. A piece of new information or a construct is regarded as "applicable" if it corresponds with the perceiver's store of knowledge (Higgins, 1996; Price & Tewksbury, 1997). An applicability effect then suggests that, among various external stimuli, news media serve to construct the connections between new and old information in order to store the new information in an individual's associative memory network. As an essential information-processing step, this connecting mechanism in the storage process—or applicability effect—should be central to any media effect, whether it is framing, priming, or agenda setting. Price and Tewksbury (1997), the scholars who distinguished the accessibility from the applicability model, also made it clear that "applicability effects should be properly seen as primary or first-order effects of stimuli" (p. 197), meaning that it actually applies to all media effect models.

Once the connection of any two constructs is stored, whether the connection is held firmly in the audience's long-term memory depends on whether the two constructs can be frequently activated in tandem by either internal or external factors such as the media coverage. Therefore, the more often two constructs are activated jointly, the greater chance these two constructs *as well as* their interrelationships can be retrieved later.

Moving to the last step of this information-processing model, individuals retrieve information by searching the associative memory network for specific information and reactivating it in the working memory (Lang, 2000). Thus, another temporary associative network (working memory) is formed. As Lang argued, the most readily retrievable information usually has the most associative links to other pieces of information in the memory network (Lang, 2000). Take the example of political communication again. When an individual retrieves an "experience" attribute to describe a certain political candidate, "experience" might be the attribute that has the most connections to other attributes. For example, the individual might associate a political candidate's experience with his leadership skills, his competence as a politician, his credibility and so forth. That is to say, "experience" becomes easily retrieved in this case not necessarily because of its importance in the attribute hierarchy in that individual's mind, but probably because of its centrality in his associative memory network. Therefore, the Network Agenda Setting Model hypothesizes that news media have the capability to construct the connections among agendas, thereby constructing the centrality of certain agenda elements in the public's mind. In other words, salience or retrieveability can also be defined as the centrality of an object or attribute on the public agenda.

Basic Network Agenda Setting Model

The Network Agenda Setting Model suggests that the news media, among other environmental factors, serve to connect new information to old information in the audience's existing associative network memory and are able to strengthen the connections by frequently activating

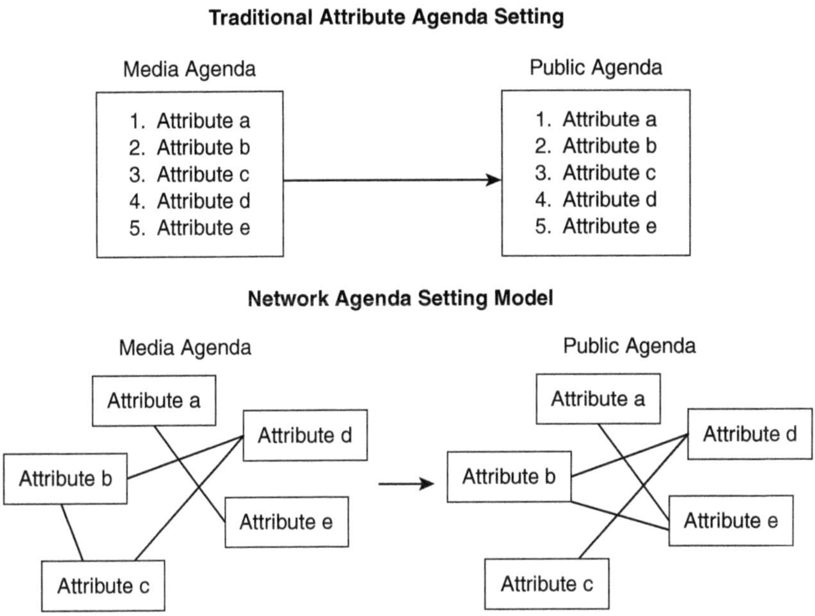

Figure 6.1 The comparison of traditional attribute agenda setting and Network Agenda Setting Model

pairs of constructs in tandem. In this way, the news media are able to construct and reconstruct the audience's associative memory network by creating new nodes to the network or altering the strength of the existing connections among different constructs (see Figure 6.1). According to the theorizations above, the central hypothesis for the Network Agenda Setting Model is that the salience of the interrelationships among constructs—or the associative network regarding a certain topic—can be transferred from the media agenda to the public agenda.

The Elaboration Factor

The Network Agenda Setting Model demonstrates that once the connection of any two constructs is stored, the more often two constructs are activated jointly by the news media,the greater chance these two constructs as well as their interconnectedness can be retrieved later. In other words, the frequency of two messages being activated in tandem during the storage process is an important independent variable that can predict the network agenda-setting effect. In fact, there is another potential independent variable: the level of elaboration.

During the storage process in Lang's (2000) model, "elaboration" is the process of connecting new information to other information stored in memory, or connecting new bits of information together in new ways (Eveland, 2001). Whether the connecting process is automatic or controlled—the two levels of "elaboration" of the interrelationships between two messages—might be another important factor that can influence the network agenda-setting effect. Specifically, two aspects can be examined for exploring the level of elaboration: the news media's elaboration and the audience's elaboration of certain messages. First whether the news media purposefully associate two messages or randomly mention them in tandem might result in different audience's perceptions of the connectedness of the two messages. On the other hand, when individuals receive messages from the news media, whether they actively elaborate on the messages or merely skim through them might also influence the network agenda-setting effect.

In the step of retrieval, more questions are then raised: Why do individuals connect messages? Do they even realize the connections they have made? As such, it is interesting to explore whether the respondents connect messages implicitly or explicitly and how the audience's explicit or implicit memories are affected by the level of elaboration. To describe, implicit memory is revealed "when previous experience facilitates performance on a task that does not require conscious or intentional recollection of those experiences," whereas explicit memory is revealed "when performance on a task requires conscious recollection of previous experience" (Schacter, 1987, p. 501).With respect to explicit memory, researchers found that elaborative processing substantially benefits this type of memory (e.g., Craik & Tulving, 1975). On the other hand, existing studies generated different results in terms of the impact of the elaboration on implicit memory. Although some researchers demonstrated that implicit memory is unaffected by elaborative versus non-elaborative processing (e.g., Carroll & Kirsner, 1982), others have suggested otherwise (e.g., Schacter & Graf, 1986). To further develop the Network Agenda Setting Model, it is important to examine how the level of elaboration on messages—news media's elaboration and individuals' elaboration—in the storage process affect the audience's implicit or explicit associations of the messages during the retrieval.

The current chapter presents the methodology and two empirical studies that tested the basic Network Agenda Setting Model. The factor of elaboration remains a future project.

Method

Applying the Network Agenda Setting Model first to the field of political communication, the following two interrelated hypotheses are presented.

H1: The salience of the relationship network of political candidate attributes on the media agenda will be positively associated with the public salience of this attribute network.

H2:The centrality of political candidate attributes in the media attribute network will be positively associated with the centrality of the attributes on the public network agenda.

Network analyses were conducted to compare the media network agendas and the public network agendas during two election periods: the 2002 Texas gubernatorial and U. S. senatorial elections, and the 2010 Texas gubernatorial elections. The 2002 elections' data were initially collected for Kim and McCombs'(2007) comprehensive analysis of attribute agenda-setting effects and their consequences for subsequent attitudes and opinions. A public survey of residents in Austin, Texas, and a content analysis of the local daily newspaper were conducted during 2002. The research strategy here is to begin the exploration of network agenda-setting effects with a data set that demonstrates strong traditional attribute agenda-setting effects. Subsequently, these network relationships were examined in an analysis of the 2010 election based on an entirely new set of content analysis and survey data. Use of data based on two election periods provides an important replication of the Network Agenda Setting Model.

Content Analysis

The *Austin American-Statesman* was used for the content analysis because it is the only local daily newspaper serving Austin residents. In addition, 83% of the respondents in the 2002 survey and 66.7% in the 2010 survey reported reading the paper.

For the 2002 election period, two sets of content analysis data are available from Kim and McCombs' (2007) project: (1) content analysis of the news coverage of Republican candidate Rick Perry and Democratic candidate Tony Sanchez, as each competed in his party's primary gubernatorial election in the spring (2002 Spring Dataset); and (2) content analysis of the news coverage of the same two candidates as well as two U. S. senatorial candidates Republican John Cornyn and Democrat Ron Kirk in the general elections in November (2002 Fall Dataset). An additional content analysis original to this study was conducted during the 2010 Texas gubernatorial election. News stories covering the incumbent Republican Governor Rick Perry and the Democrat challenger Bill White were content analyzed (see Table 6.1 for details about the datasets).

The unit of measurement in the content analysis was an assertion in an article that a candidate possessed a particular attribute describing his qualifications or character. Only attributes concerning candidate

Table 6.1 Datasets for analysis

Dataset	*Election*	*Content analysis date*	*Survey date*	*Candidates*
2002 Spring	2002 Texas primary gubernatorial election (March)	Jan. 1 – Mar. 12, 2002	Feb. 19 – Mar. 12, 2002	Rick Perry, Tony Sanchez
2002 Fall	2002 Texas gubernatorial election & U. S. senatorial election (November)	4 weeks in Sept., 2002	Sept. 26 – Oct. 11, 2002	Rick Perry, Tony Sanchez, John Cornyn, Ron Kirk
2010 Fall	2010 Texas gubernatorial election (November)	4 weeks in Oct., 2010	Oct. 26 – Nov. 2, 2010	Rick Perry, Bill White

qualifications and character were considered in this study because they were by far the most prominent attributes highlighted by the Austin media and voters (Kim & McCombs, 2007). In other words, what this study analyzed are specific sub-networks of attributes portraying the political candidates. Nine attributes were identified to define personal qualifications and character in the analyzed news stories: (1) leadership, (2) experience, (3) competence, (4) credibility, (5) morality, (6) caring about people, (7) communication skills, (8) pride in family/backgrounds, roots, and race/ethnicity, (9) non-politician[3].

Specifically, with respect to the 2002 Spring Dataset, all the news stories (N = 67) about the primary campaigns that appeared in the *Austin American-Statesmen* from January 1 through March 12, 2002 (the election day) were content-analyzed. Eight journalism graduate students in a content analysis class analyzed the media attribute agendas and achieved an average agreement of 90%. For the 2002 Fall Dataset, another group of ten student coders scrutinized all the news items (N = 107) about the four candidates during four weeks in September. The inter-coder reliability based on 25% of all assertions was 0.87, measured by Scott's pi.

A similar content analysis was performed to examine the news coverage of the two candidates for the 2010 election. Two undergraduate student coders were trained and independently identified every assertion in all the news and editorial items (N = 81) about these two candidates in the *Austin American-Statesman* during the entire month of October 2010. Cohen's kappa scores based on a random sample of 20% of all assertions was 0.75. After discussing the discrepancies of their sample coding result, the two coders continued to code all the remaining assertions separately. They then discussed their results and achieved consensus on the coding result for every assertion.

Survey

During the 2002 election period, two rounds of telephone interviews were conducted to ascertain the public's attribute agenda. From February 19 to March 12, 2002, telephone surveys were conducted with 271 randomly selected adults, 18 years or older, in the Austin metropolitan area. Another round of telephone interviews was conducted with 417 randomly selected adults between September 26 and October 11, 2002. To measure the public's attribute agenda, respondents were asked in both surveys: "Suppose that one of your friends has been away a long time and knows nothing about the political candidates. What would you tell your friend about (Cornyn, Kirk, Perry, and Sanchez)?" Responses were recorded verbatim and later coded using the same nine attribute categories in the content analysis. Inter-coder reliability based on 25% of the coding by four graduate students was 0.83, measured by Scott's pi.

However, in order to retrieve the audience's cognitive maps about the political candidates, a traditional public survey is helpful but not sufficient. Survey questions such as "what would you tell your friends about. . . (candidates)?" cannot provide researchers with direct clues with regard to the interrelationships between attribute agendas in the respondents' minds. Therefore, a new data collection method was particularly created to test the Network Agenda Setting Model. Incorporating a mind-mapping approach, this new method was used to collect the public's images of the political candidates in the 2010 election.

Specifically, mind-mapping refers to a radiant thinking approach—associative thought processes that proceed from or connect to a central point—that has already been broadly used in the advertising industry for brainstorming (Buzan & Buzan, 1996). Focusing on a certain topic, individuals are usually asked to write down the things that first come to mind and then expand outward into branches and sub-branches as fast as they can (McCutchin, 2008). Borrowing this mind-mapping concept, the new study designed a survey sheet that required respondents to fill in at most five assertions respectively describing each political candidate's (Rick Perry and Bill White) qualifications and character. After listing the attributes, they were also instructed to draw connections among each pair of attributes if they thought there were any. Figure 6.2 provides an example of a respondent's answer on the map-mapping survey sheet.

Between October 26 and November 2, 2011, these mind-mapping survey sheets were distributed to four undergraduate journalism classes at a Texas university. A total of 63 students completed the mind-mapping surveys. In this study, a convenience sample of college student subjects was mainly used to test the efficiency of the exploratory mind-mapping survey method, rather than to make generalizations to a

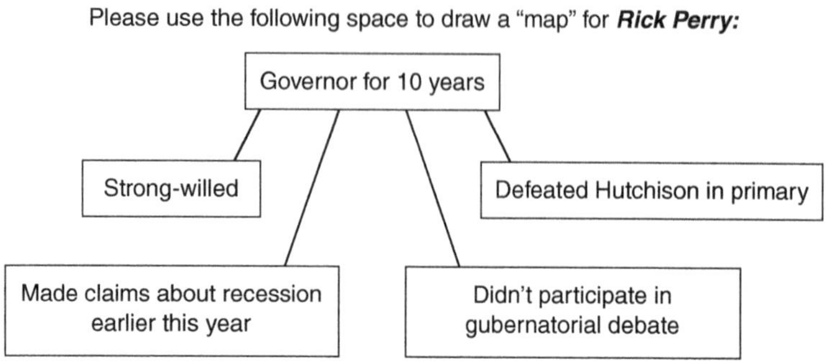

Figure 6.2 Mind-mapping survey sheet example

broader population. Students in journalism were selected because they tended to read more newspapers than the general public. Therefore, the sample should be better for finding if this new method can measure network agenda-setting effects.

Using the attribute categories from the content analysis, the two undergraduates also coded the mind-mapping survey results. The unit of analysis in the survey was an assertion in the blank filled in by the respondents to describe the two political candidates' personal qualifications and character. Based on a random sample of 20% of the total assertions, Cohen's kappa score for inter-coder reliability was 0.80. Similar to the content analysis, the two coders also independently coded all the remaining assertions in the mind-mapping survey results, compared their results afterward, and then reached agreement on every assertion.

Data Analysis

A network analysis approach was applied in this study to test the network agenda-setting effect. The content analysis data were used to define media network agendas. For this study, four subsets of content analysis data were available for use: (1) 2002 Spring Dataset; (2) 2002 Fall Dataset; (3) 2002 Combined Dataset[4]; and (4) 2010 Fall Dataset. Specifically, the media network agendas were constructed based on the interrelationships among the candidate attributes emphasized by the news coverage. Operationally, the relationship between a pair of attributes was measured according to the frequency of their co-occurrence in the same articles. As such, for each subset of data, a matrix composed of 9 rows × 9 columns was created for the network analysis (Wasserman & Faust, 1994). Each row or column represents a candidate attribute. The entry in each cell is the frequency of the co-occurrence of the

Table 6.2 Matrices of media and public agenda networks

	A	*B*	*C*	*D*	*E*	*F*	*G*	*H*	*I*
A		11	8	0	3	1	1	2	6
B	11		29	17	25	11	4	3	24
C	8	29		9	12	5	1	4	14
D	0	17	9		7	2	1	1	6
E	3	25	12	7		5	2	1	11
F	1	11	5	2	5		1	0	7
G	1	4	1	1	2	1		0	2
H	2	3	4	1	1	0	0		3
I	6	24	14	6	11	7	2	3	

2010 Fall Dataset (Content Analysis)

	A	*B*	*C*	*D*	*E*	*F*	*G*	*H*	*I*
A		6	12	2	1	3	0	3	5
B	6		31	5	1	7	2	9	29
C	12	31		6	6	7	2	7	20
D	2	5	6		2	5	2	1	8
E	1	1	6	2		2	1	0	3
F	3	7	7	5	2		1	4	9
G	0	2	2	2	1	1		0	3
H	3	9	7	1	0	4	0		3
I	5	29	20	8	3	9	3	3	

2010 Fall Dataset (Survey)

Note. A = Leadership; B = Experience; C = Competence; D = Credibility; E = Morality; F = Caring about people; G = Communication Skills; H = Pride in family/backgrounds, roots, and race/ethnicity; I = Non-politician

corresponding two attributes. The more frequently the two attributes co-occurred across news articles, the stronger their connection. Thus, the unit of analysis in this network analysis was a dyad: two attributes and their relational ties. For example, if the two attributes "leadership" and "experience" appear together in five articles, the entry is 5 in the cell corresponding to the two attributes in the matrix. Table 6.2 presents the matrix of the media network agenda portraying political candidates' attributes based on 2010 Fall Dataset.

Similarly, four subsets of survey data were available to construct the public network agendas: (1) 2002 Spring Dataset; (2) 2002 Fall Dataset;

(3) 2002 Combined Dataset; and (4) 2010 Fall Dataset. And similar matrices were created for the public network agendas regarding the political candidate attributes.

For the public network agendas based on the 2002 elections, the relationship between any two attributes was measured according to their co-occurrence in the same respondent's narrative. The more frequently the two attributes co-occurred across respondents' descriptions of the candidates, the stronger their relationship.

On the other hand, the network relationships among attributes in the 2010 public network agendas were measured in a more straightforward manner. The relationship between a pair of attributes was determined according to the explicit connections made by the respondents themselves on the mind-mapping survey sheets. Using the multiple datasets available for this study, both direct and indirect connections between attributes were coded according to network analysis theory (Wasserman & Faust, 1994). For example, if a survey respondent connects "leadership" and "experience" and at the same time connects "experience" with "competence," not only the connections between "leadership" and "experience," and that between "experience" and "competence" are built, but the connection between "leadership" and "competence" is also constructed through the mental pathways as well. Table 6.2 also presents the matrix of the public network agenda based on 2010 Fall Dataset.

It should be noted that the aim of the current study is to test the news media and the public's narratives of political candidates in general. Therefore, the study didn't distinguish between substantive attributes describing different political candidates, nor the tones of the attributes. In fact, the results obtained by Kim and McCombs (2007) are highly similar for all candidates combined, separate analyses for each of the four candidates, and separate analyses for positive and negative attribute agendas.

> **H1**: The salience of the relationship network of political candidate attributes on the media agenda will be positively associated with the public salience of this attribute network.

The QAP correlation tests on the media matrices and the survey matrices were conducted to compute the correlation between entries of four pairs of matrices: (1) 2002 Spring Dataset (content analysis) vs. 2002 Spring Dataset (survey); (2) 2002 Fall Dataset (content analysis) vs. 2002 Fall Dataset (survey); (3) 2002 Combined Dataset (content analysis) vs. 2002 Combined Dataset (survey); and (4) 2010 Fall Dataset (content analysis) vs. 2010 Fall Dataset (survey).

The QAP regression test is to regress a dependent matrix on one or more independent matrices, and assess the significance of the r-square and regression coefficient (Borgatti, Everett, & Freeman, 2002). In this analysis, the content analysis data was considered to be the independent matrices and the survey data, the dependent matrices.

> **H2:** The centrality of political candidate attributes in the media attribute network will be positively associated with the centrality of the attributes in the public network agenda.

An attribute is considered as central if it is extensively involved in relationships with other attributes (Wasserman & Faust, 1994). This study uses "degree centrality," the number of ties that a node has, which is the most direct and simplest centrality measurement. Specifically, for the attribute network matrices, the degree centrality was calculated for each attribute by computing the sums of the values of its ties with other attributes. For example, if the attribute "experience" co-occurs five times with each of the other nine attributes in the same articles, its degree centrality is 45. To compare the attribute degree centrality in the media agenda with that in the public agenda, correlation and regression tests were conducted on the four pairs of datasets.

Results

Table 6.3 reports the statistical results, which show a clear pattern in the significant degree of correspondence between the media network agendas and the public network agendas. More specifically, QAP correlation and QAP regression coefficients for all four pairs of datasets are positive and statistically significant, with QAP correlations (Pearson's r) ranging from +0.67 to +0.84 and R-square ranging from +0.45 to +0.71. Notably, the results based on the media and the public network agenda of the 2010 Texas gubernatorial election period show a very similar

Table 6.3 QAP correlation/regression and centrality correlation/regression between the media and the public attribute agenda networks

Dataset	*QAP correlation (Pearson's r)*	*QAP regression (R^2)*	*Centrality correlation (Pearson's r)*	*Centrality regression (R^2)*
2002 Spring Dataset	0.75**	0.56**	0.84*	0.70*
2002 Fall Dataset	0.67*	0.45*	0.76*	0.57*
2002 Combined Dataset	0.84**	0.71**	0.91**	0.84**
2010 Fall Dataset	0.71**	0.51**	0.81**	0.66**

Note. *$p < 0.5$, **$p < 0.01$

pattern to those from the 2002 gubernatorial and the U. S. senatorial elections. These four sets of evidence all support **H1**, that media salience of the candidate attribute networks is positively associated with the public salience of these same attribute networks.

Another finding from Table 6.3 is that the QAP correlation and regression results both demonstrate that the most significant correspondence is between the pairs of 2002 Combined Datasets: +0.84 and +0.71.

Table 6.4 details the degree centrality of the nine candidate attributes on the media and public attribute agenda networks. Pearson's r correlation and linear regression tests were conducted to test the relationship between the media and public agendas in terms of the degree centrality. In Table 6.3, the degree centrality in the media attribute agenda is significantly correlated with the degree centrality in the public agenda in all four datasets, with correlations ranging from +0.76 to +0.91. The result of the regression test, the R-square for each pair of media and public agenda, confirms the strong and positive relationships, which range from +0.57 to +0.84. Thus, the second hypothesis **H2** is well supported too. Similar to the QAP test results, the correlation and regression coefficients of the 2002 Combined Dataset are the highest among all. And the results of the 2010 election are within the range of the values computed from the 2002 election datasets.

Table 6.4 Degree centrality of candidate attributes on the media and public attribute network agenda

Attribute	*2002 spring media*	*2002 spring public*	*2002 fall media*	*2002 fall public*	*2002 combined media*	*2002 combined public*	*2010 fall media*	*2010 fall public*
A	18	7	0	26	18	33	32	32
B	39	52	27	142	66	194	123	90
C	20	37	22	126	42	163	82	91
D	24	13	24	85	48	98	43	31
E	27	20	7	71	34	91	66	16
F	6	6	18	35	24	41	32	38
G	10	10	17	4	27	14	12	11
H	9	3	0	10	9	13	14	27
I	4	9	10	42	14	51	73	80

Note. A = Leadership; B = Experience; C = Competence; D = Credibility; E = Morality; F = Caring about people; G = Communication Skills; H = Pride in family/backgrounds, roots, and race/ethnicity; I = Non-politician.

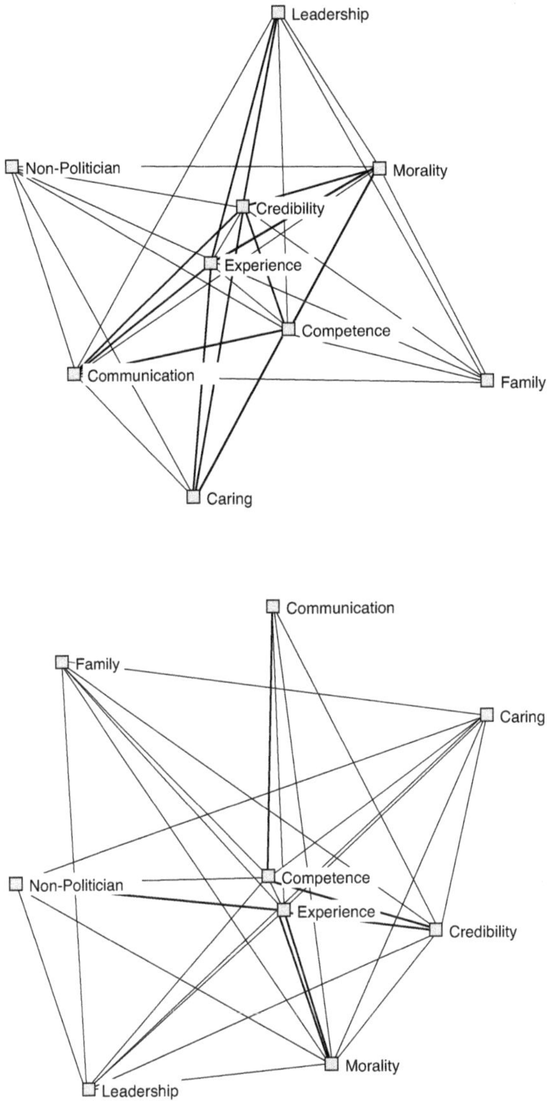

Figure 6.3 Media and public attribute agenda network (2002)

In addition, graphs were created to visualize the media and the public network agendas. Figure 6.3 shows graphs that represent the media and the public attribute agenda networks based on the 2002 Combined Dataset. Visually, these two networks are highly similar. For example, three attributes—"Experience,""Competence," and "Credibility"—are central in both the media and the public networks, and demonstrate the strongest mutual connections as shown in both graphs. The 2010 election dataset generated the media and the public attribute networks presented in Figure 6.4, which also are visually very similar.

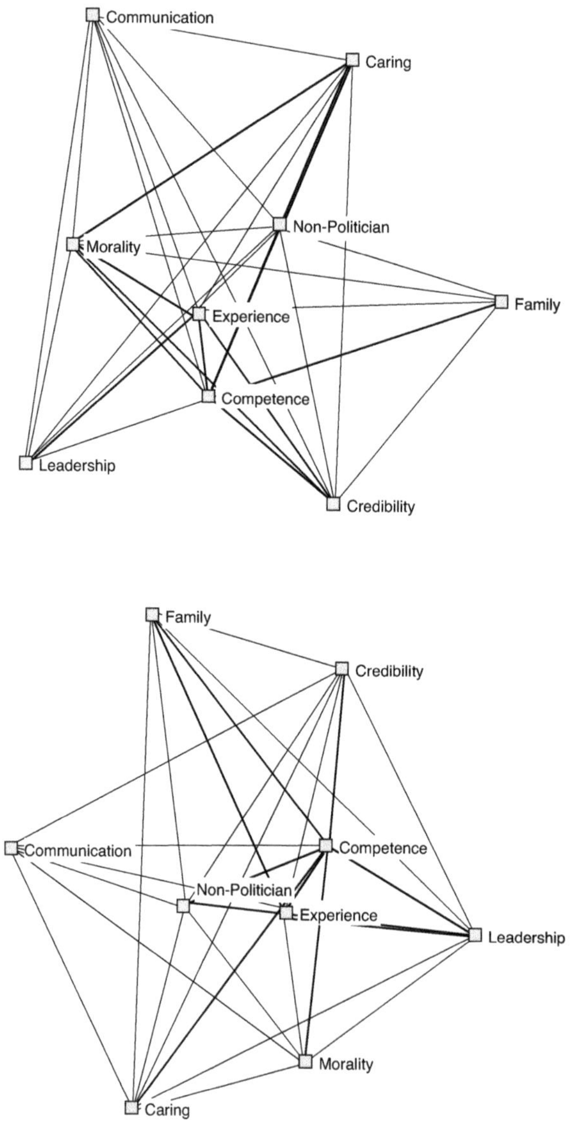

Figure 6.4 Media and public attribute agenda network (2010)

Conclusion and Discussion

This chapter describes a Network Agenda Setting Model, which hypothesizes that the news media are able to transfer the salience of the network relationships of objects and attributes to the public's mind. This new model moves beyond the two assumptions underlying the traditional approaches to agenda-setting theory, namely that the audience perceives news agenda elements in a linear logical fashion; and that objects and attributes are transferred separately. Instead, the model suggests that the news media have the capability to construct a more integrated picture of the outside world to people's heads, which moves a step closer to Walter Lippmann's (1992) picture metaphor.

In addition, two studies are presented to demonstrate that the Network Agenda Setting Model is empirically grounded. Datasets based on two election periods across eight years—the 2002 Texas gubernatorial and U. S. senatorial elections, and the 2010 Texas gubernatorial election—consistently demonstrate a significant degree of correspondence between the media network agendas and the public network agendas regarding the political candidate qualifications and character. Solid data in two studies provides firm evidence for the Network Agenda Setting Model, thereby to a large extent confirming the explanatory and predictive power of the model.

The Network Agenda Setting Model also serves to introduce the approach of network analysis to agenda-setting research. Moreover, a brand-new survey approach was created—mind-mapping—that can be used to test the network agenda-setting effect on the public agenda. Allowing respondents to make connections among attributes by themselves, this mind-mapping survey approach operationally facilitates the research by directly retrieving the audience's cognitive networks.

Some additional findings also bring insights to agenda-setting research. For example, it is noteworthy that the results for network agenda-setting effects based on the data of the 2002 election period are consistent with the attribute agenda-setting effects in Kim and McCombs'(2007) study on the same 2002 Fall Dataset. Specifically, the overall rank-order correlation coefficient (Spearman's rho =+.65) between the media and public attribute agendas corresponds well with the QAP correlation (Pearson's r =+.67) between the media and public network agendas. In other words, network agenda setting as the third level of media effects is solidly linked with the second level of agenda-setting effects.

Moreover, the evidence presented here indicates that the correlations and regression coefficients are highest in the combined datasets. This might imply that a network agenda-setting effect depends on the cumulative media coverage, rather than a small time span as frequently found in traditional agenda-setting research. In other words, it might take a longer period of time to establish the connections between the attributes than to build the salience of a sole attribute.

The studies also demonstrate that the media network agendas that describe political candidates' qualifications and character are different in the two election periods. So are the public network agendas. While the political candidates' "experience" and "competence" stayed central in the media and public agendas in both elections, the attribute "non-politician" moved from the edge in the 2002 election to the center in 2010. This finding suggests that human cognitive networks are constantly changing and the news media, among other external factors, are able to impact individuals' cognitive networks in regard to public affairs.

Since the pioneer Chapel Hill study, agenda-setting theory has been explored and developed for almost 40 years. As Graber and Smith (2005, p. 489) stated, "Agenda setting remains the predominant theoretical approach to analyzing the impact of media messages on audiences" as we entered the twenty-first century. In this chapter, the proposition of the Network Agenda Setting Model—the third level of agenda-setting theory—again proves the theory's dynamics and potential to research media effects.

Indeed, it is without doubt that the model is far from perfect, and it can be enriched and improved in many other aspects. For instance, although the research here focused specifically on the relationships between media and public networks of attributes, the definition of agenda setting from the perspective of a third level of effects is much broader: bundles of elements that are prominent on the media agenda frequently become prominent on the public agenda. These bundles, or networks, of elements can consist of attributes, objects, or combinations of objects and attributes. Therefore, future research should test the model by focusing on objects, or the mixture of different elements. Moreover, scholars should apply the model to more contexts in addition to gubernatorial elections and to various media outlets in addition to newspapers. It is also recommended that researchers examine how the aforementioned variable "elaboration"—the news media's elaboration or the audience's elaboration of how messages are associated—impacts the network agenda-setting effect. Hopefully the presentation of this model can encourage more studies to test and build on this third level of media effect.

Notes

1 This chapter is based on two network agenda-setting studies respectively presented in two conference papers: (1) Guo, L. & McCombs, M. (2011). *Network Agenda Setting: A Third Level of Media Effects*. Paper accepted for presentation to the ICA annual conference, Boston. (2) Guo, L. & McCombs, M. (2011). *Toward the Third Level of Agenda Setting Theory: A Network Agenda Setting Model*. Paper accepted for presentation to the AEJMC annual conference, St. Louis.

2 The author is pleased to acknowledge the encouragement for applying network analysis expressed by Craig Carroll and Sungsoo Bang in previous conversations with Maxwell McCombs.

3 These nine attributes were coded regardless of their tones. Specifically, (1) "leadership" refers to a person's capacity to lead. (2) "Experience" was coded if the news story mentioned the political candidate's past or current occupation, personal life achievements, or public activities. (3) "Competence" refers to the candidate's intelligence, knowledge, and qualification for being a politician. (4) "Credibility" was coded when the story mentioned whether the political candidate is trustworthy, fair, accountable, or keeps promises. (5) "Morality" refers to the quality of a person conforming to the standard of right behavior or virtue. (6) "Caring about people" was coded if the story mentioned things such as that the political candidate understands the needs of ordinary people. (7) "Communication skills" was coded if the story discussed whether the candidate is a good speaker, knows how to talk to the audience, or likes to be in the media. (8) "Pride in family/background, roots, race/ethnicity" was coded when the story covered that the candidate was proud of Texas and of his family members. (9) "Non-politician" was coded if the story mentioned that the candidate plays against established politics, or the candidate makes politically motivated choices.

4 Given that 2002 Spring Dataset and the 2002 Fall Dataset were based on the same election period, the two sets of data were combined and further analyzed as "2002 Combined Dataset."

Reference

Anderson, J. R. (1983). *The architecture of cognition*. Cambridge, MA: Harvard University Press.

Anderson, J. R., & Bower, G. H. (1973). *Human associative memory*. Washington, DC: Winston.

Armstrong, D. M. (1973). *Belief, truth, and knowledge*. London: Cambridge University Press.

Barsalou, L. (1998). Perceptual symbol systems. *Behavioral and Brain Sciences, 22*, 557–660.

Borgatti, S. P., Everett, M. G., & Freeman, L. C. (2002). *Ucinet for Windows: Software for Social Network Analysis*. Harvard, MA: Analytic Technologies.

Braddon-Mitchell, D., & Jackson, F. (2007). *Philosophy of mind and cognition* (2nd ed.). Cambridge, MA: Blackwell.

Buzan, T., & Buzan, B. (1996). *The mind map book: How to use radiant thinking to maximize your brain's untapped potential*. New York: Plume.

Campo-Flores, A. (2010, May 13). Why Americans think (wrongly) that illegal immigrants hurt the economy? *Newsweek*.

Carroll, M., & Kirsner, K. (1982). Context and repetition effects in lexical decision and recognition memory. *Journal of Verbal Learning and Verbal Behavior, 21*, 55–69.

Collins, A. M., & Loftus, E. F. (1975). A spreading activation theory of semantic processing. *Psychological Review, 82*, 402–408.

Cortese, J. (2007). *Internet learning and the building of knowledge*. Youngstown, NY: Cambria Press.

Craik, F. I. M., & Tulving, E. (1975). Depth of processing and the retention of words in episodic memory. *Journal of Experimental Psychology: General, 104*(3), 268–294.

Cummins, R. (1996). *Representations, targets, and attitudes*. Cambridge. MA: MIT Press.

Eveland, W. P., Jr. (2001). The cognitive mediation model of learning from the news: Evidence from nonelection, off-year election, and presidential election contexts. *Communication Research, 28*, 571–601.

Ghanem, S. (1997). Filling in the tapestry: The second level of agenda setting. In Maxwell McCombs, Donald Shaw, & David Weaver (Eds.), *Communication and democracy: Exploring the intellectual frontiers in agenda-setting theory* (pp. 3–14). Mahwah, NJ: Lawrence Erlbaum Associates.

Graber, D. A., & Smith, J. M. (2005). Political communication faces the 21st century. *Journal of Communication, 55*(3), 479–507.

Guo, L., & McCombs, M. (2011a). *Network agenda setting: A third level of media effects*. Paper accepted for presentation to the ICA annual conference, Boston.

Guo, L., & McCombs, M. (2011b). *Toward the third level of agenda setting theory: A network agenda setting model*. Paper accepted for presentation to the AEJMC annual conference, St. Louis.

Harris, H. R. (2011). Tavis Smiley: Obama must do more to live up to King's ideals. Retrieved from The Washington Post website: http://voices.washingtonpost.com/44/2011/01/tavis-smiley-obama-must-do-mor.html

Higgins, E. T. (1996). Knowledge activation: Accessibility, applicability, and salience. In E. T. Higgins & A. W. Kruglanski (Eds.), *Social psychology: Handbook of basic principles* (pp. 133–168). New York: Guilford.

Kaplan, S. (1973). Cognitive maps in perception and thought. In R. M. Downs & D. Stea (Eds.), *Image and Environment: Cognitive mapping and spacial behavior* (pp. 63–78). Chicago, IL: Aldine.

Kim, K., & McCombs, M. (2007). News story descriptions and the public's opinions of political candidates. *Journalism & Mass Communication Quarterly, 84*(2), 299–314.

Kiousis, S. (2005). Compelling arguments and attitude strength: Exploring the impact of second-level agenda setting on public opinion of presidential candidate images. *The Harvard International Journal of Press/Politics, 10*(2), 3–27.

Kitchin, R. M. (1994). Cognitive maps: What are they and why study them? *Journal of Environmental Psychology 14*, 1–19.

Lang, A. (2000). The limited capacity model of mediated message processing. *Journal of Communication, 50*(1), 46–71.

Lindsay, P., & Norman, D. A. (1977). *Human information processing: An introduction to psychology* (2nd ed.). New York: Academic Press.

Lippmann, W. (1922). *Public opinion*. New York: Macmillan.

McCombs, M. (2004). *Setting the agenda: The mass media and public opinion*. Cambridge, England: Polity Press.

McCombs, M., & Ghanem, S. (2001). The convergence of agenda setting and framing. In S. Reese, O. Gandy & A. Grant (Eds.), *Framing public life* (pp. 67–81). Mahwah, NJ: Lawrence Erlbaum Associates.

McCombs, M., Lopez-Escobar, E., & Llamas, J. P. (2000). Setting the agenda of attributes in the 1996 Spanish general election. *Journal of Communication, 50*(2), 77–92.
McCombs, M., & Shaw, D. L. (1972). Agenda-setting function of mass media. *Public Opinion Quarterly*, 36(2), 176–187.
McCutchin, M. (2008). *Creative problem-solving/ Ideation techniques*. PowerPoint Presentation for the Class "Creative Strategy": The University of Texas at Austin.
Monroe, B. M., & Read, S. J. (2008). A general connectionist model of attitude structure and change:The ACS (Attitudes as Constraint Satisfaction) Model. *Psychological Review, 11*(3), 733–759.
Price, V., & Tewksbury, D. (1997). News values and public opinion: A theoretical account of media priming and framing. In G. A. Barett & F. J. Boster (Eds.), *Progress in communication sciences: Advances in persuasion* (Vol. 13, pp. 173–212). Greenwich, CT: Ablex.
Rumelhart, D. E., & Norman, D. A. (1978). Accretion, tuning and restructuring: Three modes of learning. In J. W. Cotton & R. Klatzky (Eds.), *Semantic factors in cognition*. Hillsdale, NJ: Erlbaum.
Santanen, E., Briggs, R., & de Vreede, G.-J. (2000). The cognitive network model of creativity: A new causal model of creativity and a new brainstorming technique. In R. H. Spraque (Ed.), *Proceedings of the 33rd Hawaii International Conference on System Sciences*. Los Alamitos: IEEE Computer Society Press.
Schacter, D. L. (1987). Implicit memory: History and current status. *Journal of Experimental Psychology: Learning, Memory, and Cognition, 13*(3), 501–518.
Schacter, D. L., & Graf, P. (1986). Effects of elaborative processing on implicit and explicit memory for new associations. *Journal of Experimental Psychology: Learning, Memory, and Cognition, 12*(3), 432–444.
Schoenbach, K., & Semetko, H. A. (1992). Agenda-setting, agenda reinforcing or agenda-deflating? A study of the 1990 German national election. *Journalism Quarterly, 69*(4), 837–846.
Severin, W. J., & Tankard, J. W. (2001). *Communication theories: Origins, methods, and uses in the mass media* (5th ed.). New York: Longman.
Shaw, D. L., & McCombs, M. (1977). *The emergence of American political issues: The agenda-setting function of the press*. St. Paul, MN: West.
Siegel, A. W., & Cousins, J. H. (1985). The symbolizing and symbolised child in the enterprise of cognitive mapping. In R. Cohen (Ed.), *The development of spatial cognition*. Hillsdale, NJ: Lawerence Erlbaum.
Wasserman, S., & Faust, K. (1994). *Social network analysis: Methods and applications*. Cambridge: Cambridge University Press.
Weaver, D. H., Graber, D. A., McCombs, M., & Eyal, C. H. (1981). *Media agenda-setting in a presidential election: Issues, images, and interest*. New York: Praeger.
Williams, W., Jr., Shapiro, M., & Cutbirth, C. (1983). The impact of campaign agendas in perceptions of issues in the 1980 campaign. *Journalism Quarterly, 60*, 226–231.
Yioutas, J., & Segvic, I. (2003). Revisiting the Clinton/Lewinsky scandal: The convergence of agenda setting and framing. *Journalism & Mass Communication Quarterly, 80*(3), 567–582.

Chapter 7

It's Not Just a Laughing Matter

How Entertainment News Programs Influence the Transfer of the Media's Agenda to the Public's Agenda Similarly to Traditional Hard News

Jennifer Kowalewski[1]

Television viewers of the 1960s hit variety show *Rowan & Martin's Laugh In* were shocked on Sept. 16, 1968, to see a four-second snippet featuring former Vice President Richard Nixon. The 1968 Republican presidential candidate slowly faced the camera, and, in one of television history's defining moments, spouted one of the shows many catch phrases, "Sock it to me?" Decades later, show producer George Schlatter told *The New Yorker* the bit made Nixon appear like "a nice guy" (Kolbert, 2004). Schlatter posited that many people felt the appearance on the variety show was one of many reasons Nixon won the 1968 presidential election against then Vice President Hubert Humphrey, who declined to be on the show.

Although no research has shown evidence that the appearance had any impact on Nixon winning the election, we have seen more politicians turning to nontraditional news programs to reach out to potential voters. In 2000, both presidential candidates, Vice President Al Gore and Texas Governor George W. Bush, made an appearance on a primetime special of *Saturday Night Live*, in addition to their appearances on late-night comedy talk shows. In 2004, one candidate, John Edwards, announced his candidacy for president on *The Daily Show*. In 2008, Republican presidential candidate John McCain and Democratic presidential candidate Barack Obama both appeared on *The Daily Show*.

Because politicians have increasingly used these types of programs, scholars have attempted to gauge the impact these appearances might have on voters. One way to do this would be to use one of mass communication's most researched theories, agenda setting (McCombs & Shaw, 1972). Scholars need to investigate how successful these programs have been in transferring the salience of a certain issue from the media's agenda to the public's agenda. When the information is presented in an alternative format rather than the traditional hard news program, will people accept the agenda? And how do these appearances in entertainment formats impact voters? In two separate experiments, this study attempted to determine whether presentation of the information influenced the agenda-setting process.

Agenda Setting

Agenda setting remains an important area to continue to investigate because, as Camaj (Chapter 5, this volume) showed in her study, the media not only shape the public's agenda, traditional news also has shaped the political culture in which we live. Moon (Chapter 8, this volume) showed support for the idea that agenda setting remained an important area because the theory could relate to people's opinion formation and behavior. In other words, the more people focused on the media, the more involved they were politically, which showed evidence of agenda setting as a mediating variable "between media use and attitude strength" (Moon, Chapter 8, this volume). Therefore, this study expects to have similar findings to other studies.

> **H1:** Individuals exposed to information presented in a traditional news format will accept the media's agenda and cite that issue as one of the most important problems facing the nation compared to those who did not receive the issue.

We also have support, on a limited basis, to show how entertainment programs can shape the public's agenda irrespective of how the information is presented (Dearing & Rogers, 1996; Weaver, 1994). For example, when Clinton appeared on entertainment programs and spoke about certain issues important to him, individuals cited those issues as more important than other issues that he failed to mention. Due to Clinton's use of nontraditional news sources, Weaver (1994) argued, these entertainment programs had a significant agenda-setting effect on voters. Holbrook and Hill (2005) also argued that entertainment programs could impact agenda-setting effects. In their experiment, the scholars found individuals exposed to crime dramas indicated crime as an important issue in the nation, clear evidence of an agenda-setting effect. Therefore, this study expects to have similar findings.

> **H2:** Individuals exposed to information presented in an entertainment news format will accept the media's agenda and cite that issue as one of the most important problems facing the nation compared to those who did not receive the issue.

Entertainment News Versus Traditional Hard News

Zillmann (2000) argues that entertainment might influence individuals more because entertainment programs have the ability to diminish or to remove aversion to issues. Entertainment programs diminish or remove aversion to issues by making the message more entertaining. By doing so, individuals may feel more positive toward the information contained within the message; therefore, the entertaining message can be very persuasive. How and why the persuasiveness of entertainment actually occurs is a subject that many scholars have attempted to explain using a myriad of theoretical frameworks. With regard to one of the most dominant frameworks used to understand media effects, scholars have looked to dual processing models, such as the elaboration likelihood model or heuristic-systematic model, to help explain why entertainment can be more persuasive than other forms of media (Petty & Cacioppo, 1984a, 1984b; Petty, Schumann, Richman, & Strathman, 1993; Raney, 2006; Shrum, 2002; Slater and Rouner, 2002).

Petty and Cacioppo (1984a; 1984b) posited that individuals use different routes when processing information—central or peripheral. When individuals feel engaged in the message, they process the information using the central route, often counter-arguing against information contained in the message. When individuals feel less engaged in the message, they process the information using the peripheral route, often failing to counter-argue against the information contained in the message. When individuals counter-argue against the information, individuals often fail to accept the information contained in the message. They argue against the message; therefore, they dismiss the information as unimportant when they counter-argue against it.

Slater and Rouner (2002) extend the dual-processing models to entertainment programs. When the message is entertaining, although individuals use a central route to process information, they fail to counter-argue against the entertainment program because they are enjoying that program, and this makes the information more persuasive (Slater, 2002; Slater & Rouner, 2002). Shrum (2002) agreed that individuals fail to counter-argue against entertainment programs, but believed individuals exposed to entertainment programs use the peripheral route instead of the central route, because failure to counter-argue is most associated with peripheral processing. Therefore, scholars agree that individuals fail to counter-argue against entertainment programs, even if they do not agree with the premise as to why.

Besides the dual-processing models of persuasion, scholars have speculated that there are other reasons why entertainment is so persuasive. Cultivation theory helps explain why individuals accept the world of television as real, or why entertainment programs can be considered persuasive. The more individuals watch television, the more individuals accept the television world as real. This helps explain why individuals who watch entertainment programs often consider these programs as more real, or persuasive. Also, entertainment provides individuals with "a respite from the anxieties and pressures of everyday life; the opportunity to compare oneself with the demeanor, possession, and behavior of others; and a means of keeping up with what is transpiring in the world" (Comstock, 2004, p. 193). Entertainment also can be persuasive because of the idea of "willing suspension of disbelief" (Shrum, 2002). Individuals are willing to suspend their disbelief when they watch entertainment programs because they already know the information is not accurate.

However, scholars have argued that individuals do find a difference between serious messages and humorous messages. Even though individuals turn to comedy for enjoyment (Zillmann, 2000), programs containing humor are not necessarily more persuasive than programs containing serious messages (Nabi, Moyer-Gusé & Byrne, 2007). Individuals might discount the information more when it is presented in a humorous manner, making the information less persuasive in the end.

Holbert, Pillion, and Tschida (2003) have called for more research comparing entertainment to news. The authors argue "the theory of priming in political communication has been relegated to the study of news, but an argument is made for the relevance of this theory in this context to the study of entertainment television and politics" (p. 428). Since priming relates to agenda setting, especially in terms of public opinion formation, we can argue that more research needs to test entertainment news programs in relation to agenda setting. Holbrook and Hill (2005) determined that crime dramas could influence agenda setting. But little research has directly compared entertainment news programs to hard news programs. So, the question remains, will individuals accept the media's agenda more when the information is presented as traditional hard news, or as entertainment news? This study will ask the question:

RQ1: How does the presentation style of the information, entertainment versus hard news, influence individual's acceptance of the media's agenda?

Variables Influencing the Agenda-Setting Effects

After the initial agenda-setting study was published, scholars furthered the theory by investigating the magnitude of the agenda-setting effect. With regard to other factors that influence the magnitude of the agenda-setting effect, scholars have investigated how individuals have a need for orientation, or a cognitive need for information, that influences the agenda-setting process (Kim, Chapter 4, this volume; Weaver, 1980), how individuals' viewing characteristics influenced whether or not they accepted the media's agenda (Hill, 1985) regarding topics from gender and race (Miller & Wanta, 1996) to income (Shaw & Martin, 1992). Interpersonal communication (Atwater, Salwen, & Anderson, 1985; Wanta & Wu, 1992; Yang & Stone, 2003) as well as personal experience (Culbertson & Stempel, 1985; Dearing & Rogers, 1996; McCombs, 2004; Salwen, 1987) also appeared to influence the agenda-setting effects. Obtrusiveness of issues also influences the magnitude of the agenda-setting effect (Yagade & Dozier; 1990; Weaver, Zhu, & Willnat, 1992). Valenzuela (Chapter 3, this volume) talked about the variables that influenced the agenda-setting process, calling for more research into the psychological variables that might impact the effects.

One psychological variable might be the individuals' perceptions of the information,specifically, how newsworthy and how entertaining people found the news story they received, regardless of the presentation style. Participants in this study answered questions about the newsworthiness and entertainment value of the news story they received in the experiments, based on scales developed by Zillmann, Taylor, and Lewis (1998). Individuals' perceptions of the information might relate to how relevant they found the information, which is one of the two distinct characteristics that make up need for orientation (McCombs, 2004; Weaver, 1980). Individuals who find the information more relevant might accept the media's agenda more than individuals who find the information less relevant. Miller (2007) theorized that more negative news stories might influence the agenda-setting process more than more positive news stories since the negative news stories indicate the issue is more problematic. In her experiment, Miller showed individuals cited issues prominent in a negative news story more than individuals who were exposed to a positive news story about the same issue. So, Miller posited individuals adopted the issue salience by taking cues from the news media on what issues were important.

Since research indicates that many variables influence the agenda-setting effects, one could assume individuals' level of entertainment at the news stories, and how newsworthy they found the information, could influence their acceptance of the information. The study will investigate how individuals' perceptions of the information, whether they considered it newsworthy or entertaining, influenced agenda setting. Also, the study will investigate the interaction of presentation style, entertainment news versus hard news, and the influence of individuals' perception on agenda setting. Because research rarely has investigated individuals' acceptance of the media's agenda based on these scales, the study will ask the questions:

RQ2: How do individuals' feelings toward how newsworthy or how entertaining they found the information influence individuals' acceptance of the media's agenda?
RQ3: How does the presentation style, entertainment versus hard news, interact with individuals' perceptions toward the information, how newsworthy or how entertaining they found the information, influence individuals' acceptance of the media's agenda?

Method

To determine whether presentation style, namely hard news versus entertainment news, mediates the agenda-setting process, a between-subject experiment with a control group was developed. Even though many scholars have shown agenda-setting effects based off content analysis and survey research, several scholars have use experiments to find a causal relationship between exposure to issues and the salience of those issues (Iyengar & Kinder, 1986; Iyengar & Kinder, 1987). Therefore, scholars can use experiments to show agenda-setting effects. This experiment hoped to find a causal relationship between exposure to the issues in different presentation styles and the salience of those issues in the public's agenda.

Because the main component of this study was how individuals processed the presentation style differently, a pilot test was completed to assess whether the stimulus material, mainly the humor embedded in the radio address, had experimental validity. Participants (n = 40) read the news stories then answered questions about their reaction to the news story, mainly their level of enjoyment and how newsworthy they found the information, based on questions asked by Zillmann, Taylor, and Lewis (1998). Overall, the results indicated that the manipulation worked. Participants viewed the entertainment news story (M = 25.22) as more entertaining than the hard news story, M = 13.59, t-test (38df) = -3.667, $p < 0.001$.

Main Experiment Procedure

Participants first completed the pre-questionnaire, which included demographic and media reliance questions. Participants then listened to a radio address in streaming audio. After hearing the radio program, participants had a Sudoku puzzle to complete as a distraction. Participants then completed a post-questionnaire including what major issues were facing Americans, or the agenda-setting questions. After answering all questions, participants received a debriefing form.

Stimulus material: Participants heard the radio program formatted as a traditional news program or as an entertainment news program. The entertainment news program included the same information as the traditional news, with added humor (see Appendix).

Story content. All of the participants heard two buffer stories. Individuals then were randomly assigned to hear one of three issue stories created for this experiment. Each of the issue stories discussed a fictitious newlyelected senator, Joseph Beale, I-Wyo., who planned on proposing a new bill in Congress when he arrived in Washington in January. The first issue dealt with the proposal to stop Medicaid health care benefits to those individuals who were employed but made minimum wage and whose employers did not provide health care benefits. The second issue dealt with the proposal to create the Illegal Immigration Enforcement Agency, which would hunt down and deport illegal immigrants. The third issue dealt with the proposal to begin more offshore drilling, without any funding for alternative energy sources.

Both studies focused on how the presentation style, hard news versus entertainment news, influenced participants' acceptance of the information. Presentation style became the independent variable. To determine whether the presentation style influenced the agenda-setting effect, participants listed three important issues facing the nation. These are similar to questions asked both by national polls (Gallup Poll), as well as by scholars investigating agenda setting (McCombs, 2004; McCombs & Shaw, 1972; Dearing & Rogers, 1996). Responses from individuals who heard stories about health care, immigration, or offshore drilling were coded to evaluate whether these individuals cited the received issues more than individuals not exposed to stories about health care, immigration, or offshore drilling. This became the dependent variable.

The same measures used in the pilot test also were used in the main experiment to ensure that participants viewed the entertainment stories as more entertaining than the hard news stories. For each target issue story, the measures were factor analyzed using Principal Axis Factoring with Varimax rotation ($KMO = 0.83$). Two factors emerged. Four of the statements gauged how entertaining the information was: It was amusing, hilarious, enjoyable, and entertaining ($\alpha = 0.93$) (Zillmann et al., 1998). The four measures were averaged to create a variable called the entertainment scale. Five of the statements gauged how newsworthy the information was: The information was informative, authentic, credible, seemed real to me, and was newsworthy ($\alpha = .088$) (Zillmann et al., 1998). The fivemeasures were averaged to create a variable called the newsworthiness scale. The newsworthiness scale and the entertainment scale were recoded into three variables: those who scored the stories low on either scale, those who scored the stories as "only somewhat" on both scales, and those who scored the stories high on either scale. The recoding of both scales was necessary to create three groups for each scale—those who found the stories less newsworthy, somewhat newsworthy, and more newsworthy, and those who found the stories less entertaining, somewhat entertaining, and more entertaining. The scales were then analyzed to see how they impacted the acceptance of the media's agenda.

Study I

Participants. Participants were recruited from a public university. Most of the participants were female (82.6%, $n = 123$) and Caucasian (84.7%, $n = 127$). The participants ranged in age from 18 to 25, with most of them (93.3%, $n = 140$) between 20 and 22 years old. More participants considered themselves Democrat, (44.0%, $n = 66$) compared to those who considered themselves Republican (28.2%, $n = 42$). Participants appeared generally neutral in their political leaning ($M = 3.17$, $SD = 1.04$) according to their self-reports on an item ranging from a possible 1 = strongly conservative to 5 = strongly liberal.

Independent variable/presentation style. Individuals received either only the two buffer stories (control condition, $n = 78$) or the two buffer stories with one of the three issues (experimental condition, $n = 72$)—health care ($n = 23$), immigration ($n = 26$), or offshore drilling ($n = 23$)—to determine how the presentation style, hard news ($n = 77$) versus entertainment news ($n = 73$), influenced participants' acceptance of the information.

Dependent variable/agenda setting. The findings showed that individuals were more likely to cite the issue they received than those who did not receive that issue, $\chi^2(149) = 5.87$, $p < 0 .05$. The data indicated that individuals who received any of the three issues cited the particular issue they received as compared to those who did not receive that issue—health care, ($\chi^2(149) = 6.39$, $p < 0 .05$), immigration ($\chi^2(149) = 5.03$, $p < 0.05$), and offshore drilling, $\chi^2(149) = 5.87$, $p < 0.05$.

Newsworthiness/entertainment scales. Individuals exposed to the entertainment news ($M = 5.83$, $SD = 2.60$) found the issue stories more entertaining than those exposed to the hard news ($M = 3.39$, $SD = 1.65$, $t(70) = -4.76$, $p < 0.001$). Individuals indicated both the hard news ($M = 3.26$,

SD = 3.64) and entertainment news (*M* = 3.22, *SD* = 3.50) equally as newsworthy regardless of the condition they received ($t(148) = 0.07$, $p = 0.945$). Each of the issues then was analyzed separately to investigate whether individuals indicated differences between the entertainment and newsworthiness scale based on presentation style. Individuals who received the immigration story reported entertainment news (*M* = 6.70, *SD* = 2.44) more entertaining than the hard news (*M* = 2.50, *SD* = 1.29, $t(24) = -5.71$, $p > 0.001$). Individuals who received the immigration story found no difference between the hard news (*M* = 6.93, *SD* = 1.75) and the entertainment news (*M* = 6.87, *SD* = 1.23) when it came to the newsworthiness scale ($t(24) = .10$, $p = .923$). Individuals who received the health care story did not perceive a difference between the hard news (*M* = 3.96, *SD* = 1.48) and entertainment news (*M* = 4.96, *SD* = 2.25) when it came to the entertainment scale ($t(21) = 0.10$, $p = 0.220$). Individuals who received the health care story found the hard news (*M* = 7.43, *SD* = 1.30) more newsworthy than the entertainment news (*M* = 5.31, *SD* = 1.88, $t(21) = 3.16$, $p < 0.01$). Individuals who received the offshore drilling story found no difference between the hard news (*M* = 4.11, *SD* = 1.85) and entertainment news (*M* = 5.82, *SD* = 2.90) when it came to the entertainment scale ($t(21) = -1.57$, $p = 0.130$). Individuals who received the offshore drilling story found no difference between the hard news (*M* = 6.46, *SD* = .79) and the entertainment news (*M* = 7.23, *SD* = 1.46) when it came to the newsworthiness scale ($t(21) = -1.45$, $p = 0.162$). Although for health care and offshore drilling, the entertainment version failed to be more entertaining than the hard news version based on the results from the entertainment scale, the data appeared to be trending in that direction. All three variables were kept in further analysis.

Study 2

Participants: Participants were recruited from both a private university and the local community for the second experiment. Most of the participants were female (73.4%, *n* = 141) and Caucasian (83.3%, *n* = 160). The participants ranged in age from 18 to 84 with individuals almost evenly split between younger participants, 18 to 24 (52.1%, n = 100) and older participants, 25 and older (47.9%, n = 92). More participants considered themselves Democrat (53.1%, *n* = 102) compared to those who considered themselves Republican (25.0%, *n* = 48). Participants appeared generally neutral in their political leaning (*M* = 2.68, *SD* = 1.08) according to their self-reports on an item ranging from a possible 1 = strongly conservative to 5 = strongly liberal.

Independent variable/presentation style. All participants (*n* = 192) received all three news stories including the buffer stories of abstinence and video games. Participants were randomly assigned to receive one of the issue stories—health care ($n = 61$), immigration ($n = 70$), and offshore drilling ($n = 61$)—presented as either entertainment news ($n = 95$) or traditional hard news ($n = 97$).

Dependent variable/agenda setting. The findings showed that individuals were more likely to cite the issue they received than those who did not receive that issue, $\chi^2(192) = 13.78, p < 0.05$. The data indicated that individuals cited the three issues more when they received those particular issues—health care ($\chi^2(192) = 7.41, p < 0.01$), immigration ($\chi^2(192) = 10.24, p < 0.001$), and offshore drilling, $\chi^2(192) = 17.69, p < 0.001$.

Newsworthiness/entertainment scales. Individuals exposed to the entertainment news ($M = 5.77, SD = 2.56$) found the issue stories more entertaining than those exposed to the hard news ($M = 3.28, SD = 1.73, t(190) = -7.89, p < 0.001$). Individuals found the hard news ($M = 6.16, SD = 1.79$) more newsworthy than the entertainment news ($M = 5.51, SD = 1.90, t(190) = 2.42, p < 0.05$). Each of the issues then was analyzed separately to investigate whether individuals indicated differences between the entertainment and newsworthiness scale based on presentation style. Individuals who received the health care story indicated the entertainment news ($M = 6.58, SD = 2.55$) more entertaining than the hard news ($M = 3.94, SD = 1.90, t(59) = -4.60, p < 0.001$). Individuals who received the health care story indicated the hard news ($M = 6.17, SD = 2.05$) more newsworthy than the entertainment news ($M = 5.16, SD = 1.75$) when it came to the newsworthiness scale ($t(59) = 2.07, p < 0.05$). Individuals who received the immigration story found the entertainment news ($M = 5.82, SD = 2.55$) more entertaining than the hard news ($M = 3.24, SD = 1.79$) when it came to the entertainment scale ($t(68) = -4.91, p < 0.001$). Individuals who received the immigration story found no difference between the hard news ($M = 6.17, SD = 1.64$) and entertainment news ($M = 5.75, SD = 1.97$) on the newsworthiness scale ($t(68) = .97, p = 0.338$). Individuals who received the offshore drilling story indicated the entertainment news ($M = 4.92, SD = 2.40$) more entertaining than the hard news ($M = 2.65, SD = 1.16$) when it came to the entertainment scale ($t(59) = -4.68, p < 0.001$). Individuals who received the offshore drilling story found no difference between the hard news ($M = 6.14, SD = 1.73$) and the entertainment news ($M = 5.60, SD = 1.98$) when it came to the newsworthiness scale ($t(59) = 1.14, p = 0.260$). The findings indicated individuals who received the entertainment version, regardless of the issue, indicated the entertainment version more entertaining as compared to those who received the hard news version. Overall, the manipulation worked.

Results

To answer the hypotheses, Chi-square tests were performed on all three issues combined, then all three issues separately to determine if those who received an issue were more likely to cite the issue as one of the most important facing the nation. Two analyses were performed—one on those who received traditional hard news and one on those who received entertainment news.

In study 1, individuals in the hard news condition were more likely to cite the issue when they received that issue than those who did not receive the issue, $\chi^2(76) = 6.87$, $p < 0.01$. When looking at the individual issues, an interesting pattern emerged. Individuals in the health care or offshore drilling conditions were more likely to cite the issue when they received the hard news, $\chi^2(45) = 5.01$, $p < 0.05$ and $\chi^2(45) = 7.25$, $p < 0.01$ respectively. However, individuals in the immigration condition were not more likely to cite the issue when they received the hard news, $\chi^2(25) = .81$, $p = 0.368$. In study 2, individuals in the hard news condition were more likely to cite an issue that they received than those who did not receive the issue, $\chi^2(192) = 9.43$, $p < 0.01$. When looking at the individual issues, individuals in each condition were more likely to cite the issue they received—health care ($\chi^2(192) = 6.08$, $p < 0.05$), immigration ($\chi^2(192) = 4.37$, $p < 0.05$), offshore drilling, $\chi^2(192) = 9.25$, $p < 0.01$. Therefore, the results mostly supported the first hypothesis in that individuals in the hard news condition were more likely to cite the issue they received as a major problem compared to those who did not receive the issue.

In study 1, individuals in the entertainment news condition were not more likely to cite the issue they received than those who did not receive the issue when investigating all issues combined, $\chi^2(72) = .71$, $p = 0.401$. When looking at each issue separately, individuals in the health care or offshore drilling condition were not more likely to cite the issue when they received the entertainment news, $\chi^2(71) = 1.71$, $p = 0.191$ and $\chi^2(72) = .85$, $p = 0.358$ respectively. However, individuals in the immigration condition were more likely to cite the issue when they received the entertainment news, $\chi^2(72) = 5.63$, $p < 0.05$. In study 2, individuals in the entertainment news condition were not more likely to cite the issue they received than those who did not receive the issue when investigating all issues combined, $\chi^2(192) = 4.90$, $p = 0.086$. However, the results indicated the entertainment condition was trending in that general direction. When looking at each issue separately, individuals who received the health care piece were not more likely to cite the issue when they received the entertainment news, $\chi^2(192) = 1.89$, $p = 0.169$. However, individuals in the immigration and offshore drilling condition were more likely to cite the issue when they received the entertainment news, $\chi^2(192) = 5.97$, $p < 0.05$ and $\chi^2(192) = 8.23$, $p < 0.01$ respectively. Therefore, the results indicated partial support for the second hypothesis that individuals in the entertainment condition would cite the issue they received more than individuals who did not receive the issue (see Table 7.1).

Table 7.1 Cross-tabulation of presentation style and issue received on whether individuals cited that issue as one of the most important problems facing the country

Issue received	*Presentation style*	*Did they receive the issue*	*Study 1* Did they cite the issue they received as the MIP			*Study 2* Did they cite the issue they received as the MIP		
			No	*Yes*	c^2	*No*	*Yes*	χ^2
Health care	Hard news	No	39 (0.6)	26 (-0.7)	5.01*	39 (1.0)	27 (-1.0)	6.08*
		Yes	3 (-1.4)	9 (1.5)		10 (-1.4)	21 (1.4)	
	Entertainment News	No	41 (0.3)	21 (-0.4)	1.71	38 (0.5)	27 (-0.6)	1.89
		Yes	5 (0.3)	6 (1.0)		13 (-0.8)	17 (0.8)	
	Total	No	80 (0.6)	47 (-0.8)	6.39*	77 (1.1)	54 (-1.1)	7.41*
		Yes	8 (-1.5)	15 (1.8)		23 (-1.6)	38 (1.6)	
Immigration	Hard news	No	55 (0.1)	7 (-0.4)	.811	53 (0.6)	8 (-1.1)	4.37*
		Yes	12 (-0.3)	3 (0.8)		25 (-0.7)	11 (1.5)	
	Entertainment News	No	56 (0.3)	6 (-0.9)	5.63*	55 (0.6)	6 (-1.3)	5.97*
		Yes	7 (-0.8)	4 (2.0)		24 (-0.8)	10 (1.8)	
	Total	No	111 (0.3)	13 (-0.9)	5.03*	108 (0.8)	14 (-1.7)	10.24*
		Yes	19 (-0.7)	7 (1.9)		49 (-1.1)	21 (2.3)	
Offshore drilling	Hard news	No	64 (0.3)	4 (-0.9)	7.25*	57 (0.8)	10 (-1.5)	9.25*
		Yes	6 (-0.8)	3 (2.4)		17 (-1.2)	13 (2.2)	
	Entertainment News	No	55 (0.1)	4 (-0.4)	.85	53 (0.9)	11 (-1.4)	8.43*
		Yes	12 (-0.2)	2 (0.8)		17 (-1.2)	14 (2.0)	
	Total	No	119 (0.3)	8 (-0.9)	5.87*	110 (1.2)	21 (-2.1)	17.69*
		Yes	18 (-0.7)	5 (2.1)		34 (-1.7)	27 (3.0)	
All issues	Hard news	No	21 (1.4)	20 (-1.1)	6.87*			
		Yes	8 (-1.5)	28 (1.2)		52	45	9.43*
	Entertainment News	No	18 (0.4)	19 (-0.4)	.71			
		Yes	14 (-0.4)	22 (0.4)		54	41	4.90
	Total	No	39 (1.3)	39 (-1.1)	5.87*			
		Yes	22 (-1.3)	50 (1.1)		106	86	13.76*

Note: * $p < 0.05$. Adjusted standardized residuals appear in parentheses after group frequencies.

To answer the first research question, Chi Squares tested whether individuals in the hard news condition were more likely to cite the issue they received, as compared to those in the entertainment news condition. For both studies, the Chi Square was not significant, showing individuals in the entertainment news condition were just as likely to cite the issue they received as a major issue as compared to those in the hard news condition, (study 1: $\chi^2(150) = .24$, $p < 0.628$; study 2: $\chi^2(192) = 0.20$, $p < .652$) The results answered the first research questions, in that individuals in the entertainment condition indicated an agenda-setting effect similar to individuals in the hard news condition.

When answering the second and third research questions, an ANOVA was performed to determine if individuals' perceptions of how newsworthy or how entertaining they found the information influenced their acceptance of the media's agenda. For study 1, an interaction occurred between presentation style and how newsworthy individuals found the information, $F(71) = 4.26$, $p < 0.05$. Results indicated when individuals felt the information was more newsworthy; they cited the issue more when they received the hard news as compared to the entertainment news. Unlike study 1, the results for study 2 showed no interaction between the presentation style and individuals' perceptions, or how newsworthy or entertaining they found the stories, on whether

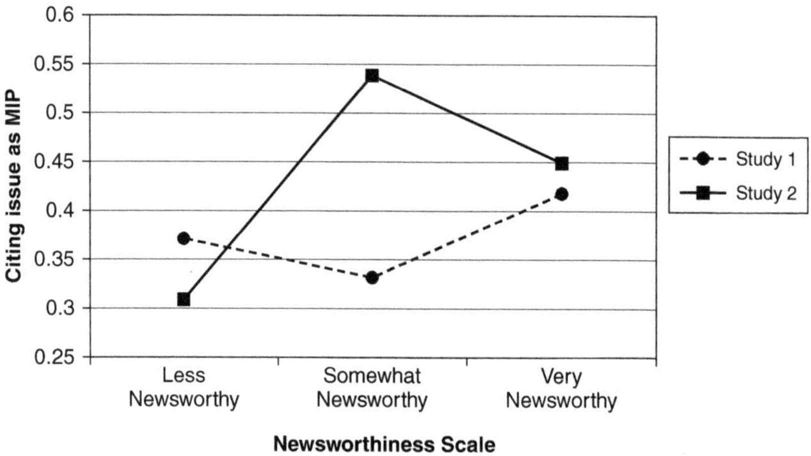

Figure 7.1 How newsworthy individuals felt the stories were and how that impacted the transfer of issue salience from the media's agenda to the participants' agenda. The graph shows the simple slope for each study

The Y-axis shows the newsworthiness scale, which was based on three points the mean of those who considered the story less newsworthy, the mean of those who considered the story somewhat newsworthy; and the mean of those who considered the story very newsworthy. The X-axis shows the likelihood individuals cited the issue they received as the most important problem facing the nation.

Table 7.2 ANOVA analysis of presentation style and newsworthiness scale on whether individuals cited the issue they received as the most important issue

	Study 1				Study 2			
	df	*F*	η^2	*p*	*df*	*F*	η^2	*p*
Presentation	1	0.39	0.01	0.536	1	0.00	0.00	0.984
Scale	2	.015	0.00	0.863	2	3.28	0.30	0.040
Presentation * Scale	2	4.27	0.11	0.018	2	0.50	0.10	0.605
Error	66		.		86			

Note: The ANOVA analysis of presentation * scale refers to the interaction of presentation style and response on the newsworthiness scale to the dependent variable.

individuals accepted the media's agenda. However, results indicated the newsworthiness scale (F(191) = 3.28, $p < 0.05$) impacted individuals' acceptance of the media's agenda, regardless of presentation (see Figure 7.1). When individuals felt the information more newsworthy, they cited the issue more than those who felt the information was less newsworthy, answering the first part of the second research question (see Table 7.2).

When investigating how entertaining individuals found the information, the results showed no interaction between individuals' perceptions on the entertainment scale and the presentation style on influencing the agenda-setting effects, answering the third research question(see Table 7.3). The results indicated, though, that individuals' perceptions influenced their acceptance of the media's agenda, regardless of presentation style, for both study 1 (F(71) = 3.26, $p < 0.05$) and study 2

Table 7.3 ANOVA analysis of presentation style and entertainment scale on whether individuals cited the issue they received as the most important issue

	Study 1				Study 2			
	df	*F*	η^2	*p*	*df*	*F*	η^2	*p*
Presentation	1	1.82	0.03	0.182	1	1.20	0.01	0.274
Scale	2	3.26	0.09	0.045	2	3.04	0.03	0.050
Presentation * Scale	2	0.18	0.01	0.837	2	1.23	0.01	0.296
Error	66		.		186			

Note: The ANOVA analysis of presentation * scale refers to the interaction of presentation style and response on the entertainment scale to the dependent variable.

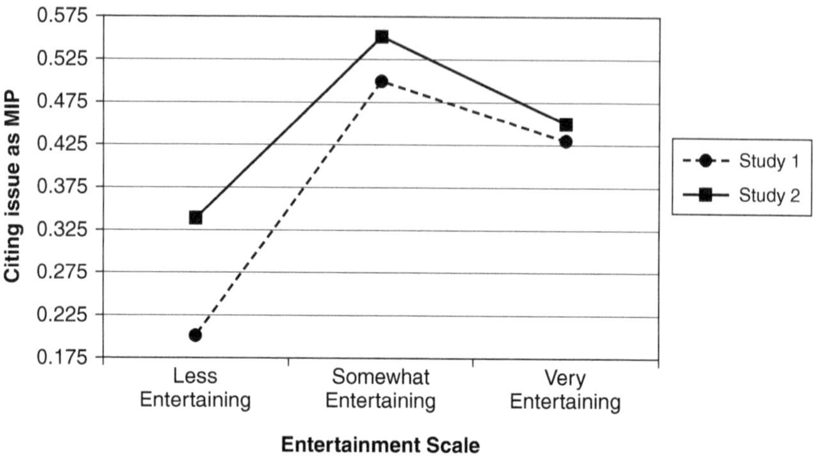

Figure 7.2 How entertaining individuals felt the stories were and how that impacted the transfer of issue salience from the media's agenda to the participants' agenda

The graph shows the simple slope for each study. The Y-axis shows the entertainment scale, which was based on three points: the mean of those who considered the story less entertaining, the mean of those who considered the story somewhat entertaining; and the mean of those who considered the story very entertaining. The X-axis shows the likelihood individuals cited the issue they received as the most important problem facing the nation.

($F(191) = 3.04$, $p < 0.05$). When individuals felt the information was more entertaining, whether they received the entertainment or the hard news version, they indicated more acceptance of the media's agenda than those who felt the information was less entertaining, answering the second research question (see Figure 7.2).

Discussion

The results indicated, overall, that individuals failed to react differently to the information presented as either the traditional hard news or the entertainment news. When investigating each presentation style separately, results indicated individuals in the hard news condition were more likely to cite the issue they received as more of a problem facing the nation than individuals who did not receive the issue, showing an agenda-setting effect. When the news media increase focus on certain issues, those issues become more critical to people; therefore, individuals cite those issues more as the most important problems facing the nation. The results indicated overall support for the first hypothesis; however, when investigating each issue individually, the findings were somewhat problematic. Individuals in the hard news condition who received the immigration story during the first study apparently failed to think of that issue as a major problem facing the nation. However, everyone does not automatically discuss an issue because the media discusses that issue. For example, obtrusive issues are more likely to influence the agenda of an individual if those issues are obtrusive in his or her own life. In this case, for instance, immigration is more prevalent for individuals living in certain areas than for individuals from other areas. Participants in the first study lived in an area with a large immigration population. Recent press coverage has discussed how the rising number of immigrants has led to an expansion of local schools (Maguire, 2006). Regardless of the reason for individuals to fail to react to the immigration hard news story in the first study, though, the findings for the first hypothesis indicate support for the extensive research from past studies on agenda setting (Dearing & Rogers, 1996; McCombs, 2004; McCombs & Shaw, 1972).

The study also found partial support for an agenda-setting effect for those who received the entertainment version. In the first study, individuals in the entertainment condition failed to cite the issue they received as a problem facing the nation. However, the findings were trending in the general direction to support the hypothesis in the second study. Individuals who received the entertainment news story apparently accepted the issue as the most important problem facing the nation in the second study. When each issue was investigated separately, individuals who received the immigration news story presented as entertainment cited the issue as a major problem facing the nation in both studies. And in the second study, individuals who received the offshore drilling news story presented in the entertainment format cited the issue more than those individuals who failed to get the issue. But individuals who received the entertainment health care story did not show an agenda-setting effect in either study. And individuals who received the entertainment offshore drilling story failed to list the story as an important issue. Nevertheless, the results indicate partial support for the scant amount of research into how the entertainment programs influence the agenda-setting process (Holbrook & Hill, 2005; Weaver, 1994).

When comparing those in the hard news versus those in the entertainment news condition, the results indicated no difference in the agenda-setting effect. In other words, individuals in either condition were more likely to cite the issue they received as one of the most important facing the nation. The findings indicate individuals react similarly to the information, even when it is presented in two different formats. In other words, the findings here indicate that entertainment programs could set the public's agenda. The results indicate traditional hard news programs do not have a monopoly on influencing the public's agenda. However, the traditional hard news appeared to have a stronger influence on the agenda-setting process in that the results indicated more support for the first hypothesis and only partial support for the second hypothesis.

But individuals' perception of the information may have impacted the agenda-setting effects, answering the second research questions. Overall, the studies found that individuals who felt the information was more newsworthy or more entertaining were more likely to cite the issue they received as a major problem facing the nation. However, only one study indicated an interaction effect between presentation style and individuals' perception on agenda setting, answering the third research question. A slight interaction occurred for individuals in study 1 who received the hard news, because they cited the issue more when they felt the information was newsworthy, as compared to those who received the entertainment news. No interaction effect occurred between presentation style and the entertainment scale, as well as between presentation style and the newsworthiness scale in study 2.

One interesting thing to note with the main effects of the newsworthiness and entertainment scales is that individuals' perceptions of the information apparently reached saturation at some point. Individuals who found the information very entertaining or very newsworthy were not more likely to cite the issue they received as compared to those who found the information somewhat entertaining or somewhat newsworthy. In fact, when individuals felt the information very entertaining or very newsworthy, they apparently began to dismiss the issue as being of major concern. Individuals who consider something very entertaining appear to dismiss the information, as other research has argued (Nabi, Moyer-Gusé & Byrne, 2007), because the program did not take the issue seriously. Why individuals appeared to dismiss the information when they considered it very newsworthy is a somewhat perplexing finding. Individuals may have reached the saturation point of tying their individual perceptions to their answers to the most important problem facing the nation.

However, individuals' perceptions of how newsworthy the information was in study 1 did not act like those in study 2. Individuals who found the information very newsworthy were more likely to cite the issue than those who listed the information as less newsworthy or somewhat newsworthy. The results may be because of the small sample size in study 1. Individuals in the entertainment condition apparently had a similar relationship to the other individuals' perceptions in that individuals who listed the information as somewhat newsworthy were more likely to cite the issue than those who felt the information was less newsworthy or very newsworthy. The anomaly occurred with those in the hard news condition. Individuals who felt the information was somewhat newsworthy were less likely to cite the issue than those who felt the information was less newsworthy or very newsworthy.

When trying to determine why the issues appeared to react differently based on the two hypotheses, the reason behind the different findings may be related to individuals' perceptions of the information. Scholars have argued that individuals' perceptions, whether they find the information obtrusive (Weaver, 1980), or they have past experience (Culbertson & Stempel, 1985; Salwen, 1987), might influence the agenda-setting process. With the studies here, individuals may have reacted differently to the issues because of how newsworthy or how entertaining they found the information. Valenzuela (Chapter 3, this volume) argued for research into the psychological variables influencing agenda setting, which might explain why individuals' perceptions impacted the results. Individuals' perceptions might relate to how relevant they found the information. The more relevant individuals found the information, the more newsworthy or the more entertaining they found the information. When they found the information less relevant, individuals felt the information was not as newsworthy or as entertaining; therefore, the psychological variables had an impact on agenda setting.

In the first study, individuals who received the health care or offshore drilling stories failed to find the entertainment version as statistically more entertaining that the hard news version. Therefore, because they were not entertained by the information, they apparently felt the information was not as important an issue facing the nation. And in both studies, individuals who received the traditional hard news version of the health care story indicated they felt the stories were more newsworthy than those who received the same information in the entertainment condition. When individuals view the information as more newsworthy, their perception of the issue should cause them to accept the media's agenda more than someone who may not believe the information was newsworthy. Therefore, because they found the hard news version so much more newsworthy than the entertainment version, the hard news version of the health care story should have more impact on people's acceptance of the media's agenda. When it came to the immigration news stories, individuals in study 1 who received the hard news story failed to indicate an agenda-setting effect. Again, when investigating individuals' perceptions of the information, individuals felt the entertainment news story was much more entertaining than the hard news story. Individuals may have not been entertained by the hard news version of the story. And the more entertaining someone found the information, the more likely there was an agenda-setting effect. Apparently, when they were not entertained by the information, even somewhat, they dismissed it. So individuals' perceptions have a relationship with the agenda-setting effect,as similar research has shown (McCombs, 2004; Weaver, 1980).

Despite the promising results, the study had several limitations. Study 1 had a low number of participants (n = 72). The results would have been stronger if more participants had been in the experimental condition; however, showing any statistically significant findings should indicate some very important results here. Another limitation would have been the time frames for the two studies. Study 1 occurred in January 2009; study 2 took place from July 2010 through March 2011. Different issues were of major concern during the two time periods, which might have influenced the results. For example, in January 2009, the media had focused attention on the economic problems, which many participants listed as a major concern. For the second study, the media focused on other issues, such as the BP oil spill. Despite these limitations, the studies indicated some very important findings for researchers.

Conclusion

What does this mean for journalism today? People do not discount fake news just because we have labeled it as such. Scholars have shown that although Jon Stewart claims his program is not real news, his television audience does consider it real news (Baym, 2007; McKain, 2005). *The Daily Show* even parodies nightly news programs stylistically, with Jon Stewart playing the anchor, and his correspondents bringing the audience the "news" (Fox, Koloen, & Sahin, 2007). Research has even shown that individuals who watch *The Daily Show* or *The Colbert Report* had similar political knowledge to those who watched hard news programs, such as CNN (Kohut, 2004; 2007; Xenos & Becker, 2009).

Traditional news media have struggled with losing audiences to alternative sources. Young people have turned more to programs like *The Daily Show* or *The Colbert Report* to obtain political information, as well as to be entertained (Kohut, 2004, 2007). The news media struggle with this. We have seen typical hard news networks incorporating humor into their delivery of the news. For example, CNN anchor Anderson Cooper attempts humor in his program *AC360* with his "Ridiculist" segment that has focused on some person or group acting "ridiculous." But traditional journalists need to take heed because if individuals find their information to be not entertaining, they may discount the information. On the other hand, if individuals find the information too entertaining, they may do likewise.

More research needs to be done determine why individuals who feel very entertained by the information apparently discounted the information more than those who feel somewhat entertained by the information. Future studies could attempt to understand this impact by creating information with several different levels of humor, such as no humor, somewhat humorous, and very humorous to determine if individuals discount the seriousness of the issue if they have a very humorous news story. Studies may also examine content from actual programs, both serious and entertaining. Overall, though, this study indicated entertainment news programs have been shown to influence the transfer of issue salience from the media's agenda to the public's agenda. As we move into the next 40 years of agenda-setting research, we need to determine how these different programs influence the process, and why.

Notes

1 Funding for this research, study 1, was due in part to Roy H. Park Fellow at the University of North Carolina at Chapel Hill. Funding for this research, study 2, was due in part to the Research and Creative Activities Fund and the Junior Faculty Summer Research Program at Texas Christian University. Funding for study 2 also was due in part to the College of Communication at Texas Christian University.

References

Atwater, T., Salwen, M. B., & Anderson, R. B. (1985). Interpersonal discussion as a potential barrier to agenda-setting. *Newspaper Research Journal,* 37–47.

Baym, G. (2007). Crafting new communicative models in the televisual sphere: Political interviews on The Daily Show. *Communication Review, 10*(2), 93–115.

Bryant, J., & Miron, D. (2002). Entertainment as media effect. In J. Bryant & D. Zillmann (Eds.), *Media effects: Advances in theory and research (*2nd ed.) (pp. 549–582). Mahwah, NJ: Lawrence Erlbaum Associates, Inc.

Comstock, G. (2004). Paths from television violence to aggression: Reinterpreting the evidence. In L. J. Shrum (Ed.), *The psychology of entertainment media: Blurring the lines between entertainment and persuasion* (pp. 193–212). Mahwah, NJ: Lawrence Erlbaum Associates, Inc.

Culbertson, H. M., & Stempel, G. H. III, (1985). Possible barriers to agenda setting in medical news. *Newspaper Research Journal*, 53–60.

Dearing J. W. & Rogers E. M. (1996). *Agenda-setting.* Thousand Oaks, CA: Sage Publications.

Fox, J. R., Koloen, G., & Sahin, V. (2007). No Joke: A comparison of substance in The Daily Show with Jon Stewart and broadcast network television coverage of the 2004 presidential election campaign. *Journal of Broadcasting & Electronic Media, 51*(2), 213–227.

Hill, D. B. (1985). Viewer characteristics and agenda setting by television news. *Public Opinion Quarterly, 49,* 340–350.

Holbert, R. L, Pillion, O., & Tschida, D. A. (2003). The West Wing as endorsement of the U. S. presidency: Expanding the bounds of priming in political communication. *Journal of Communication, 53*(3), 427–443.

Holbrook, R. A., & Hill, T. G. (2005). Agenda-setting and priming in prime time television: Crime dramas as political cues. *Political Communication 22,* 277–295.

Iyengar, S., & Kinder, D. R. (1986). More than meets the eye: TV news, priming, and public evaluations of the president. In G. Comstock (Ed.),*Public Communication and Behavior, Volume 1* (pp. 135–163). Syracuse, NY: Academic Press, Inc.

Iyengar, S., & Kinder, D. R. (1987). *News that matters: Television and American opinion.* Chicago: The University of Chicago Press.

Kolbert, E. (2004, April 19). Stooping to conquer: Why candidates need to make fun of themselves. *The New Yorker.*

Kohut, A. (2004, January 11). Perceptions of partisan bias seen as growing—Especially by Democrats: Cable and Internet loom large in fragmented political news universe. Pew Research Center for People and the Press. Retrieved from http://people-press.org/reports/pdf/200.pdf

Kohut, A. (2007 April 15). What Americans know: 1989–2007: Public knowledge of current affairs little changed by news and information. Retrieved from http://people-press.org/reports/display.php3?ReportID = 319

Lippmann, W. (1922). *Public opinion.* New York: Harcourt Brace.

Maguire, M. (2006, February 27). Schools bear burden of immigration. *News & Observer,* A1. Retrieved from http://www.newsobserver.com/1155/story/412207.html

McCombs, M. (2004). *Setting the agenda: The mass media and public opinion.* Cambridge, England: Polity Press.

McCombs, M. E., & Shaw, D. L. (1972). The agenda-setting function of mass media. *Public Opinion Quarterly, 36*(2), 176–187.

McKain, A. (2005, December). Not necessarily not the news: Gatekeeping, remediation, and The Daily Show. *Journal of American Culture*, 415–430.

Miller, J. M. (2007). Examining the mediators of agenda setting: A new experimental paradigm reveals the role of emotions. *Political Psychology, 28*(6), 689–717.

Miller, R. E. & Wanta, W. (Winter 1996). Race as a variable in agenda setting, *Journalism & Mass Communication Quarterly, 73*(4), 913–925.

Nabi, R. L., Moyer-Gusé, E., & Byrne, S. (2007). All joking aside: A serious investigation into the persuasive effect of funny social issue messages. *Communication Monographs, 74*(1), 29–54.

Petty, R. E., & Cacioppo, J. T. (1984a). The effects of involvement on responses to argument quantity and quality: Central and peripheral routes to persuasion. *Journal of Personality and Social Psychology, 46*, 69–81.

Petty, R. E., & Cacioppo, J. T. (1984b). Source factors and the Elaboration Likelihood Model of Persuasion. *Advances in Consumer Research, 11*, 668–672.

Petty, R. E., Schumann, D. W., Richman, S. A., & Strathman, A. J. (1993). Positive mood and persuasion: Different roles for affect under high- and low-elaboration conditions. *Journalof Personality and Social Psychology, 64*, 5–20.

Raney, A. A. (2006). The psychology of disposition-based theories of media enjoyment. In J. Bryant & P. Vorderer (Eds.), *Psychology of Entertainment* (pp. 137–150). Mahwah, NJ: Lawrence Erlbaum Associates, Inc.

Salwen, M. B. (1987). Mass media issue dependency and agenda setting. *Communication Research Reports, 4*(1), 26–31.

Shaw, D. L. & Martin, S. E. (Winter 1992). The function of mass media agenda setting. *Journalism Quarterly, 69*(4), 902–920.

Shrum, L. J. (2002). Media consumption and perception of social reality: Effects and underlying processes. In J. Bryant & D. Zillmann (Eds.), *Media effects: Advances in theory and research,* (2nd ed.) (pp. 69–96). Mahwah, NJ: Lawrence Erlbaum Associates, Inc.

Slater, M. D. (2002). Entertainment education and the persuasive impact of narratives. In M. C. Green, J. J. Strange, & T. C. Brock (Eds.), *Narrative impact:*

Social and cognitive foundations (pp. 157–181). Mahwah, NJ: Lawrence Erlbaum Associates, Inc.

Slater, M. D., & Rouner, D. (2002). Entertainment-education and elaboration likelihood: Understanding the processing of narrative persuasion. *Communication Theory, 12*(2), 173–191.

Wanta, W., & Wu, Y. (Winter 1992). Interpersonal communication and the agenda-setting process. *Journalism Quarterly, 69*(4), 847–855.

Weaver, D. (1980). Audience need for orientation and media effects. *Communication Research, 7*(3), 361–377.

Weaver, D. (1994). Media agenda setting and elections: Voter involvement or alienation? *Political Communication, 11*, 347–356.

Weaver, D. H., Zhu, J., & Willnat, L. (1992). The bridging function of interpersonal communication in agenda-setting. *Journalism Quarterly, 69*(4), 856–867.

Xenos, M. A., & Becker, A. B. (2009). Moments of Zen: Effects of The Daily Show on information S\seeking and political learning. *Political Communication, 26*(3), 317–332.

Yagade, A., & Dozier, D. M. (1990). The media agenda-setting effect of concrete versus abstract issues. *Journalism Quarterly, 67*(1), 3–10.

Yang, J. & Stone, G. (2003). The powerful role of interpersonal communication in agenda setting. *Mass Communication & Society, 6*(1), 57–74.

Zillmann, D. (2000). Humor and comedy. In D. Zillmann, & P. Vorderer (Eds.), *Media entertainment: The psychology of its appeal* (pp. 37–58). Mahwah, NJ: Lawrence Erlbaum Associates, Inc.

Zillmann, D., Taylor, K., & Lewis, K. (1998). News as nonfiction theater: How dispositions toward the public cast of characters affect reactions. *Journal of Broadcasting & Electronic Media, 42*(2), 153–170.

Appendix

Stimulus Material

Hard news treatment story about immigration policy:	**Entertainment news treatment story about immigration policy:**
There's a new congressman in town and he's got a plan to "clean up" America. . .but at what cost?	There's a new congressman in town and he's got a plan to "clean up" America. . .but at what cost?
Newly-elected Senator Joseph Beale, an independent, plans to propose a new anti-immigration policy with a big price tag when he arrives in Washington this January.	Newly-elected Senator Joseph Beale, an independent, plans to propose a new anti-immigration policy with a big price tag when he arrives in Washington this January. You know, outside of the killings, Washington, D. C., has one of the lowest crime rates in the country.
That new policy will include creating a special branch of the Justice Department called the Illegal Immigration Enforcement Agency, and Beale refused to comment on what this could cost taxpayers.	That new policy will include creating a special branch of the Justice Department called the Illegal Immigration Enforcement Agency, and Beale refused to comment on what this could cost taxpayers. I am going out on a limb here and saying this is more expensive than their first plan to put a 500-foot-long neon "No Vacancy" sign at each of our borders.
The agency's only mission would be to find and return undocumented immigrants to their own countries, even if they fled their own country because of the dangerous environment.	The agency's only mission would be to find and return undocumented immigrants to their own countries, even if they fled their own country because of the dangerous environment, like the legendary Chupacabra, the goatsucking vampire creature that we all must fear.
Beale says an agency like this is needed if we are to stop the influx of people entering the country illegally.	Beale says an agency like this is needed if we are to stop the influx of people entering the country illegally. Measures include frowning exercises to show those foreigners how unwelcoming America really is.
The law could hurt people like Cuban-born Maria Gonzales and her three innocent children.	The law could hurt people like Cuban-born Maria Gonzales and her three innocent children.

(Continued)

Gonzales came to America after her husband was killed for speaking out against Fidel Castro and has been unsuccessful in her attempts to get a Green Card.	Gonzales came to America after her husband was killed for speaking out against Fidel Castro and has been unsuccessful in her attempts to get a Green Card.
She fears what will happen to her and her children if she is forced to return to Cuba.	She fears what will happen to her and her children if she is forced to return to Cuba. Those waters get rough during hurricane season.
According to the U. S. Committee for Refugees and Immigrants, an estimated 14 million people worldwide fled from their homes because of war and persecution last year. The list includes countries from all continents.	According to the U. S. Committee for Refugees and Immigrants, an estimated 14 million people worldwide fled from their homes because of war and persecution last year. The list includes countries from all continents.
Immigration advocates like Sam Brown say instead of forcing these people out, America should make an effort to assimilate them into society and culture.	Immigration advocates like Sam Brown say instead of forcing these people out, America should make an effort to assimilate them into society and culture.
He also reminds people that America was founded and built by immigrants.	He also reminds people that America was founded and was built by immigrants, which reminds us that illegal immigration is nothing new. In fact, I believe the Indians had a special name for it. They called it 'white people.'
Beale's proposed law comes just months after Congress passed a bill to beef up security by building a 700-mile fence along the Texas–Mexico border to stop undocumented workers from searching out a better life for themselves and their families.	Beale's proposed law comes just months after Congress passed a bill to beef up security by building a 700-mile fence along the Texas–Mexico border to stop undocumented workers from searching out a better life for themselves and their families. And the materials to build that fence? 1200 miles of pure Angus beef. Delicious.
Words: 290 No humor interjections	Words: 415 (290+125) 7 humorous interjections
Both articles are against the proposal by Beale; anti-immigration policy, pro acculturation	

Chapter 8

From What the Public Thinks About to What the Public Does

Agenda-Setting Effects as a Mediator of Media Use and Civic Engagement

Soo Jung Moon

Although considerable research has demonstrated a significant positive relationship between civic engagement and media use, especially information-seeking in the news media, previous studies have provided only limited explications of the process that underlies this relationship. This study seeks to fill that void by addressing these questions:

- How does news use contribute to civic engagement?
- Why do some people exhibit higher levels of civic energy than others even though all use the same amount of news?

To answer these questions, this study examines the role of agenda-setting effects as a mediator between media use and civic engagement. Discussions that link agenda setting and behavioral outcomes such as civic engagement begin with the selective attention of the news media toward public issues. Subsequently, public consensus emerges about the importance of those issues, and people form opinions about them. In turn, the public participates to varying degrees in various forms of civic engagement that respond to these issues.

These links between agenda-setting effects, attitude strength, and behavior are grounded in previous studies in the fields of communication and psychology. The literature shows, simply stated, that people devote more thought to those objects that they regard as important (Tesser, Martin, & Mendolia, 1995), and the greater the amount of thought, the stronger the attitudes. In turn, strong attitudes operate as strong predictors of behaviors. This study expands the behavioral consequences of agenda setting to include not only voting, but also various other types of civic engagement such as political discussion, donations, and community participation.

The key elements of the theoretical models investigated here are:

> Attention to media → Agenda-setting effects → Opinion formation → Civic behaviors

To explore attitudinal and behavioral consequences of agenda setting, content analysis and survey data analysis are employed. To measure media salience, news stories from the *New York Times* and NBC's Nightly News are analyzed. For survey data, the American National Election Studies (ANES), conducted during the 2004 presidential election, are relied upon.

This research concentrates on the individual level, and as such, differs from most other agenda-setting studies that investigate aggregate data. Previous studies of individual-level analysis with multiple issues are relatively rare. However, aggregate-level analyses tend to ignore individual characteristics such as demographic variables and party identification that have substantial influence on media use and its consequences (McCombs, Lopez-Escobar, & Llamas, 2000).

The theoretical contribution of this study extends the concept of agenda-setting effects beyond salience to encompass its consequences for attitudes and, even, behaviors. This study perceives people as active information processors rather than as passive stimulation receivers.

From a broader perspective, this study considers the role of the news media in a democratic society. Information and engagement are regarded as a single, indivisible concept for the functioning of a good citizenry. Normative discourse about an "informed citizenry" gains substantive power through examination of the sequence of events that begins with media use and leads to public engagement.

Theoretical Background

Media and civic engagement. The news media encourage civic participation through providing information that leads to discussion and deliberation (Beaudoin & Thorson, 2004; Camaj, Chapter 5, this volume). In general, civic engagement can be divided into two categories: political participation and civic participation (Verba & Nie, 1972). Political participation usually means participation in the electoral process. Civic participation, on the other hand, is defined as an "organized voluntary activity focused on problem solving and helping others" (Zukin, Keeter, Andolina, & Jenkins, 2006, p 7).

Putnam (1995) ignited debate about the role of media by arguing that civic engagement has drastically declined since the 1960s or 1970s and by blaming TV viewing as the "single most consistent" predictor of decline. However, many communication scholars point out that the relationship between media use and civic engagement varies according to differences in content, genres, and usage patterns (Kang & Kwak, 2003; Norris, 2002). Generally, information seeking has been found to have a positive impact on civic engagement while entertainment or recreational user viewing is observed to have a negative effect (Shah,

McLeod, & Yoon, 2001). For that reason, different measures of TV use lead to varying results. Norris (1996), for instance, included three variables to measure TV use: news attention, current affairs program attention, and total hour consumption. Among those variables, viewer attention to news and current affairs programs showed a positive relationship with viewers' informal community participation yet did not have a direct relationship with conventional political participation. In terms of viewers' total hours of TV consumption, Norris found a negative relationship with six of the seven participation indicators and a significantly negative correlation relative to voting and informal community activity. While this finding seems to support Putnam's (1995) argument, Norris emphasized the positive impact of news and current affairs programs, such as "Nightline" and "60 Minutes," relative to the health of a democratic society.

Scholars rarely doubt the positive function of newspaper readership, recognizing especially that public affairs news use has a positive relationship with various types of engagement. Almost all studies repeatedly support the contribution of newspaper readership to enhanced participation (Beaudoin & Thorson, 2004; Moy, Torres, Tanaka, & McCluskey, 2005; Norris, 2000; Shah, 1998; Viswanath, Finnegan, Rooney, & Potter, 1990).

In addition to media use, some scholars have introduced other conditional variables, such as community context (Beaudoin & Thorson, 2004; McLeod & Scheufele, 1999), knowledge (Moy, Torres, Tanaka, & McCluskey, 2005), and interpersonal communication (Kang & Kwak, 2003; Paek, Yoon, & Shah, 2005). Norris (1996) examined a broad range of civic participation activities ranging from informal community involvement to protesting, and found that newspaper readership had a significant correlation with most indicators of participation, especially voting. From the perspective of civic engagement at the community level, local newspaper readership was found to be the strongest predictor of civic participation in an index that included attending club meetings, church attendance, volunteer work, and working on community projects (Paek et al., 2005).

Unlike the findings on newspaper readership, studies on TV news viewing have presented mixed results. Kang and Kwak (2003) and Shah et al. (2001) reported that TV hard news viewing, both local and national news programs or local public affairs programs, was positively associated with civic engagement whereas the total amount of viewing time resulted in a negative relationship. The positive relationship between civic engagement and TV news viewing has been supported in Europe and Asia, as well (Inoguchi, 2002; Norris, 2002). In contrast, McLeod and Scheufele (1999) reported no direct impact of TV news use on voting and contacting officials. Scheufele and Shah (2000) and

Uslaner (1998) found also that TV hard news use did not affect various forms of engagement.

Agenda setting. Three key words link agenda setting with democracy and engagement: attention, consensus, and problem solving. Attention refers to a set of thoughts and opinions about social and political life and the object of attention can be anything belonging to the public affairs arena in the widest possible sense (Inoguchi, 2002; Offe & Fuchs, 2002). Simply put, "Democracy means paying attention."(Bellah, 1991, p. 254)

In general, issues are social problems that involve conflicted interests: problems that need to be fixed. In that respect, agenda setting is an intrinsically political process within democracy (Dearing & Rogers, 1996). The term *agenda* has been defined as a set of issues or a list of problems to be concerned about and solved (Cobb & Elder, 1983; Dearing & Rogers, 1996; Kingdon, 1995). Conventional agenda-setting research, therefore, has paid attention to the agenda of public issues mainly by asking "Most Important Problem" (MIP) survey questions. Once the public has started to pay attention to a particular issue and consensus about its importance has started to build, the public then begins to make a decision about where to start: which aspect of the problem needs to be solved first. In that respect, agenda setting is an intrinsically political process within democracy (Dearing & Rogers, 1996). The media launches a process that might lead to policy change through agenda-setting effects. Dearing and Rogers (1996) summarized that agenda setting presents an explanation of "why information about certain issues, and not other issues, is available to the public in a democracy; how public opinion is shaped; and why certain issues are addressed through policy actions while other issues are not." (p. 1)

Empirical evidence exists to support the role of consensus building in agenda setting (Higgins, Chapter 11, this volume). Studies that have examined the correlation within the demographics of gender, age, and race found that the more people read newspapers, the more they agreed on important public issues (Shaw & Martin, 1992).They found that among infrequent newspaper readers, the rank-order correlation between men's and women's issues was 0.55, while among regular readers there was a perfect correlation of 1.0. Lopez-Escobar et al. (1998) found results similar to those of Shaw and Martin (1992) in terms of gender. In addition to gender, there were differences in issue salience between readers and non-readers of newspapers in terms of educational subgroups that was not supported in the previous research, and ideology that was a newly added factor.

Agenda-setting effects are not limited to salience. As Entman (1989) noted, "If the media (or anyone) can affect what people think about—the information they process—the media (or anyone) can affect their attitudes" (p. 76). Priming is the most prominent topic with regard to these consequences of agenda setting. Iyengar and Kinder (1987) observed that "By calling attention to some matters while ignoring others, television news influences the standards by which governments, presidents, policies, and candidates for public office are judged." (p. 63).

Although priming is the most popular concept in studying consequences of agenda setting, there have been several other studies that looked at the influence of agenda-setting effects. Kiousis and McCombs (2004) found a strong correlation between public salience and attitude strength toward 11 public figures. Respondents recognized the names of public figures mentioned more frequently by media, and respondents' feeling thermometers toward the salient figures were more dispersed. Attitude dispersion was measured by the proportion of respondents who showed non-neutral feeling.

Another study about political socialization of adolescents supported the linkage between agenda setting and attitudes (Kiousis, McDevitt, & Xu, 2005). Students' perceived issue salience affected their opinion extremity about certain issues such as gambling and class sizes. In the study, issue salience, worked as mediator between news attention and opinion strength.

Several differences exist between priming studies and other agenda-setting consequences research. First, the main interest of priming research is cognitive evaluation,whereas other recent studies have sought to analyze the affective dimension of attributes. Some studies have looked at sentiment and feeling as well as cognitive evaluations. Coleman and Wu (2010) found that media's emotional comments about candidates affect the public's feelings about those candidates. Furthermore, there were stronger agenda-setting effects on voters' feelings than on voters' cognitions. Visual images of candidates on network TV also influenced audiences' feelings toward the candidates (Coleman & Banning, 2006). Results showed the U. S. public perceived foreign nations negatively if coverage about the nations was negative (Kiousis & Wu, 2008; Wanta, Golan, & Lee, 2004).

Second, priming studies talk about the direction of evaluation, yet Kiousis and his colleagues focused on strength of evaluation. Third, as Kiousis and McCombs (2004) pointed out, "One underdeveloped area in priming research has been empirical work investigating how media attention toward political figures themselves influences public attitudes toward those same figures" (p. 37).

In addition to the political arena, scholars have also looked at issues in the realm of economic news (Ragas, Chapter 12, this volume). This area is particularly rich for the study of attitudinal effects because of the availability of various indices that measure people's attitudes over long periods of time, such as the Conference Board's Consumer Confidence Survey and the Michigan Consumer Sentiment Index (MCSI).

Hester and Gibson (2003) and Blood and Phillips (1995) found that the negative tone of news stories, for example, articles about a recession, significantly influenced peoples' confidence or sentiment about the nation's economic performance. The more the media wrote or talked about bad news, the more the public expressed deep concerns and negative outlooks about the economy. Hetherington (1996) reported that negative reporting of economic performance during an election year negatively affected voters' perception of the economy; moreover, negative perceptions were significantly related to voter choices.

Although studies of economic news produced empirical evidence to support observations that new stories have an effect on the public's attitudes, strictly speaking, those studies missed a certain key link with regard to agenda setting in terms of whether the attitudinal effects were because of the media's salience, on one hand, or due to the public's salience, on the other. Blood and Phillips (1995), Goidel and Langley (1995), and Hester and Gibson (2003) went directly to *attitudes* of the public without examining the *salience* of the public (recognition by the public that the issue was of importance). That oversight occurred because the economic indices relied upon do not offer items that are appropriate for measuring public salience. Hetherington's (1996) study began with the assumption that general economic news is negative without conducting content analysis that would have allowed measurements of the media's salience.

Beyond attitudinal consequences, a few studies have reported the behavioral effects of agenda setting. Kepplinger and Roth (1979) showed the media's influence on the purchase of petroleum products during an oil crisis in Germany. Advertising is an especially rich field associated with behavioral affects of agenda setting. Sutherland and Galloway (1981) noticed that the concept of salience in agenda setting operated in parallel with a strategy of advertising so that issue salience could be applied to an advertising strategy based on brand salience. "That is, perceived popularity functions as an intervening variable that is inferred from salience or prominence and thereby links agendas with behavioral outcomes" (pp. 27–28). In the study, consumers were asked, for example: "What attribute or product feature is the first one you think of when I say toothpaste?" The questions were similar to survey questions that readers/listeners are asked in second-level agenda-setting studies.

Based on their study, Ghorpade (1986) proposed a two-stage model: a first stage that refers to the transfer of salience from advertising to the public mind, and a second stage that indicates the transfer of salience from the public mind to a behavioral outcome. Ghorpade's (1986) two-stage model was empirically supported by Roberts (1992). Her investigation of the 1990 Texas gubernatorial election found that 70%

of the voters were correctly classified in their reported vote for governor according to their level of issue concern. Weaver (1991) looked at diverse political behaviors, for example, signing petitions, attending meetings, writing letters and voting, in which he explored the relationship between the salience of the federal budget deficit issue and public knowledge, opinion, and behavior. Weaver found a significant correlation between issue salience and all the dependent variables, concluding that the greater importance people assigned to the deficit problem, the more knowledge they sought about causes and solutions relative to the problem, and the stronger their opinions were about the topic. He also found that the more concerned people were about the deficit issue, the more likely they were to engage politically through various behaviors. Stroud and Kenski (2007) applied agenda-setting effects to a unique behavior: participation in survey. They found a negative relationship between 2004 presidential campaign news and refusal rate of taking a survey. In other words, the more the media covered the election, the more the respondents agreed to answer questions of an election survey. Recently, Kiousis and McDevitt (2008) have suggested a more elaborate model to explain the influence of agenda setting on behavior using panel data. In their study, agenda setting was conceptualized as an intrinsic process within political socialization that impacts actual political participation through opinion extremity and political ideology. The sequence of cognitive (Iraq issue importance), affective (feelings about the U. S. government's actions about Iraq), and behavioral outcomes (voting) was statistically supported.

Hypotheses. Figure 8.1 presents the links from media use to agenda setting and to civic engagement. The important aspect of this model is that

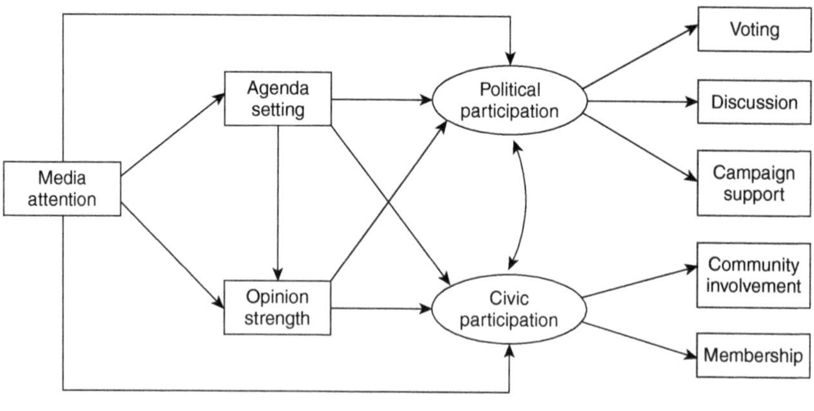

Figure 8.1. Conceptual framework

agenda setting is the "general" learning process of individuals rather than being limited to specific issues or specific electoral contexts. That is the rationale for including civic participation in the model. In addition, empirical studies demonstrate that political and civic participation are distinct yet strongly correlated concepts (Delli Carpini & Keeter, 2006; Verba & Nie, 1972; Wilkins, 2000; Zukin et al., 2006).

Despite rare consensus on the precise definition of attitude, scholars agree that evaluation is an essential part of attitude (Ajzen, 1980; Millar & Tesser, 1986; Oskamp, 1991). Eagly and Chaiken (1995) pointed out that "Attitude is a psychological tendency that is expressed by evaluating a particular entity with some degree of favor or disfavor . . . attitudes are people's evaluations of attitude objects." (p. 414). Among the various viewpoints about attitudes, this study employs the definition that attitudes are multi-component entities that encompass cognition and affect (Millar & Tesser, 1986; Sherif, Sherif, & Nebergall, 1965).

The models relied on in this study regard attitude strength rather than attitude change as a consequence of agenda setting. According to Krosnick and Petty (1995), attitude strength is defined as "the extent to which attitudes manifest the qualities of durability and impactfulness" (p. 3), and attitude strength is measured in terms of extremity, intensity, magnitude, or polarity (Prislin, 1996).

The link between agenda-setting effects and attitude strength can be understood in terms of a simple logic: "Since the mass media (from an agenda-setting perspective) tend to stimulate more thinking and learning about objects and attributes in people's minds, one might consequently expect that this increased thinking would lead to strengthened attitudes "(Kiousis, 2005, p. 7). Tesser, Martin, and Mendolia (1995) noted that thinking operates as an antecedent of attitude strength and concluded: "thought, then, tends to make evaluations more extreme, more accessible and more enduring" (p. 75). Tesser and his colleagues repeatedly demonstrated that during experiments participants showed more extreme evaluations about the attitude object when they were directed to take some time to think about the object (Tesser & Conlee, 1975; Tesser & Cowan, 1977).

With regard to the link between attitudes and behaviors, numerous studies have explicated the A (attitude) to B (behavior) relationship and A-B consistency (Ajzen & Fishbein, 1977; Bentler & Speckart, 1979; Petty & Krosnick, 1995). Behavior refers to any act, verbal or nonverbal, that individuals generally assume to require real commitment (Schuman & Johnson, 1976). The interests in attitudes of scholars results partially from the consistency that has been found to exist between attitudes and behavior. In other words, attitude frequently is a predictor of behavior and, consequently, attitude change produces behavioral change (McGuire, 1985). Oskamp (1991) summarized the relationship as follows: "an attitude is a theoretical construct which is not observable in itself, but which mediates or helps to explain the relationship between certain observable stimulus events and certain behavioral responses" (p. 14). Therefore, "the stronger an attitude, the more likely it will drive behavior" (Tesser et al., 1995, p. 75). Extreme attitudes have been shown to be more consistent with behaviors than are less extreme attitudes (Fazio & Zanna, 1978; Judd & Brauer, 1995). For example, Kallgren and Wood (1985) reported that college students who expressed more favorable attitudes about environmental preservation participated more actively in pro-environmental works, such as signing petitions and recycling projects.

In the electoral context, Crano (1995) noted, "people are more likely to vote in accord with their attitudes as perceived issue importance increases." (p. 135). If agenda setting influences voter turnout, then it is logical to assume that it may also affect other types of campaign participation such as attending campaign meetings and making financial contributions to parties or candidates.

Based on the previous literature, this study posits the following hypotheses:

H1. The more people pay attention to news, the higher the positive relationship between media salience and public salience on the issues.

H2. Agenda-setting effects serve as a mediator in the relationship between media use and opinion strength.

H3. Agenda-setting effects influence civic engagement through opinion strength.

Structural Equation Modeling (SEM) was used to test these hypotheses. SEM is useful not only to show specific relationships between variables presented by each arrow but also to examine whether the overall model is retainable using model fit statistics.

Method

Content analysis. Stories from the *New York Times* and NBC's Nightly News were chosen for analysis, using keyword searches in LexisNexis Academic (http://www.lexisnexis.com). In the United States, the *New York Times* is regarded as a major prestige newspaper and is commonly used for media research (Kiousis, 2004; Tedesco, 2005). During the 2004 presidential campaign, NBC was reported to have more viewers than its competitors (Associated Press, 2004). The sampling of this study also relies upon the concept of intermedia agenda setting,in which a high level of homogeneity existing among various media outlets resulted from a role of elite media as agenda setters (Dearing & Rogers, 1996; McCombs, 2004; Reese & Danielian, 1989).

To measure public salience, data collected as part of the American National Election Studies (ANES) survey during the 2004 presidential campaign were analyzed. The issues for the content analysis drawn from the ANES survey questions were diplomacy, government spending, defense spending, health insurance, jobs, aid to Blacks, environment, gun access, and women's equal role. The wording of the ANES questions was: "How important is this issue to you personally?" While the question wording is different from traditional agenda-setting studies' Most Important Problem (MIP) questions, a recent study found little difference between responses to MIP and personal context questionnaires (Min, Ghanem, & Evatt, 2007).

News stories were deleted if at least one paragraph did not address one of the issues. This filtering process resulted in a final sample of 1,737 articles from the *New York Times* and 287 stories from NBC for August and September of 2004. The unit of analysis for this study was the news story, meaning that only one main issue per story was coded as "1" and other issues were coded as "0."

News stories about the Iraqi war were usually coded as "diplomacy" following ANES's codebook that classifies "Iraq war" as "foreign affairs." However, when stories focused mainly on the financial cost of the conflict to the U. S. government or the war's effects on national safety, the Iraqi war issue was coded as "defense." Terrorism and 9/11 were generally coded as "defense" but when the articles talked about those issues in the context of foreign policy, those issues were coded as "diplomacy." (Additional details of the content analysis coding are in Appendix A.)

This study includes two timeframes based on the dates of the ANES interviews. When respondents answered in September, the content analysis of August was matched. When interviews were conducted from October 1 to November 1, news stories from September were matched. The sample included 1,066 respondents who participated in both pre-election and post-election interviews. Inter-coder reliability for the content analysis was determined by two coders using a random sub-sample of 10 days' stories. The overall agreement was 96%. All the values of Scott's pi for each issue were acceptable and higher than 0.70 (Riffe, Lacy, & Fico, 2005): diplomacy (0.82), defense spending (0.73), government spending (0.88), health insurance (0.88), environment (0.96), aid to Blacks (0.72), gun control (0.80), and women's role (0.73).

Individual agenda-setting index. Wanta (1997) and Wanta and Hu's (1994) methodology was used to create an individual agenda-setting index. Creation of this index involves a series of steps. First, through content analysis of the news coverage, issues in the media are divided into two groups: major issues and minor issues. In order to establish a weighted measurement system, the percentage of frequency for each issue in the media is calculated. This weighting system is determined across all issues, not across major and minor issues separately, in order to assure that a minor issue group has equal weight as a major issue group. Next, the weight is multiplied by the survey respondents' answers measured on a 5-point scale. Finally, the sum of the minor issues is subtracted from the sum of the major issues so that the weights of minor issues are inversed (percentage of 1/frequency).

Opinion strength and other variables. The opinion strength index is the sum of the extremity points regarding major issues. This study investigated not the direction, but the extent of dispersal, so that the 7-point scale was folded as follows: 1 and 7 into 4, 2 and 6 into 3, 3 and 5 into 2, 4 into 1. If the respondent answered, "Haven't thought much about this," his or her score was 0.

Media use was measured in terms of attention, a measurement commonly used in media effects studies (Chaffee and Schleuder, 1986; Drew and Weaver, 1990). Various political behaviors were considered, including voting, campaign support, political discussion, and campaign donations. The voting index was an additive score with a possible range of 0 to 3. Respondents were asked if they voted for candidates for president, the U. S. House of Representatives, and the U. S. Senate. Each question was scored "1" ("voted") or "0" ("did not vote"). If respondents voted in all three elections, they received "3," and if there was no senatorial election, those cases were omitted from analysis. To avoid losing data, this study used pairwise deletion rather than casewise deletion.

Campaign support included attending a meeting, putting up campaign signs for a particular party or candidates, and campaign donations for an individual candidate, political party, or any other group that supported or opposed the candidates. All these items were dummy-coded as "1" ("Yes") or "0" ("No") and the possible range was 0 to 5. Political discussion was measured by the number of days in a week respondents reported discussing politics with family or friends.

Civic participation was measured by the community involvement index and by organizational memberships. Brady (1999) categorized these works as conventional and nonelectoral activities, while Putnam (2000) called it cooperative activities. Community involvement was a three-item measure with a score ranging from 0 to 3. The questions asked if respondents had worked with other people to deal with some issue facing their community, attended a meeting about an issue facing the community, or volunteered. Membership was indicated by the number of organizations the respondents participated in. (See Appendix B for the details of measures.)

Four demographic variables and party identification from the 2004 ANES data were used. Overall, 51% were females. Age was measured in years (M = 46.4, SD = 17.4). Education was measured on a 7-point scale (Median = 4, more than 12 years of schooling, no higher degree), with responses ranging from "8 grades or less and no diploma or equivalency" to "post BA degree, including LLB." Income was measured on a 23-point scale (Median = 11, $22,000-$24,999), with possible responses ranging from "none or less than $2,999" to "$120,000 or over." Party identification was folded depending on its strength: strong democrat/strong republican (4); weak democrat/weak republican (3); independent-democrat/independent-republican (2); independent-independent (1).

Results

In both media, diplomacy and defense were dominant issues. The *New York Times* published 650 articles on diplomacy, and 416 articles on defense. Diplomacy (150 articles) and defense (96 articles) represented 86% of the sampled NBC news stories. Table 8.1 presents the number of each story and its weight for calculation of individual agenda-setting index. The issues of the *New York Times* and NBC showed significant rank-order correlations with a value of 0.82 (August, $p < 0.01$) and 0.98 (September, $p < 0.01$). The newspaper and TV's agenda-setting effects were also highly correlated (.81, $p < 0.01$).

Figure 8.1 shows the results for the hypothesized Structural Equation Modeling (SEM) . The model assumes a direct as well as an indirect relationship between media use and engagement. Exogenous variables were omitted from the figure in order to highlight the relationships of primary interest in this study. This study used a trimming method that is based on path coefficients of demographic variables based on a rationale: namely, a combination of empirical and theory-driven modeling. The paths between the endogenous variables were not deleted from the final models even though the coefficients were not significant. As Schumacker and Lomax (2004) suggested, "If a parameter is not significant but is of sufficient substantive interest, then the parameter should probably remain in the model." (p. 71)

The statistical analysis of the current research relied on weighted data to adjust for the difference between the sample and national demographic characteristics. Appendix C displays the correlation matrix used for the SEM analysis.

Although the model yielded a significant model χ^2 (109.61, df = 37, $p < 0.001$) indicating that the model does not reflect the data, other fit statistics for this model support a different interpretation. The

Table 8.1. The issue grouping and weight

		August			*September*		
		Issue	*No. story*	*Weight*	*Issue*	*No. story*	*Weight*
New York Times	*Major*	Diplomacy	264	0.38	Diplomacy	386	0.52
		Defense	255	0.36	Defense	161	0.22
		Jobs	106	0.15	Jobs	120	0.16
		Environment	76	0.11	Insurance	73	0.10
		N	701	1.00	N	740	1.00
	Minor	Gun control	4	0.72	Blacks	9	0.41
		Women	28	0.10	Gun control	10	0.37
		Blacks	31	0.09	Women	27	0.14
		Budget	33	0.09	Budget	43	0.09
		N	96	1.00	N	89	1.00
NBC	*Major*	Diplomacy	73	0.54	Diplomacy	77	0.56
		Defense	50	0.37	Defense	46	0.34
		Budget	6	0.04	Insurance	8	0.06
		Insurance	5	0.04	Jobs	6	0.04
		N	134	1.00	N	137	1.00
	Minor	Blacks	1	0.37	Women	1	0.48
		Gun control	1	0.37	Blacks	2	0.24
		Women	2	0.19	Environment	3	0.16
		Environment	5	0.07	Gun control	4	0.12
		N	9	1.00	N	10	1.00

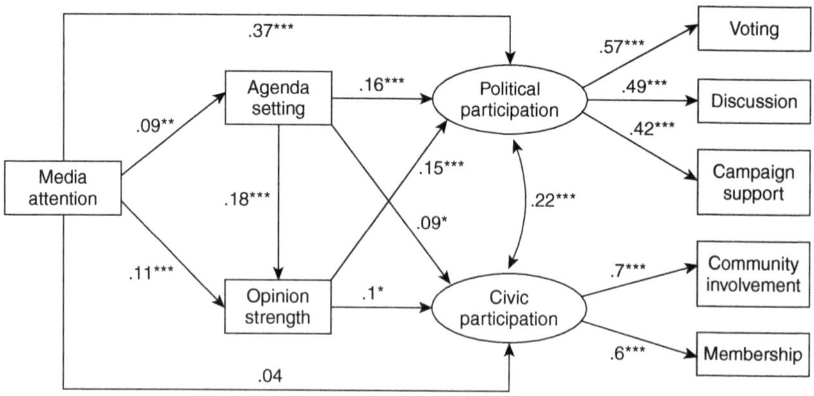

Figure 8.2. Result of hypothesized SEM

normed χ^2 (χ^2/ df) was 2.96, which is a reasonable value, and other fit statistics revealed satisfactory results: GFI (Goodness-of-fit) of 0.98, CFI (Comparative Fit Index) of 0.96, RMSEA (Root Mean Square Error of Approximation) of 0.04 and SRMR(standardized root mean square residual) of 0.03. The overall set of statistics indicates this model is retainable. Hu and Bentler (1999) suggested two kinds of combined criteria to retain a model, such as (a) CFI > 0.96 and SRMR < .1, (b) RMSEA < 0 .06 and SRMR < 0.10. This model meets both criteria.

The effect sizes indicate that news attention had a significant relationship with agenda-setting effects, political participation, and opinion strength, but not with civic participation. Agenda-setting effects had a significant relationship with all variables of opinion strength, political participation, and civic participation. Opinion strength showed a stronger relationship with political participation than with civic participation.

The squared multiple correlation for political participation was 0.61 and for civic participation 0.27 in this model. This means that 61% of the variance in the latent variable political participation was explained by all predictors, including exogenous variables.

Hypotheses Testing

H1. The more people pay attention to news, the higher the positive relationship between media salience and public salience on the issues.

The standardized coefficient of the path from media attention to the agenda-setting effect was 0.09 ($p < 0.01$). H1 was supported.

H2. Agenda-setting effects serve as a mediator in the relationship between media use and opinion strength.

If agenda setting operates as a mediating variable, then the indirect effect of media use on opinion strength must be significant. To test the significance of indirect effects, this study used two methods. First, Sobel's test (1982) was employed to produce a critical ratio of statistical significance. The indirect effect (0.02) of media attention on opinion strength through agenda setting was significant at the 0.01 level (critical ratio = 2.88).

Next, Baron and Kenny's (1986) method using three OLS regression analyses was used controlling for demographic variables and party identification strength. The first two regressions, the independent variable, media attention, had significant effects on agenda setting ($\beta = 0.1$, $p < 0.01$) and on opinion strength (β = = 0.12, $p < 0.001$). In the third regression, the coefficient of media attention was reduced to 0.09 ($p < 0.05$) while agenda setting was a significant independent variable ($\beta = .19$, $p < 0.001$). Therefore, H2 about the mediating role of agenda setting was supported.

H3. Agenda-setting effects influence civic engagement through opinion strength.

Table 8.2 summarizes the effects of key variables on civic engagement. As hypothesized, agenda setting was significant across all the engagement variables. The coefficient of agenda setting's direct effects on political participation was 0.16 ($p < 0.001$) and the total effect was 0.18 ($p < 0.001$). Agenda setting also affects political participation through opinion strength (0.02, $p < 0.01$).With regard to civic participation, the total effect (0.11), direct effect (0.09), and indirect effect (0.02) of agenda setting were all significant at the 0.05 level. Opinion strength was a significant predictor for participations.

In addition, the influence of agenda setting on specific indicators was examined. It was found to significantly affect all measurement variables, including voting (0.11, $p < 0.001$), campaign support (0.08, $p < 0.001$) and political discussion (0.09, $p < 0.001$). The influence was stronger on voting and discussion than on campaign support activities. Agenda setting's influence on organizational membership (0.06) and community involvement (0.07) also was significant at the 0.05 level.

Influence of exogenous variables. Table 8.3 demonstrates the influence of the four demographic variables along with party identification strength.

Although the income and age variables were relatively weak, other variables showed influence generally on endogenous variables. The strongest demographic variable in terms of direct effect was education, significant across all the endogenous variables with the exception

Table 8.2 Effects decomposition

Causal Variables		*Endogenous Variables*			
	Effects	*Agenda setting*	*Opinion Strength*	*Political participation*	*Civic participation*
Media attention	*Direct*	0.09**	0.10***	0.37***	0.04ns
	Indirect		0.02**	0.03***	0.02**
	Total	0.09**	0.12***	0.40***	0.06**
Agenda setting	*Direct*		0.18***	0.16***	0.09*
	Indirect			0.02**	0.02*
	Total		.018***	0.18***	0.11*
Opinion strength	*Direct*			0.15***	0.1*
	Indirect				
	Total			0.15***	0.1*

Note. * p < 0.05, **p < 0.01, ***p < 0.001, ns not significant

Bootstrapping was used to test significance of the *standardized* coefficients.

Table 8.3. Influence of exogenous variables

Paths		*Direct*	*Total*
Age			
→	News attention	0.25***	0.25***
→	Agenda setting	–	0.02**
→	Opinion strength	–	-0.06*
→	Political participation	–	0.09***
→	Civic participation	–	0.01[ns]
Gender[a]			
→	News attention	-0.07*	-0.07**
→	Agenda setting	-0.15***	-0.15***
→	Opinion strength	-0.14***	-0.17***
→	Political participation	0.09*	0.01[ns]
→	Civic participation	0.15***	0.12**
Education			
→	News attention	0.2***	0.2***
→	Agenda setting	0.11***	0.13***
→	Opinion strength	–	0.04***
→	Political participation	0.24***	0.35***
→	Civic participation	0.40***	0.43***
Income			
→	News attention	–	–
→	Agendasetting	–	–
→	Opinion strength	–	–
→	Political participation	0.15**	0.15**
→	Civic participation	0.11**	0.11**
Party Strength			
→	News attention	0.15***	0.15***
→	Agendasetting	–	0.01**
→	Opinion strength	.010**	0.11***
→	Political participation	0.36***	0.43***
→	Civic participation	–	0.02*

+ $p < .01$,* $p < 0.05$, **$p < 0.01$, ***$p < 0.001$, ns not significant

[a]. male = 1; female = 2

- paths were constrained to zero

Bootstrapping was used to test the significance of the standardized coefficients.;

of opinion strength. In general, the better-educated respondents gave more attention to news and revealed more agenda-setting effects. These respondents were also more actively engaged in politics and in their communities. The standardized total effect size of education on civic participation was more than 0.3, thus indicating a greater role than media strength. Total effects of education on all variables were significant at 0.001 level. Gender effect was relatively widespread. In general, older male respondents paid more attention to the media. However, female respondents were more engaged in their communities and campaign.

Party strength significantly ($p < 0.001$) affected political participation, as expected. If respondents indicated that they were Strong Republican or Strong Democrat, they reported being more actively engaged in the political process. Party identification had a significant relationship with media attention and opinion strength, as well.

Discussion and Conclusion

On balance, the fit statistics indicated that the SEM model was retainable, meaning that the hypothesized sequence reflects the data reasonably well. As previous studies found, the more people use news media, the more they participated in political and community activities. The statistical test for agenda setting as a mediator supported the hypothesis that agenda-setting effects work as mediating variables between media use and attitude strength. The model supported the information processing of agenda-setting effects as a "learning process"(Takeshita, 2006; Wanta, 1997). People are active problem solvers, not submissive followers led by media. Once people recognize particular issues as important, they are willing to solve problems through participation. Forms of participation vary, ranging from voting to joining organizations.

When we understand agenda setting as a learning process in terms of individual autonomy, *relevance* might be one of the keys leading to agenda-setting effects. Although this study did not include that variable, the perceived relevance of an issue, a candidate, elections in general, or a community can be increased by attention to the news. As a result, relevance becomes the main antecedent variable for agenda-setting effects and a variety of consequences as the concept of need for orientation has discussed. The effects of agenda setting should be evaluated from a broader context beyond the specific impact of the media on issues or candidates. Agenda setting enhances general salience of public issues in electoral and non-electoral contexts and makes the issues relevant to an individual's life.

Agenda setting affects engagement, in general, and specifically demonstrates stronger influence on political participation. The overall model better explained the latent variable of political participation than civic participation. The square multiple correlations (R^2) of political participation was 0.61 whereas civic participation was only 0.27. The difference between political and civic participation can be explained by the electoral context of the survey: The media content analysis focused on national political issues. Using community issues with local-level surveys would be helpful to obtain a deeper understanding of media effects on civic engagement.

Noteworthy, however, is the observation that although agenda setting affects civic participation, media attention does not. In other words, even if people watch TV news or read newspapers with attention, that particular type of media use, in itself, does not encourage audiences to become involved in community activities. Instead, only those audiences who, themselves, recognize the importance of issues highlighted in the news are willing to join community meetings and other organizations. This finding appears to underscore the importance of agenda setting with reference to civic participation. As previous studies discussed, agenda setting is a deliberative process of considerable awareness that promotes civic participation (Kiousis & McDevitt, 2008; Kiousis, McDevitt, & Xu, 2005).

It is important to point out that two issues included in the agenda-setting index had a different status in other public opinion research. Although government spending/federal budget deficit is typically a major political topic for Americans (Weaver, 1991), that issue was classified as a minor one in the newspaper. By contrast, the environment, which typically has not been considered an important issue, emerged as a major concern of the *New York Times*. This unusual grouping of issues contributed to the capture of real agenda-setting effects in real-time based on the importance of the environment issue at the time of the study. For example, if a respondent answered that government spending was important because that issue had been important in the past and said that the environment was not important because it had not been important previously, he/she is assumed to have not given attention to news over a certain period of time. To paraphrase Druckman (2005): to avoid a bias requires studying issues that are available in the given medium, but that are not too conventional (p. 466).

This study relied on an individual agenda-setting index to measure the effects on each of the individual survey respondents, proposed by Wanta (1997). The index was developed through rigorous means that employed several steps to prevent inflation of effects. The individual level of analysis provided the opportunity to examine the influence of demographic variables. Education, along with party identification, was found to be one of the strongest variables that affected all other endogenous variables.

The main drawback of this study is that it is a one-time cross-sectional analysis. This study posits Structural Equation Modeling, regarded as a useful tool for demonstrating causal relationships (Kline, 2005) but the findings must be interpreted cautiously. The shortcoming can be overcome only by a panel study. While the fit statistics indicate the models were "retainable," that does not mean there is no need to consider alternative models that further explicate in greater detail the consequences of agenda-setting effects on political and civic engagement. Despite statistical significance of all expected relationships, the coefficient values of each path were very small.

As the result of hundreds of studies, we know that the media set the agenda. A compelling question that remains unanswered is: What can the media do through these effects? From a journalistic viewpoint, agenda setting is able to function as a stimulator for engagement. One of the critical concerns surrounding journalism is oftentimes connected to concerns about young people's indifference to news and politics (Mindich, 2005). From the perspective of agenda setting, the media can invite young people to become involved by setting an agenda that appeals to the "dot net" generation. In that respect, generational effects on agenda setting and subsequent engagement are worthy of future research.

References

Ajzen, I. (1980). *Understanding attitudes and predicting social behavior*. Englewood Cliffs, NJ: Prentice-Hall.

Ajzen, I., & Fishbein, M. (1977). Attitude-behavior relations: A theoretical analysis and review of empirical research. *Psychological Bulletin, 84*(5), 888–918.

Associated Press (October 21, 2004, PM cycle). NBC's Nightly News Dominates Ratings.

Baron, R. M., & Kenny, D. A. (1986). The moderator-mediator variable distinction in social psychological research: Conceptual, strategic, and statistical considerations. *Journal of Personality and Social Psychology, 51*(6), 1173–1182.

Beaudoin, C. E., & Thorson, E. (2004). Social capital in rural and urban communities: Testing differences in media effects and models. *Journalism & Mass Communication Quarterly, 81*(2), 378–399.

Bellah, R. N. (1991). *The Good society*. New York: Knopf : Distributed by Random House.

Bentler, P. M., & Speckart, G. (1979). Models of attitude-behavior relations. *Psychological Review, 86*(5), 452–464.

Blood, D. J., & Phillips, P. C. B. (1995). Recession headline news, consumer sentiment, the state of the economy and presidential popularity: A time series analysis 1989–1993. *International Journal of Public Opinion Research*, 7(1), 2–22.

Brady, H. E. (1999). Political participation. In J. P. Robinson, P. R. Shaver, & L. S. Wrightsman (Eds.), *Measures of political attitudes* (pp. 737–801). New York: Academic Press.

Chaffee, S. H., & Schleuder, J. (1986). Measurement and effects of attention to media news. *Human Communication Research, 13*(1), 76–107.

Cobb, R. W., & Elder, C. D. (1983). *Participation in American politics: The dynamics of agenda-building.* Baltimore: Johns Hopkins University Press.

Coleman, R., & Banning, S. (2006). Network TV news' affective framing of the presidential candidates: Evidence for a second-level agenda-setting effect through visual framing. *Journalism & Mass Communication Quarterly, 83*(2), 313–328.

Coleman, R., & Wu, H. D. (2010). Proposing emotion as a dimension of affective agenda setting: separating affect into two components and comparing their second-level effects. *Journalism & Mass Communication Quarterly, 87*(2), 315–327.

Crano, W. D. (1995). Attitude strength and vested interest. In R. E. Petty & J. A. Krosnick (Eds.), *Attitude strength: Antecedents and consequences* (pp. 131–157). Mahwah, NJ: Lawrence Erlbaum Associates.

Dearing, J. W., & Rogers, E. M. (1996). *Agenda-setting.* Thousand Oaks, CA: Sage.

Delli Carpini, M. X., & Keeter, S. (1996). *What Americans know about politics and why it matters.* New Haven, CT: Yale University Press.

Drew, D., & Weaver, D. (1990). Media attention, media exposure, and media effects. *Journalism Quarterly, 67*(4), 740–748.

Druckman, J. (2005). How newspapers and television news cover campaigns and influence voters. *Political Communication, 22*, 463–481.

Eagly, A. H., & Chaiken, S. (1995). Attitude strength, attitude structure, and resistance to change. In R. E. Petty & J. A. Krosnick (Eds.), *Attitude strength: Antecedents and consequences* (pp. 413–432). Mahwah, NJ: Lawrence Erlbaum Associates.

Entman, R. M. (1989). *Democracy without citizens: Media and the decay of American politics.* New York: Oxford University Press.

Fazio, R. H., & Zanna, M. P. (1978). On the predictive validity of attitudes: The roles of direct experience and confidence. *Journal of Personality, 46*(2), 228–243.

Ghorpade, S. (1986). Agenda setting: A test of advertising's neglected function. *Journal of Advertising Research, 26*(4), 23–27.

Goidel, R. K., & Langley, R. E. (1995). Media coverage of the economy and aggregate economic evaluations: Uncovering evidence of indirect media effects. *Political Research Quarterly, 48*(2), 313–328.

Hester, J. B., & Gibson, R. (2003). The economy and second level agenda setting: a time-series analysis of economic news and public opinion about the economy. *Journalism & Mass Communication Quarterly, 80*(1), 73–90.

Hetherington, M. J. (1996). The Media's Role in Forming Voters' National Economic Evaluations in 1992. *American Journal of Political Science, 40*(2), 372–395.

Hu, L., & Bentler, P. M. (1999). Cutoff criteria for fit indexes in covariance structure analysis: Conventional criteria versus new alternatives. *Structural Equation Modeling: A Multidisciplinary Journal, 6*(1), 1–55.

Inoguchi, T. (2002). Broadening the basis of social capital in Japan. In R. D. Putnam (Ed.), *Democracies in flux: the evolution of social capital in contemporary society.* New York: Oxford University Press.

Iyengar, S., & Kinder, D. R. (1987). *News that matters: Television and American opinion*. Chicago: University of Chicago Press.
Judd, C. M., & Brauer, M. (1995). Repetition and evaluative extremity. In R. E. Petty & J. A. Krosnick (Eds.), *Attitude strength: Antecedents and consequences* (pp. 43–71). Mahwah, NJ: Lawrence Erlbaum Associates.
Kallgren, C. A., & Wood, W. (1985). Access to attitude-relevant information in memory as a determinant of attitude-behavior consistency. *Journal of Experimental Social Psychology, 22*(4), 328–338.
Kang, N., & Kwak, N. (2003). A multilevel approach to civic participation: Individual length of residence, neighborhood residential stability, and their interactive effects with media use. *Communication Research, 30*(1), 80–106.
Kepplinger, H. M., & Roth, H. (1979). Creating a crisis: German mass media and oil supply in 1973–74. *Public Opinion Quarterly, 43*(3), 285–296.
Kingdon, J. W. (1995). *Agendas, alternatives, and public policies*. New York: Longman.
Kiousis, S. (2004). Explicating media salience: A factor analysis of New York Times issue coverage during the 2000 U. S. Presidential election. *Journal of Communication, 54*(1), 71–87.
Kiousis, S. (2005). Compelling arguments and attitude strength: Exploring the impact of second-level agenda setting on public opinion of presidential candidate images. *Harvard International Journal of Press/Politics, 10*(2), 3–27.
Kiousis, S., & McCombs, M. E. (2004). Agenda-setting effects and attitude strength: Political figures during the1996 presidential election. *Communication Research, 31*(1), 36–57.
Kiousis, S., & McDevitt, M. (2008). Agenda setting in civic development: Effects of curricula and issue importance on youth voter turnout. *Communication Research, 35*(4), 481–502.
Kiousis, S., & Wu, X. (2008). International agenda-building and agenda-setting: Exploring the influence of public relations counsel on US news media and public perceptions of foreign nations. *International Communication Gazette, 70*(1), 58–75.
Kiousis, S., McDevitt, M., & Xu, W. (2005). The genesis of civic awareness: Agenda setting in political socialization. *Journal of Communication, 55*(4), 756–774.
Kline, R. B. (2005). *Principles and practice of Structural Equation Modeling*. New York: Guilford Press.
Krosnick, J. A., & Petty, R. E. (1995). Attitude strength: An overview. In R. E. Petty & J. A. Krosnick (Eds.), *Attitude strength: Antecedents and consequences* (pp. 1–24). Mahwah, NJ: Lawrence Erlbaum Associates.
Lopez-Escobar, E., Llamas, J. P., & McCombs, M. E. (1998). Agenda setting and community consensus: First and second level effects. *International Journal of Public Opinion Research, 10*(4), 335–348.
McCombs, M. E. (2004). *Setting the agenda: The mass media and public opinion*. Cambridge, UK: Polity.
McCombs, M., Lopez-Escobar, E., & Llamas, J. P. (2000). Setting the agenda of attributes in the 1996 Spanish general election. *Journal of Communication, 50*(2), 77–92.

McGuire, W. J. (1985). Attitudes and attitude change. In G. Lindzey & E. Aronson (Eds.), *Handbook of social psychology* (3rd ed., pp. 233–346). Hillsdale, NJ: Random House; Distributed exclusively by L. Erlbaum Associates.

McLeod, J. M., & Scheufele, D. A. (1999). Community, communication, and participation: The role of mass media and interpersonal discussion. *Political Communication, 16*(3), 315–336.

Millar, M. G., & Tesser, A. (1986). Effects of affective and cognitive focus on the attitude-behavior relation. *Journal of Personality and Social Psychology, 51*(2), 270–276.

Min, Y., Ghanem, S. I., & Evatt, D. (2007). Using a split-ballot survey to explore the robustness of the 'MIP' question in agenda-setting research: A methodological study. *International Journal of Public Opinion Research, 19*(2), 221–236.

Mindich, D. T. Z. (2005). *Tuned out: Why Americans under 40 don't follow the news*. New York: Oxford University Press.

Moy, P., Torres, M., Tanaka, K., & McCluskey, M. R. (2005). Knowledge or trust? Investigating linkages between media reliance and participation. *Communication Research, 32*(1), 59–86.

Norris, P. (1996). Does television erode social capital? A reply to Putnam. *PS: Political Science and Politics, 29*(3), 474–480.

Norris, P. (2000). *A virtuous circle: Political communications in postindustrial societies*. Cambridge, UK: Cambridge University Press.

Norris, P. (2002). Editorial: Social capital and the news media. *Harvard International Journal of Press/Politics*, 7(1), 3–8.

Offe, C., & Fuchs, S. (2002). A decline of social capital? The German Case. In R. D. Putnam (Ed.), *Democracies in flux: the evolution of social capital in contemporary society*. New York: Oxford University Press.

Oskamp, S. (1991). *Attitudes and opinions* (2nd ed.). Englewood Cliffs, NJ: Prentice Hall.

Paek, H. J., Yoon, S. H., & Shah, D. V. (2005). Local news, social integration, and community participation: Hierarchical linear modeling of contextual and cross-level effects. *Journalism & Mass Communication Quarterly*, *82*, 587–606.

Petty, R. E., & Krosnick, J. A. (1995). *Attitude strength: Antecedents and consequences*. Mahwah, NJ: Lawrence Erlbaum Associates.

Prislin, R. (1996). Attitude stability and attitude strength: One is enough to make it stable. *European Journal of Social Psychology, 26*(3), 447–477.

Putnam, R. D. (1995). Tuning in, tuning out: The strange disappearance of social capital in America. *PS: Political Science and Politics, 28*(4), 664–683.

Putnam, R. D. (2000). *Bowling alone: The collapse and revival of American community*. New York: Simon & Schuster.

Reese, D. S., & Danielian, H. L. (1989). Intermedia influence and the drug issue: Converging on cocaine. In P. J. Shoemaker (Ed.), *Communication campaigns about drugs: Government, media, and the public* (p. x). Hillsdale, NJ: L. Erlbaum Associates.

Riffe, D., Lacy, S., & Fico, F. (2005). *Analyzing media messages: Using quantitative content analysis in research* (2nd ed.). Mahwah, NJ: Lawrence Erlbaum.

Roberts, M. S. (1992). Predicting voting behavior via the agenda-setting tradition. *Journalism Quarterly, 69*(4), 878–892.

Sobel, M. E. (1982). Asymptotic confidence intervals for indirect effects in structural equation models. In S. Leinhart (Ed.), *Sociological methodology 1982* (pp. 290–312). San Francisco: Jossey-Bass.

Scheufele, D. A., & Shah, D. V. (2000). Personality strength and social capital. *Communication Research, 27*(2), 107–131.

Schumacker, R. E., & Lomax, R. G. (2004). *A beginner's guide to structural equation modeling* (2nd ed.). Mahwah, NJ: Lawrence Erlbaum Associates.

Schuman, H., & Johnson, M. P. (1976). Attitudes and behavior. *Annual Reviews in Sociology, 2*(1), 161–207.

Shah, D. V. (1998). Civic engagement, interpersonal trust, and television use: An individual-level assessment of social capital. *Political Psychology, 19*(3), 469–496.

Shah, D. V., McLeod, J. M., & Yoon, S.-H. (2001). Communication, context, and community: An exploration of print, broadcast, and Internet influences. *Communication Research, 28*(4), 464–506.

Shaw, D. L., & Martin, S. E. (1992). The function of mass media agenda setting. *Journalism Quarterly, 69*(4), 902–920.

Sherif, C. W., Sherif, M., & Nebergall, R. E. (1965). *Attitude and attitude change: The social judgment-involvement approach*. Philadelphia: Saunders.

Stroud, N. J., & Kenski, K. (2007). From agenda setting to refusal setting. *Public Opinion Quarterly, 71*(4), 539–559.

Sutherland, M., & Galloway, J. (1981). Role of advertising: Persuasion or agenda setting? *Journal of Advertising Research, 21*(5), 25–29.

Takeshita, T. (2006). Current critical problems in agenda-setting research. *International Journal of Public Opinion Research, 18*(3), 11–32.

Tedesco, J. C. (2005). Issue and strategy agenda setting in the 2004 presidential election: Exploring the candidate-journalist relationship. *Journalism Studies, 6*(2), 187–201.

Tesser, A., & Conlee, M. C. (1975). Some effects of time and thought on attitude polarization. *Journal of Personality and Social Psychology, 31*(2), 262–270.

Tesser, A., & Cowan, C. L. (1977). Some attitudinal and cognitive consequences of thought. *Journal of Research in Personality, 11*(2), 216–226.

Tesser, A., Martin, L., & Mendolia, M. (1995). The impact of thought on attitude extremity and attitude-behavior consistency In R. E. Petty & J. A. Krosnick (Eds.), *Attitude strength: Antecedents and consequences* (pp. 1–24). Mahwah, NJ: Lawrence Erlbaum Associates.

Uslaner, E. M. (1998). Social capital, television, and the "Mean world": Trust, optimism, and civic participation. *Political Psychology, 19*(3), 441–467.

Verba, S., & Nie, N. H. (1972). *Participation in America: Political democracy and social equality*. New York: Harper & Row.

Viswanath, K., Finnegan Jr., J. R., Rooney, B., & Potter, J. (1990). Community ties in a rural Midwest community and use of newspapers and cable television. *Journalism Quarterly, 67*(4), 899–911.

Wanta, W. (1997). *The public and the national agenda: How people learn about important issues*. Mahwah, NJ: Lawrence Erlbaum Associates.

Wanta, W., & Hu, Y-W. (1994). The effects of credibility, reliance and exposure on media agenda-setting: A path analysis model. *Journalism Quarterly, 71*(1), 90–98.

Wanta, W., Golan, G., & Lee, C. (2004). Agenda setting and international news: Media influence on public perceptions of foreign nations. *Journalism & Mass Communication Quarterly, 81*(2), 364–377.

Weaver, D. (1991). Issue salience and public opinion: Are there consequences of agenda-setting? *International Journal of Public Opinion Research, 3*(1), 53–68.

Wilkins, K. G. (2000). The role of media in public disengagement from political life. *Journal of Broadcasting & Electronic Media, 44*(4), 569–580.

Zukin, C., Keeter, S., Andolina, M., Jenkins, K., & Delli Carpini, X. M. (2006). *A new engagement? : Political participation, civic life, and the changing American citizen.* New York: Oxford University Press.

Appendix A. Codebook

V1. Coder ID
V2. Article ID
V3. Publication 1. New York Times 2. NBC
V4. Published Date
V5. Foreign Affairs
Relations with other countries or international organizations
Foreign involvement (except War in Iraq)
Aid to Israel/Arab states
Firmness in foreign policy; maintenance of position of military/diplomatic strength
Prevention of war; establishment of peace
Obligation to take care of problems at home before helping foreign countries
V6. Defense
Defense budget; level of spending on defense
Disarmament; arms race
Space race; space program
Weapons development; missile program; bomb testing
General or specific references to functioning and performance of defense; waste, inefficiency
Nuclear war; the threat of nuclear war; nuclear proliferation
V7. Terrorism; War on Terror-Foreign Affairs
V8. Terrorism; War on Terror-Defense
V9. War in Iraq-Foreign Affairs
V10. War in Iraq—Defense
V11. 9/11- Foreign Affairs
V12. 9/11- Defense

Note. V7–12. Terrorism, War in Iraq, and 9/11 will be coded differently depending on the context. These three issues should be included as keywords for sampling because they were most common terms in the 2004 election (Tedesco, 2004). The problem is that the issues contain multiple dimensions.

News stories about Iraq war, for example, are usually coded as "foreign affairs" but if the story focuses on financial cost of the U. S. government, then it would be coded as "defense" or "defense spending."

Terrorism and 9/11 are usually coded as "defense" but if the articles talk about the issue in the context of foreign policy such as cause of the events related to international relationship and establishment of national safety through military involvement, it would be coded as "foreign affairs."

V13. Government spending
Balancing of the budget
Tax cut
Against government stimulation of the economy
The size of the budget deficit
Mention of "twin deficit"
Government spending on education, health care, and other social welfare systems
The size of federal government

Note: I included education, health care, and social welfare because the NES asked that "Some people think the government should provide fewer services even in areas such as health and education in order to reduce spending. . ."

To prevent overlap of V13 and V15, we need criteria: if health care is talked about solely in the article (paragraph), then it will be coded as V18. If it was mentioned with other issues, it belongs to V16.

V14. Jobs
The number of people with jobs
Unemployment rate
Create jobs
Recruiting
V15. Medical Insurance
Medial cost and hospital cost
Medicare; prescription drug program
Health insurance program
Medicaid
V16. Environment
Conservation of natural resources
Protecting the environment/endangered species
Regulating growth or land development
Pollution, global warming, climate change
Development of natural resources/energy sources
V17. Aid to Blacks
Civil rights/racial problems
Programs to enable Blacks to gain social/economic/educational/political equality
Relations between Blacks and Whites
V18. Gun control
Unregistered ownership of guns
Legislative control of guns
Right to have guns
V19. Women's rights

Women's rights
References to women's issues
Economic equality for women

Appendix B. ANES Questionnaires of Measured Variables

Media Attention

- How much attention do you pay to newspaper articles about the campaign for president
- How much attention do you pay to news on national news shows about the campaign for president

A GREAT DEAL, QUITE A BIT, SOME, VERY LITTLE, OR NONE?

Agenda Setting

How important is this issue to you personally?
EXTREMELY important, VERY important, SOMEWHAT important, NOT TOO important, or NOT AT ALL important

- Interventionism/diplomacy
- Services/spending
- Defense spending
- Government insurance plan
- Guaranteed jobs
- Aid to Blacks
- Environment/jobs
- Women's role
- Whether buying gun should be easier/more difficult

Opinion Strength

-Some people believe the United States should solve international problems by using diplomacy and other forms of international pressure and use military force only if absolutely necessary. Suppose we put such people at "1" on this scale. Others believe diplomacy and pressure often fail and the United States must be ready to use military force. Suppose we put them at number 7. And of course others fall in positions in-between, at points 2, 3, 4, 5, and 6.

Where would you place YOURSELF on this scale, or haven't you thought much about this?

-Some people think the government should provide fewer services even in areas such as health and education in order to reduce spending. Suppose these people are at one end of a scale, at point 1. Other people feel it is important for the government to provide many more services even if it means an increase in spending. Suppose these people are at the other end, at point 7. And, of course, some other people have opinions somewhere in between, at points 2, 3, 4, 5, or 6.

-Some people believe that we should spend much less money for defense. Suppose these people are at one end of a scale, at point 1. Others feel that defense spending should be greatly increased. Suppose these people are at the other end, at point 7. And, of course, some other people have opinions somewhere in between, at points 2, 3, 4, 5, or 6.

-There is much concern about the rapid rise in medical and hospital costs. Some people feel there should be a government insurance plan that would cover all medical and hospital expenses for everyone. Suppose these people are at one end of a scale, at point 1. Others feel that all medical expenses should be paid by individuals through private insurance plans like Blue Cross or other company paid plans. Suppose these people are at the other end, at point 7. And, of course, some other people have opinions somewhere in between, at points 2, 3, 4, 5, or 6.

-Some people feel the government in Washington should see to it that every person has a job and a good standard of living. Suppose these people are at one end of a scale, at point 1. Others think the government should just let each person get ahead on their own. Suppose these people are at the other end, at point 7. And, of course, some other people have opinions somewhere in between, at points 2, 3, 4, 5, or 6.

-Some people feel that the government in Washington should make every effort to improve the social and economic position of Blacks. (Suppose these people are at one end of a scale, at point 1.) Others feel that the government should not make any special effort to help Blacks because they should help themselves. (Suppose these people are at the other end, at point 7.) And, of course, some other people have opinions somewhere in between, at points 2, 3, 4, 5, or 6.

-Some people think it is important to protect the environment even if it costs some jobs or otherwise reduces our standard of living. (Suppose these people are at one end of the scale, at point number 1.) Other people think that protecting the environment is not as important as maintaining jobs and our standard of living. (Suppose these people are at the other end of the scale, at point number 7.) And of course, some other people have opinions somewhere in between, at points 2, 3, 4, 5, or 6).

-Recently there has been a lot of talk about women's rights. Some people feel that women should have an equal role with men in running business, industry, and government. (Suppose these people are at one end of a scale, at point 1.) Others feel that a woman's place is in the home. (Suppose these people are at the other end, at point 7.) And, of course, some other people have opinions somewhere in between, at points 2, 3, 4, 5, or 6.

-Do you think the federal government should make it MORE DIFFICULT for people to buy a gun than it is now, make it EASIER for people to buy a gun, or keep these rules ABOUT THE SAME as they are now?

1. More difficult
3. Make it easier
5. Keep these rules about the same
8. Don't know
9. Refused
0. NA

Voting

-How about the election for president? Did you vote for a candidate for PRESIDENT?

1. Yes, voted for president
5. No, didn't vote for president
8. Don't know
9. Refused
0. NA

-How about the election for the HOUSE OF REPRESENTATIVES in Washington. Did you vote for a candidate for the U. S. House of Representatives?

1. Yes, voted for House of Representatives
5. No, didn't vote for House of Representatives
7. R VOLUNTEERS: voted but names are not correct for R's district
8. Don't know
9. Refused
0. NA

-If senate election in state:
Vote for a candidate for the U. S. SENATE?

1. Yes, voted for Senate
5. No, didn't vote for Senate
8. Don't know
9. Refused
0. NA

Campaign Discussion

-How many days in the past week did you talk about politics with family or friends?

0–7. days in the past week
=88. Don't know
89. Refused
99. NA

Campaign Support

-Did you go to any political meetings, rallies, speeches, dinners, or things like that in support of a particular candidate?

-Did you wear a campaign button, put a campaign sticker on your car, or place a sign in your window or in front of your house?

-Did you give money to an INDIVIDUAL CANDIDATE running for public office?

-Did you give money to a POLITICAL PARTY during this election year?

-Did you give any money to ANY OTHER GROUP that supported or opposed candidates?

1. Yes
5. No
8. Don't know
9. Refused
0. NA

Community Involvement

-During the PAST 12 MONTHS, have you worked with other people to deal with some issue facing your community?

-During the PAST 12 MONTHS, did you attend a meeting about an issue facing your community or schools?

-Many people say they have less time these days to do volunteer work. What about you, were you able to devote any time to volunteer work in the last 12 months or did you not do so?

1. Yes
5. No
8. Don't know
9. Refused
0. NA

Organization Membership

-How many organizations are you currently a member of?

0. No organizations
1. One organization
2–96. Organizations
97. 97 or more organizations
88. Don't know
89. Refused
99. NA

Appendix C

Appendix C. Correlation Matrix of the Observed Variables

	1	*2*	*3*	*4*	*5*	*6*	*7*	*8*	*9*	*10*	*11*	*12*	*13*
1. Gender													
2. Age	0.055												
3. Education	0.012	-0.087											
4. Income	-0.310	0.109	0.421										
5. Party strength	0.121	0.121	0.128	0.013									
6. Media use	-0.037	0.251	0.194	0.152	0.199								
7. AS effects	-0.150	0.033	0.129	0.136	0.033	0.120							
8. Opinion	-0.160	-0.058	0.109	0.120	0.094	0.127	0.214						
9. Voting	0.042	0.147	0.313	0.256	0.308	0.296	0.188	0.122					
10. Discuss	-0.001	0.032	0.153	0.068	0.232	0.234	0.086	0.104	0.193				
11. Campaign support	-0.007	0.021	0.183	0.109	0.185	0.275	0.127	0.220	0.302	0.240			
12. Community	0.082	0.008	0.329	0.149	0.057	0.100	0.135	0.120	0.245	0.228	0.293		
13. Membership	0.022	0.015	0.296	0.215	0.058	0.100	0.056	0.074	0.213	0.225	0.193	0.409	
N	1,066	1,063	1,066	963	1,049	1,065	1,017	1,066	720	1,066	1,060	1,066	1,065
Mean	1.506	46.430	4.053	10.605	2.849	5.162	0.317	17.074	2.165	0.775	2.622	0.944	0.824
SD	0.500	17.398	1.648	6.488	0.993	2.253	1.581	6.125	1.250	1.268	2.553	1.065	1.397

Chapter 9

The Public Agenda Along the Life Span

Testing the Life-Cycle Effect of Age on the Agenda-Setting Process

Jae Kook Lee

Scholarly inquiries into media effects are often framed as an interaction between individual audience members and the media as an environment. Thus, it is natural that theories of media effects have tried to explain the roles played by an audience's individual characteristics. Researchers conceived that the process of media effects starts from media exposure. Audience characteristics have usually been considered contingent conditions or moderating variables in the process. As a basic demographic variable, age has been studied as such a condition in media effects research.

In this sense, it is worthwhile to see if age functions as a contingent condition in agenda setting, one of the most widely studied media effects theories. One particular question is whether or not agenda-setting effects of the media vary by people's location in their life span. If people are exposed to the media in differing patterns along their life cycle, it is possible that agenda-setting effects will also show variation. In theories on the role of age, life-cycle effects represent the amount of time that has elapsed since birth, and explain how an individual's position in his or her life span affects attitudes toward public affairs (Danowski & Ruchinskas, 1983; Eisenstadt, 1956; Glenn, 2005; Hellevik, 2002; Klecka, 1971).

Applied to the agenda-setting process, the life-cycle hypothesis states that the location of an individual in his or her life span affects exposure to media content and leads to the formation of different agendas at different ages. There is evidence that an audience is variably exposed to the news media depending on their location in the life span (Robinson, Skill, & Turner, 2004). However, it is still a matter of debate whether such variance in news media exposure actually leads to differing perception of issue importance. Hence, the purpose of the present study is in testing the hypothesis with a longitudinal analysis design. This project seeks to understand how differently or similarly people come to form ideas about issue importance by investigating the life-cycle effects of age on agenda setting.

Although many studies have explored the contingent conditions that regulate agenda-setting effects, there is a paucity of research on age as an intervening variable. This study will contribute to the understanding of public opinion formation by using longitudinal data to explore the relationship between aging and agenda setting. The present study first investigates the stability of the public agenda along the life span of audience members, and then explores its relationship with the media agenda. Three generational age brackets are used in analysis of the public opinion survey for more detailed documentation of variance in the audience agenda. Also, secondary content analysis data of *The New York Times Index* is used to investigate the correspondence between the media and the public agenda.

Literature Review

Age and Agenda Setting

Since the introduction of agenda setting (McCombs & Shaw, 1972), studies have examined contingent conditions that moderate the process. Research has identified audience characteristics such as the need for orientation (Kim, Chapter 4, this volume; Valenzuela, Chapter 3, this volume; Weaver, 1977) and issue obtrusiveness (Zucker, 1978) among the conditions that weaken or strengthen agenda-setting effects. Research has explored age as a contingent condition of agenda setting, but the function of age is not clearly understood. Younger and older people were found to have increasingly similar issue agendas as both groups read local newspapers more frequently (Shaw & Martin, 1992). A study in Spain found a curvilinear relationship of attention to political information with consensus between younger people's issue and attribute agendas and those of older people (Lopez-Escobar, Llamas, & McCombs, 1998). The effects of age on agenda setting were mediated by political interest in a study, which included age as an independent variable (Wanta, 1997). The findings indicated that, compared to younger generations, older people were more likely to be influenced by agenda setting, because they were more interested in public affairs and exposed to more news. However, a direct investigation of age found that agenda setting occurred regardless of generational differences in media use (Coleman & McCombs, 2007). There were no significant differences in rank-order correlations of agendas among three different generations.

In sum, age has at least some effects on the agenda-setting process, but the influence is largely regulated by individual interest in public affairs. However, the difference between age groups in terms of agenda-setting effects is mostly minimal and not captured in correlation analysis. The inconsistent results call for more research. Previous studies are based on cross-sectional data and did not examine possible changes in audience in relation to the media over a long period of time. To address the issue, this study employs a theoretical framework of the life-cycle effect of age, and investigates whether individuals maintain or change their perception of issue importance as they age. In so doing, the study sheds more light on the roles played by age in the agenda-setting process.

Effects of Aging

People change their perceptions of issues and their attitudes toward them as they age, and this individual phenomenon inevitably plays an important role in society. Eisenstadt (1956, p. 21) noted, “age and differences of age are among the most basic and crucial aspects of human life and determinants of human destiny.” This claim should also be applicable to research in media effects and the public opinion process.

It is natural that individuals go through changes in their biophysical and socio-psychological experience, such as physical aging and shifts in social roles, as they move on along their life span. Literature has identified the effects of such movements on various attitudinal and behavioral variables as life-cycle effects among other influences of age (Wimmer & Dominick, 2006). The life-cycle effect explains that differences between generations can be found if people are influenced by changes in the processes and content of socialization as one becomes older (Klecka, 1971).

Among many communication variables, media use is particularly important in examination of the life-cycle effects of age on the agenda-setting process, because agenda setting basically assumes that media effects originate in audience media exposure. A person’s age, or location in their life span, also has been found to influence media use and political communication. Young, middle-aged, and older persons are different from each other in their media use, and older people in particular are more reliant on television for information (Stephens, 1981). Such reliance on television by older people is explained by process gratification rather than by content gratification (Cutler & Danowski, 1980). Older people do not care much about what is on television, but care about whether television is involved in their communication. Also, older people are likely to spend more time on reading the newspaper. Newspaper readership and newspaper satisfaction have both been linked to increased age (Burgoon & Burgoon, 1980; Schoenbach, Lauf, McLeod, Scheufele, 1999). Recently, a European study found that age has a curvilinear relationship with newspaper readership, indicating that older people are more likely to read newspapers, except those who are very old (Elvestad &Blekesaune 2008).

Literature of uses and gratifications offers an explanation of the life-cycle effect on media use by using the concept of need structure (Dimmick, McCain, & Bolton, 1979). The hierarchy of individual needs may alter over time, as people move through their life span, which obviously has an impact on their need for information. Once a person is retired from his or her long-time workplace, he or she is not likely to maintain the same level of appetite for news or information. The need structure has just been re-organized at the time of retirement. The re-organization of the need structure leads to media use patterns different from before. As people move their location from young adults to middle-aged to older, they may need a varying amount of news. It may also influence the content of the media they need. Middle-aged people are more likely than young adults to view more informational shows on television (Dimmick et al., 1979).

As a consequence of the change in media use patterns, the life-cycle effects will influence the agenda-setting effects of the media. People may read or watch different news reflecting their changing interests by the location in their lifetime. This may result in longitudinal differences in the agenda for a group of people as they age. For instance, middle-aged people who are likely to have children in grade school may be more prone to think of education as an important issue, because their position as young parents influences their perceptions. When these same people become seniors, however, health issues such as the universal health care bill and Medicare may become more important to them than education.

The life-cycle effect is based on the assumption that people's different situations in their life span, or individual age, affects their use of media, which leads to different perceptions of issue importance by people of different ages. Although Coleman and McCombs (2007) found a positive relationship between the media agenda and the youngest generation's agenda, it did not investigate the possibility of a life-cycle effect. They surmised that slight differences in agendas between the youngest generation and the two older ones were based on life-cycle differences; for example, the youngest among them were more interested in education than health care because they or their children were more likely to be in college or K-12 schools; however, the study did not try to find direct evidence for this. Thus, the existence of a life-cycle effect is worth testing.

One of the main arguments for the life-cycle effect on agenda setting is based on different media use by people as they age. Extant agenda-setting research on contingent conditions offers an explanation to the differential media use by aging, which possibly leads to a variation in the agenda in one's life span. As indicated in research of need for orientation (Kim, Chapter 4, this volume; Matthes, 2006; Valenzuela, Chapter 3, this volume; Weaver, 1977), personal relevance of certain issues may an have effect on the individual need for orientation, which subsequently influences the magnitude of the agenda-setting effects. The explanation of mediated relationship between life-cycle effect and agenda setting via personal relevance is consistent with the findings of aforementioned research of media use patterns along the life span in the uses and gratifications tradition. If personal relevance based on one's location in the life span causes significant differences in the issue agenda, it will be evidence for the life-cycle effect of age.

The present study tests this hypothesis in two ways for three different generational groups, using a longitudinal research design. First, this study will test the stability of the public agenda along the life span, conducting correlation analysis between the agendas at different time points. Second, the public agenda will be compared with the media agenda at different years to see if there is a variation of correspondence between them. The utility of three age groups is discussed below.

Age Breaks

The purpose of this study is to test if the life-cycle effect has a significant impact on the agenda-setting process. Hence, the older generation is the main focus of the analysis, because the members of the age group have already lived along their life span to almost the end of it. However, analysis of young and middle-aged generations can still be used as a comparison to see if there is any distinct pattern (Cutler & Danowski, 1980).

Studies have used different age breaks to address their unique research questions. For instance, a study on the Vietnam War used the "Munich generation" to describe people who lived through the 1930s and remembered the Munich Conference of 1938 as a symbol of American isolationism, and the "Vietnam generation" designating those who experienced the Vietnam War to investigate effects of age (Holsti & Rosenau, 1980). Recent research has often used a three-generation grouping of Civic Generation, Baby Boomers, and Generation X (e.g. Shah, Kwak, & Holbert, 2001).

The current study employed one of the most widely used three-generation groupings, designated by the Civic Generation, Baby Boomers, and the youngest generation, which incorporates both Gen X and Gen Y, following Coleman and McCombs (2007). For longitudinal analyses exploring the life-cycle effect, these three groups are defined as those who were born between 1926 and 1945, between 1946 and 1964, and 1965 or after. The generations are called the Civic Generation, Baby Boomers, and youngest generation, respectively, as in previous studies (Coleman & McCombs, 2007; Shah et al., 2001).

Method

Data

This study used data from the American National Election Studies (ANES) from 1960 to 2004 to measure the public agenda. The test of life-cycle effects on agenda setting requires comparisons of the agendas of a generation or age cohort over a long period of time. Specifically, public opinion data from the Civic Generation, or those born in or before 1945, are necessary to compare agendas of that generation with their agendas at later time points. ANES data are available from 1948, meeting the conditions for this study. The *New York Times Index* data provided by Policy Agendas Project are used to measure the media agenda.[1] The number of available years from both datasets totals 27. If correlation tests were conducted for all of the years, it may include many results significant only by chance. Nine years were randomly selected out of the 27 to reduce such risks. However, a condition of one or two years in each decade was applied to the selection process to address the purpose of the study, which compares a generation's agendas along the life span of its members.

The main variables in this study are the public agendas of generational subgroups and the media agenda represented by the *New York Times Index*. For measurement of the agenda, we followed extant literature (McCombs & Shaw, 1972) using the Most Important Problem (MIP) questions. A typical wording of the MIP question is: "What do you think is the most important problem facing the country today?" There is much variation in MIP questions, suggesting the possibility of differing measurements of opinion. However, a recent study found that there are strong correlations between agendas measured by various forms of the MIP question (Min, Ghanem, & Evalt, 2007).

Following previous agenda-setting research (e.g. McCombs & Shaw, 1993), an agenda was conceptualized as a ranked list of issues based on frequencies of respondents mentioning those issues. The MIP question has been included in the ANES since 1960, but the coding strategy of the open-ended question has changed over time. The original number of issue categories was eight in 1960, but later increased to 11, excluding a category of nonpolitical problems. To address the overarching question in this study, comparisons of public agendas in different years are necessary. For this reason, the number of categories was adjusted to eight for all datasets from later years. ANES has often created a new issue category by dividing one into several. For instance, an original issue category of "racial and public order problems" in 1960 was later divided into "racial problems" and "public order problems." In this study, these two new categories were combined back into the original category to compare agendas over time. The eight issue categories are social welfare, agriculture and natural resources, labor and union-management relations, race and public order, economy, foreign affairs, national defense, and functioning of the government. On the media side, the Policy Agendas Project uses 27 different categories of major topics in coding the *Times* stories. To compare the media agenda with the public agenda, major topics of the *Times* stories were recoded, following the ANES master code of the MIP question. For instance, topics of macroeconomics and foreign trade were recoded into economy. With the recoding process, it was possible to measure the *New York Times*' agenda using eight issue categories.

Analysis

Data analysis used the rank-order correlation of issue agenda, following previous agenda-setting literature (McCombs & Shaw, 1972). To test the life-cycle effect hypothesis, the issue agendas of three generations at certain years were compared with their own agendas at later time points. For example, the issue agendas of the Civic Generation in the 1960s were compared with their agendas in the 1970s, 1980s, 1990s, and 2000s. Also, rank-order correlation coefficients were estimated between agendas of the three generations and the *New York Times* agenda each year, to see the correspondence between the media and the public agenda. For instance, the Baby Boomers' agenda was compared with the *Times* agenda in each year of 1980, 1986, 1990, 1996, 2000, and 2004. Comparisons were conducted for more years for older generations and for fewer years for younger generations because of the difference in the respective length of the life span.

Results

Table 9.1 illustrates the media agendas represented by the *New York Times Index* for nine years from 1960 to 2004. Out of eight issue categories, economy, foreign affairs, and the functioning of the government have consistently appeared in the top three spots. Economy was mentioned by the *Times* most frequently in all nine years, except 1976, when it was the second most frequently mentioned. The issue of foreign affairs was ranked second for five years, third for three years, and fourth for one year. The functioning of the government was the most frequently covered problem in two years, second in one year, and third in four years. Issues other than these three appeared in the top three ranks only in three years. The relative stability of the media agenda was also found in correlation analyses. The analyses generated 36 Spearman's rho correlation coefficients for all agendas in nine years. Thirty coefficients are statistically significant ($p < 0.05$). The median value of the coefficients is +0.81.

The public agendas for the same years are shown in Table 9.2. Compared to the media agenda, the public agenda captured by the ANES survey shows more variation in terms of most frequently mentioned issues. However, three issues of foreign affairs, social welfare, and the economy appeared at the top three spots in the agendas. They

Table 9.1. Agendas of the *New York Times* by year: Frequencies (and ranks)

	1960	1968	1976	1980	1986	1990	1996	2000	2004
Economy	109 (1)	165 (1)	174 (2)	225 (1)	244 (1)	229 (1)	193 (1)	282 (1)	300 (1)
Foreign affairs	103 (2)	97 (3)	130 (3)	74 (3)	66 (4)	69 (2)	80 (2)	129 (2)	117 (2)
Functioning of the government	96 (3)	165 (1)	209 (1)	137 (2)	68 (3)	30 (6)	69 (3)	109 (3)	88 (4)
Social welfare	54 (4)	84 (4)	94 (4)	50 (7)	70 (2)	55 (3)	52 (4)	76 (4)	68 (6)
National defense	52 (5)	40 (6)	40 (7)	74 (3)	44 (6)	35 (5)	11 (7)	6 (8)	97 (3)
Racial and public order	30 (6)	83 (5)	71 (5)	58 (6)	49 (5)	54 (4)	45 (5)	63 (5)	80 (5)
Agriculture & natural resources	20 (7)	23 (7)	63 (6)	60 (5)	33 (7)	26 (7)	20 (6)	25 (6)	34 (7)
Labor	13 (8)	19 (8)	22 (8)	25 (8)	12 (8)	8 (8)	9 (8)	7 (7)	26 (8)
Total	476	724	803	668	586	506	479	697	799

were mentioned as the most important problems in the ANES survey in three years. Further, at least two of the three issues appeared in the top three spots in the agenda in all years. Out of 36 Spearman's rho correlation coefficients, 26 are statistically significant ($p < 0.05$). The median value of the coefficients is +.71.

To test the life-cycle effect hypothesis, this study first examined stability of the public agenda. As aforementioned, correlation analysis was performed for three age groups, the Civic Generation, Baby Boomers, and the youngest generation. The hypothesis predicted that people change their need structure as they age because of changes in personal relevance to issues, which leads to differential use of media contents, leading to fluctuation of the issue agenda.

The issue agendas of each generation in earlier years were compared with the issue agendas of the same generation in different years. Correlation analyses for the Civic Generation were conducted for eight datasets from 1968 to 2004. Because the age of the cohort members ranged from 23 to 42 in 1968, their issue agenda was captured in 1968, and compared with their agendas in the 1970s, 1980s, 1990s, and 2000s. The tests for Baby Boomers were performed for six datasets from 1980 to 2004. The age of the cohort ranges from 18 to 34 in 1980, and

Table 9.2. The public agendas measured in ANES by year: Frequencies (and ranks)

	1960	*1968*	*1976*	*1980*	*1986*	*1990*	*1996*	*2000*	*2004*
Foreign affairs	592 (1)	785 (1)	61 (5)	320 (2)	279 (4)	605 (1)	21 (5)	59 (4)	271 (2)
Social welfare	155 (2)	207 (3)	606 (2)	200 (3)	555 (1)	323 (3)	294 (1)	375 (1)	75 (4)
Economy	87 (3)	102 (4)	773 (1)	626 (1)	530 (2)	447 (2)	157 (3)	133 (3)	156 (3)
Agriculture & natural resources	68 (4)	17 (6)	39 (6)	34 (5)	82 (6)	120 (5)	20 (6)	23 (7)	1 (7)
National defense	66 (5)	16 (7)	28 (7)	113 (4)	279 (4)	15 (7)	15 (7)	39 (6)	26 (6)
Racial & public order	58 (6)	449 (2)	150 (3)	18 (7)	287 (3)	213 (4)	241 (2)	192 (2)	478 (1)
Labor	23 (7)	9 (8)	8 (8)	1 (8)	2 (8)	1 (8)	0 (8)	1 (8)	0 (8)
Functioning of the government	4 (8)	27 (5)	66 (4)	33 (6)	76 (7)	49 (6)	29 (4)	45 (5)	31 (5)
Total	1053	1612	1731	1345	2090	1773	777	867	1038

thus the issue agendas of this group can be obtained. The analyses for the youngest generation were executed for five datasets from 1986 to 2004. Members of the cohort were age 21 or younger in 1986, so their issue agendas can be measured and compared with their agendas in the 1990s and 2000s.

Table 9.3 shows the Spearman's rho correlation coefficients estimated in the analyses to test stability of the public agenda over the life span. As suggested by literature, if the life-cycle effect works in the agenda-setting process, insignificant correlations will prevail. The left-hand section of Table 9.3 reports all 28 of Spearman's rho correlation coefficients of longitudinal comparisons of the agendas of the Civic Generation for nine datasets from the 1968 to 2004 elections. Out of the 28 comparisons, all are positive, and 23 are statistically significant. The median value is +0.78. The first row of Table 9.3 illustrates the comparisons of the 1968 agenda of the Civic Generation with their agendas in later years. Eight of the nine coefficients generated in the analysis are statistically significant and positive. The median value is +0.79 for all nine comparisons. The findings indicated that age does not have a life-cycle effect on the agenda-setting process.

Table 9.3. Correlations between the different years' agendas of three generations

	The Civic Generation						
	1976	*1980*	*1986*	*1990*	*1996*	*2000*	*2004*
1968	0.64*	0.41	0.64*	0.83**	0.81**	0.79*	0.95**
1976		0.55	0.81**	0.74*	0.91**	0.86**	0.73*
1980			0.88**	0.74*	0.45	0.50	0.37
1986				0.79*	0.76*	0.81**	0.64*
1990					0.76*	0.71*	0.81**
1996						0.98**	0.83**
2000							0.80**
N	8	8	8	8	8	8	8

	Baby Boomers					*Youngest Generation*			
	1986	*1990*	*1996*	*2000*	*2004*	*1990*	*1996*	*2000*	*2004*
1980	0.67*	0.20	0.26	0.45	0.45				
1986		0.45	0.64*	0.81**	0.62	0.80**	0.59	0.86**	0.72*
1990			0.50	0.45	0.33		0.33	0.71*	0.77*
1996				0.88**	0.59			0.74*	0.34
2000					0.81**				0.80**
N	8	8	8	8	8	8	8	8	8

** $p < 0.01$; * $p < 0.05$

N = number of issue categories

If it follows the life-cycle effect hypothesis, the Baby Boomers' agenda in the earliest year will have positive correlations with their own agendas in later years. The findings were much different from those for the Civic Generation. In its middle section, Table 9.3 also reports all Spearman's rho correlation coefficients of longitudinal comparisons of Baby Boomers' agendas for six datasets from 1980 to 2004. All 15 coefficients are positive. Out of these, only five are statistically significant. The median value is +0. 50. As found in Table 9.3, Baby Boomers' agenda in 1980 only correlates with the 1986 agenda. The findings indicated that the life-cycle effect influenced the agenda-setting process for Baby Boomers.

The right-hand section of Table 9.3 reports all 10 of Spearman's rho correlation coefficients of longitudinal comparisons of the youngest generation's agendas for five datasets from the 1986 to 2004 elections. First, all coefficients are positive. Out of the 10 comparisons, seven are statistically significant. The median value is +0.73. Further, the first row in the section of Table 9.3 illustrates the comparisons of the 1986 agenda of the youngest generation with their agendas in later years. Three of the four coefficients generated in the analysis are statistically significant and positive. The median value is +0.76, for all four comparisons. However, it should be noted that the youngest generation has lived a relatively short life in their life span. Hence, the results should be interpreted in a very limited context.

This study next turns to examining the variation of correspondence between the agenda of each generation and the media agenda. The three generation agendas were compared with the *New York Times* agendas in different years. Correlations between the media agenda and the agendas of the three generations were calculated in nine different years from 1968 to 2004 for the Civic Generation, in six from 1980 to 2004 for Baby Boomers, and in five from 1986 to 2004 for youngest generation. All correlation coefficients were positive and were generally statistically significant, except for Baby Boomers. Findings are reported in Table 9.4.

If the life-cycle effect works, there should not be significant correlations between the media agenda and the generation agendas. The first row of Table 9.4 shows the nine comparisons between the *Times* agenda and the Civic Generation agenda from 1968 to 2004. All of the coefficients are positive. Five of the nine coefficients generated in the analysis are statistically significant. The median value is +0.68, for all comparisons. The result is not clear, but the fluctuation of the correlation showed some sign of life-cycle effect on the agenda-setting process.

If the case of Baby Boomers follows the life-cycle effect hypothesis, there will be little correlation between their agenda and the *New York*

Table 9.4. Correlations between the *New York Times* agenda and agendas of three generations in each year

	1968	*1972*	*1976*	*1980*	*1986*	*1990*	*1996*	*2000*	*2004*
The Civic Generation	0.42	0.67*	0.71*	0.52	0.83**	0.91**	0.62	0.45	0.68*
Baby Boomers				0.52	0.74*	0.36	0.52	0.57	0.74*
Youngest Generation					0.74*	0.87**	0.26	0.57	0.74*
N	8	8	8	8	8	8	8	8	8

** $p < 0.01$; * $p < 0.05$

N = number of issue categories

Times agenda over their life span. If the other scenario is true, researchers will find more significant correlations. The second row of Table 9.4 reports the six comparisons between Baby Boomers' agenda and the *New York Times* agenda from 1980 to 2004. Out of the six comparisons, two are statistically significant. The median value of the coefficients is +0.55. In general, the results support the life-cycle effect hypothesis.

The third row of Table 9.4 illustrates the five correlations between the *Times* agenda and the young age group's agenda. All of the five coefficients are positive. Three of the coefficients are statistically significant. The median value is +0.74. The findings indicated that life-cycle effect is working to some extent. Again, the results for the youngest generation should be interpreted with caution.

Discussion and Conclusion

This study explores age as a contingent condition of agenda-setting effects by a mechanism such that personal relevance varying with individual life cycle eventually influences the media effects process. The life-cycle effect hypothesis states that as people age, they naturally change their media use patterns and the actual news content they are exposed to, which results in a differing agenda along an individual life span. The conditioning effect of personal relevance is considered an antecedent of the life-cycle effect of age. To test the hypothesis, correlation analysis is performed for the Civic Generation, and also for Baby Boomers and the youngest generation for the purpose of comparison.

Findings indicated that the agenda of the Civic Generation is in general stable along the life span. The agenda did not change much as its members became older. In 28 longitudinal comparisons of the Civic Generation agendas, the median value of the correlations is +0.78. The generation had significant and strong correlations on the issues in 23 of 28 pairwise year comparisons. Similar results were found for the youngest generation—7 out of 10 pairwise year comparisons were significant with a median correlation of +0.73. The finding indicated that there is little life-cycle effect on the agenda-setting process. However, Baby Boomers showed much variation within their generation across the years—in 15 pairwise year comparisons, 10 did not reach traditional significance levels. The findings for Baby Boomers support the life-cycle effect hypothesis.

The test of correspondence between the generations and the media agenda showed more variations in correlation tests. The Civic Generation's agendas were found close to the *New York Times*' agenda in each year. In nine comparisons between the Civic Generation and the *Times*' agenda, five were significant with a median correlation of +0.71. For Baby Boomers, only two correlations out of six comparisons reached the traditional level of statistical significance. For the youngest generation, three correlations out of five comparisons were significant with a median value of +0.74. From the findings for the Civic Generation, it can be concluded that the public agendas at different points of audience life span are more similar to one another.

However, the findings should be interpreted with caution before they can be generalized to all populations. The correlation analysis revealed that there *is* significant difference between agendas of Baby Boomers along their life span. The finding indicated a high probability of the life-cycle effect of aging on the agenda-setting process for Baby Boomers. It is notable that correlations are significantly weak between the Baby Boomer agenda and the media agenda. The weak relationship made it possible to speculate that the variation in the Baby Boomers' agendas along their life cycle is somehow related to the media's influence on the demographic group.

On another front, the varying agenda along life span suggests that Baby Boomers are intrinsically different from the preceding generation, the Civic Generation. This is an argument for generation effect, which can be observed when generational differences found in a cross section are affected by exposure to historical events. It is not clear that the difference found in the agenda originated in the life-cycle effect or in generation effect, due to limitation of current data. Unlike the Civic Generation, Baby Boomers still have many years to live. It is possible that the Baby Boomer agenda will show more similarity to their agenda of the past in years to come.

The nature of this longitudinal project leaves some points untouched. The most important thing to note is that this study cannot consider other factors that may influence the relationship between aging and agenda setting. It is possible that individual characteristics such as political knowledge play important roles as conditioning variables in the relationship. The limit stems from the fact that this study is focused on generations and thus used aggregate-level analysis. Future research can address this problem by employing individual-level analysis. Also, the characteristics of issues are not taken into consideration in this study. By looking at the relationship between aging and issue characteristics such as obtrusiveness in the agenda-setting process, future research may be able to provide fuller explanation.

Despite the obvious assertion that age is one of the most crucial aspects of human life, and that the evidence of age reveals differences in many areas—ideology; civic engagement; attention to, knowledge of, and interest in politics, among others—this does not translate clearly into the life-cycle effect on agenda setting. The longitudinal stability of the Civic Generation agenda illustrated this point. However, relatively more fluctuating relationship between the generation agenda and the media agenda along their life span provides some, albeit limited, hints of the life-cycle effect. The case of Baby Boomers showed more leaning toward life-cycle effect hypothesis. The generation agenda showed much variance along their life span as well as its relationship with the media agenda. Again, it should be noted that younger generations still have many years to live, unlike the Civic Generation. Further, consideration of other kinds of age influence such as generation effect will contribute to better understanding of age as a contingent condition of the agenda-setting effects.

Note

1 The data used here were originally collected by Frank R. Baumgartner and Bryan D. Jones, with the support of National Science Foundation grant number SBR 9320922, and were distributed through the Department of Government at the University of Texas at Austin and/or the Department of Political Science at Penn State University. Neither NSF nor the original collectors of the data bear any responsibility for the analysis reported here.

References

Burgoon, J. K., & Burgoon, M. (1980). Predictors of newspaper readership. *Journalism Quarterly, 57*(4), 589–596.

Coleman, R., & McCombs, M. E. (2007). The young and agenda-less? Exploring age-related differences in agenda setting on the youngest generation, Baby Boomers, and the Civic Generation. *Journalism & Mass Communication Quarterly, 84*(3), 495–508.

Cutler, N. E., & Danowski, J. A. (1980). Process gratification in aging cohorts. *Journalism Quarterly, 57*(2), 269–276.

Danowski, J. A., & Ruchinskas, J. E. (1983). Period, cohort, and aging effects: A study of television exposure in presidential election campaigns, 1952–1980. *Communication Research, 10*(1), 77–96.

Dimmick, J. W., McCain, T. A., & Bolton, W. T. (1979). Media use and the life span. *American Behavioral Scientist, 23*(1), 7–31.

Eisenstadt, S. N. (1956). *From generation to generation: Age groups and social structure* (3rd ed.). New Brunswick, NJ: Transaction.

Elvestad, E., & Blekesaune, A. (2008). Newspaper readers in Europe. *European Journal of Communication, 23*(4), 425–447.

Glenn, N. D. (2005). *Cohort analysis* (5th ed.). Thousand Oaks, CA: Sage.

Hellevik, O. (2002). Age differences in value orientation: Life cycle or cohort effects? *International Journal of Public Opinion Research, 14*(3), 286–302.

Holsti, O. R., & Rosenau, J. N. (1980). Does where you stand depend on when you were born? The impact of generation on post-Vietnam foreign policy beliefs. *Public Opinion Quarterly, 44*(1), 1–22.

Klecka, W. R. (1971). Applying political generations to the study of political behavior: A cohort analysis. *Public Opinion Quarterly, 35*(3), 358–373.

Lopez-Escobar, E., Llamas, J. P., & McCombs, M. E. (1998). Agenda setting and community consensus: First and second level effects. *International Journal of Public Opinion Research, 10*(4), 335–348.

Matthes, J. (2006). The need for orientation towards news media: Revising and validating a classic concept. *International Journal of Public Opinion Research, 18*(4), 422–444.

McCombs, M. E., & Shaw, D. L. (1972). The agenda-setting function of mass media. *Public Opinion Quarterly, 36*(2), 176–187.

McCombs, M. E., & Shaw, D. L. (1993). The evolution of agenda-setting research: 25 years in the marketplace of ideas. *Journal of Communication, 43*(2), 58–67.

Min, Y., Ghanem, S. I., & Evatt, D. (2007). Using a split-ballot survey to explore the robustness of the "MIP" question in agenda-setting research: A methodological study. *International Journal of Public Opinion Research, 19*(2), 221–236.

Robinson, J. D., Skill, T., & Turner, J. W. (2004). Media usage patterns and portrayals of seniors. In J. F. Nussbaum & J. Coupland (Eds.), *Handbook of communication and aging research* (2nd ed., pp. 423–450). Mahwah, NJ: Lawrence Erlbaum Associates.

Schoenbach, K., Lauf, E., McLeod, J. M., & Scheufele, D. A. (1999). Distinction and integration: Sociodemographic determinants of newspaper reading in the USA and Germany, 1974–96. *European Journal of Communication, 14*(2), 225–239.

Shah, D. V., Kwak, N., & Holbert, R. L. (2001). "Connecting" and "disconnecting" with civic life: Patterns of Internet use and the production of social capital. *Political Communication, 18*(2), 141–162.

Shaw, D. L., & Martin, S. E. (1992). The function of mass media agenda setting. *Journalism Quarterly, 69*(4), 902–920.

Stephens, N. (1981). Media use and media attitude changes with age and time. *Journal of Advertising, 10*(1), 38–45.

Wanta, W. (1997). *The public and the national agenda: How people learn about important issues*. Mahwah, NJ: Lawrence Erlbaum Associates.

Weaver, D. H. (1977). Political issues and voter need for orientation. In D. L. Shaw & M. E. McCombs (Eds.), *The emergence of American political issues* (pp. 107–119). St. Paul: West.

Wimmer, R. D., & Dominick, J. R. (2006). *Mass media research: An introduction*. Belmont, CA: Wadsworth Pub Co.

Zucker, H. G. (1978). The variable nature of news media influence. In B. D. Ruben (Ed.), *Communication Yearbook* (Vol. 2, pp. 225–240). New Brunswick, NJ: Transaction Books.

Chapter 10

Online Agenda Setting

A New Frontier for Theory Development

Hai Tran

At the turn of the twenty-first century, researchers prophesized about the end of agenda-setting effects due to the rise of the Internet. They predicted that the decentralized characteristics of the new media communication would make it less possible for the media to tell people what to think about. However, online agenda setting has become a new frontier as media scholars seek to examine the relevance of the theory in the Web 2.0 world.

According to PEW surveys, the U. S. online news audience grew from 2% in 1995 to 74% in 2010 (Jansen, 2010). The spread of the Internet has changed the way information is disseminated, sought, and received. The abundance of divergent agendas on the Web allows people to mix them in a highly individualized manner, which may create a fragmenting media landscape (Maier, 2010; Shaw, Hamm, & Knott, 2001). On the other hand, there is a process of convergence between various media in the online environment (Strömback & Kiousis, 2010). Therefore, the concept of media salience as a whole remains relevant despite differential media use (Coleman & McCombs, 2007). Though dramatic changes brought by the Internet may not necessarily lead to an entire elimination of a common agenda in the focal attention of the public, there is no question that they are altering the way effects occur (McCombs, 2005; Meraz, Chapter 1, this volume; Sayre, Bode, Shah, Wilcox, & Shah, 2010). This brings both challenges and opportunities to the work of broadening the applicability of agenda-setting theory.

As Web-based mass communication became a ubiquitous phenomenon, scholars have moved research beyond the traditional media context and taken up the task of testing the agenda-setting hypothesis with newer media platforms. In the past 10 years or so, several studies examined the specific effects of agenda setting as it occurs online. It is both timely and possible now to conduct a review of this newly formed, coherent literature.

The current study looks at journal articles as a barometer of online agenda-setting scholarship. Relevant studies might also be found in books, book chapters, or at refereed conferences, but peer-reviewed journals remain the main channel for reporting cutting-edge research

(Kamhawi & Weaver, 2003). A comprehensive search in two EBSCOhost databases, *Academic Search Premier* and *Communication and Mass Media Complete*, yielded an adequate number of research articles that meet the criteria of a formal review. These published studies utilize conceptualizations of agenda setting and empirical methods to investigate the transfer of salience that involves the online media and online audience. This sample, though it may not provide an exhaustive list of all existing research, is representative of the overall effort by academics in advancing our knowledge of agenda setting on the Web.

The present status report offers a close reading of relevant studies. The ultimate goal is to identify dominant trends, areas of strengths and weaknesses, thereby pointing to opportunities for theory development. In subsequent sections, the chapter provides (a) an overview of relevant research; (b) an in-depth review of key approaches and findings; (c) a discussion of research methodology; and (d) a list of questions for future consideration.

Overview

Online agenda-setting research, as found in this sample, made its way to academic journals in 2002, when PEW Internet & American Life Project released statistics showing the emergence of the Internet as a mainstream information tool (Rainie & Horrigan, 2002). Since then, at least 26 primary research articles examining the agenda-setting function of online media have appeared in 16 different academic outlets. Though the frequency of relevant studies does not increase steadily from year to year, there is a surge in publication volume between 2007 and 2011[1]. Roughly 73% of selected articles were published within this time interval. The data reported here indicate a sharp turn to strong interest in online agenda setting in the last five years.

Online agenda-setting research is characterized by a high degree of concentration. Half of the studies examine issues pertaining to elections and politics. This reflects a clear topical preference as often found in traditional agenda-setting literature (Martin, Chapter 2, this volume; McCombs, 2005). Content analysis is the most common choice, with 61.5% of the articles solely using this method. Online agenda-setting effects are most likely to be measured by aggregate data (88.4%), which provide evidence of the shifting salience of a particular issue over time or of an array of issues competing for attention. According to McCombs (2004), such aggregate measures are the foundations of the agenda-setting tradition. In terms of geography, while more than three quarters of the publications (77%) focus on U. S. context, online agenda setting has also been studied internationally.

Evidence of Effects Online

Looking at agenda setting as a political process, Dearing and Rogers (1996) identified three traditions of research: media agenda setting, public agenda setting, and policy agenda setting. Although agenda-setting studies initially emphasized the role of the media in influencing public opinion, the question of how the media's agenda is shaped started to intrigue scholars in the 1980s (McCombs, 2004). Online agenda-setting research has a heavy focus on the transfer of salience among media agendas. Though the extant literature does explore public agenda setting and effects on attitudes and behaviors, a majority of published studies (69.2%) address interactions among various online media.

Intermedia Agenda Setting

Antecedents of the media agenda are described by McCombs (2004) as concentric layers of an onion. These include sources and their own agendas, other media's agendas, and professional norms. The Internet and its diverse mix of news and information create an ideal environment for interactions among numerous media (mass, niche, social). A comparison of 18 months of news content generated by major news websites (Yahoo News, MSNBC, CNN.com. AOL News, Google News) and legacy media (newspapers, network television, cable television, and radio) shows a 59% commonality in news agenda and a similar depth of coverage (Maier, 2010). Nevertheless, online news does not simply follow the lead of traditional news. News sites offer a news mix with a wider range of issues. Tracking news links that appeared in blogs and social media in 2009, Maier (2010) found stark differences between the issue agenda of these platforms and that of mainstream media—including news websites. Less than 30% of the top story topics in blogs and social media match with professional media top issues. When their agendas do overlap, blog and social media content is slightly more extensive. This indicates a tendency to diverge from the mainstream news agenda. In cyberspace, there are many external sources beyond the mainstream channels for the studying of intermedia agenda setting (Meraz, Chapter 1, this volume).

Political blogs. One of the most researched areas is a possible agenda-setting relationship between the blogosphere and the news media. Anecdotal evidence indicates that political blogging, on several occasions, put pressure on the mainstream media, forcing them to pay more attention to certain issues (Tomaszeski, Proffitt, & McClung, 2009). As blogs appear to play an influential role in the media landscape, scholars have empirically investigated who sets whose agenda under what circumstances (McCombs, 2005).

In various publications, Meraz (2009, 2011a, 2011b) analyzed different levels of agenda setting in a sample of 18 top independent political blogs, two traditional media outlets (the *New York Times* and the *Washington Post*), and their 11 political newsroom blogs. The author focused on three issues that provided notable discussion across all media networks over two months in 2007: the Alberto Gonzales hearings on the NSA wiretapping; the Larry Craig sex scandal; the David Petraeus report on Iraq and the MoveOn.org "Betray us" ad. Several findings are noteworthy. First, a two-way relationship exists between the mainstream media and the blogosphere. The traditional media set the agenda for almost all blog networks (except for left-leaning blogs). On the other hand, all blog networks had success at setting the traditional media's agenda. Moreover, political blog platforms set each other's agenda. Meanwhile, mainstream newsroom blogs were less of an influencer and more influenced by their news sites and political blogs (Meraz, 2011a). In terms of attribute agendas, partisan blogs (left vs. right) differed on two out of the three issues. Both moderate and left-leaning blog networks shared strongly correlated attribute agendas with the mainstream media, but the right-leaning blogs showed little agreement with elite media agendas at the attribute level. In the online, networked environment, partisan bloggers can handpick, bypass, or even resist elite media's attribute agenda-setting influence to best suit their own ideologies (Meraz, 2011b). Moreover, hyperlink analysis shows that while traditional elite media maintained relative dominance as top sources within each blog network for two of the three issues, they no longer exerted a singular, universal influence (Meraz, 2009). Across three issues, political blogs also linked to citizen media. There was some evidence, though modest, that elite traditional media entities paid attention to independent blogs run by news professionals. Overall, the mainstream media remain a driving force in influencing the blogosphere, but political blogs are redistributing power between elite media and citizen media (Meraz, 2009).

Along a similar line, Messner and DiStaso (2008) examined intermedia agenda setting between the traditional media and the blogosphere by studying the process through which one becomes a legitimate source for the other. The research first looked at the use of blogs as sources in the *New York Times* and the *Washington Post* over a six-year period (2000–2005). As shown in the findings, the number of articles mentioning the blogosphere increased steadily over time and the traditional media increasingly legitimized blogs as credible sources. The blogosphere was initially used in news coverage as a secondary mention, then as a news source, and eventually as a primary news topic. The two newspapers used blogs primarily as sources for political and business news.

A separate analysis of the top 120 blogs in 2005 found that the blogosphere relied heavily on the traditional media as sources, especially for public affairs information. This interdependence between the traditional mass media entities and the blogosphere creates a news source cycle with agendas being passed back and forth in the online environment.

A complex, bidirectional relationship between the mainstream media agenda and the blog agenda was also documented during the 2004 U. S. presidential elections. Tracking media coverage and blog discussion of 35 issues during the campaign, Wallsten (2007) demonstrates how political bloggers and journalists interacted, followed, and responded to each other's coverage, thereby setting each other's agenda. An in-depth analysis of interactions among 10 A-list blogs, 50 less popular blogs, and the *New York Times* reveals mutual influence between the newspaper and top political blogs on 19 issues and two-way relationships between the newspaper and less popular blogs on 22 issues. This study indicates that the agenda-setting relationship between mainstream media and political blogs is "a high-speed, two-way street rather than a slow-moving, one-way road leading from media coverage to blog discussion or vice versa"(Wallsten, 2007).

In some cases, political bloggers have an intense desire to push an issue agenda into the mainstream media. Schiffer (2006) focused on the Downing Street Memo controversy in 2005 to explore blogswarm agenda setting, which refers to the ability of the blogosphere to force mainstream media coverage of ignored issues. The salience of this issue in 10 elite blogs (five from each side of the political spectrum), the Daily Kos diaries, broadcast content (NBC, CBS, ABC, CNN, Fox News), and newspaper content (28 national and big-city dailies) over two months was examined. As liberal blogs pressed the issue, newspapers and CNN gave considerable coverage to it. The conservative blogs, broadcast networks, and Fox appear to have ignored the story. In terms of directional relations, newspaper and broadcast coverage largely followed official action. Though blogswarm did not impact news content, it had an easier time penetrating the opinion pages of newspapers.

Online campaigning tools. Increasing the salience of candidates and their issue agenda in media coverage and public attention is an important goal of political communication efforts (Kiousis, Mitrook, Wu, & Seltzer, 2006). Since the 1996 election campaigns, the Internet has been used as a tool to convey political messages directly to voters. Web-based communication strategies such as candidate blogs, campaign websites, and online political advertisements have become new subjects for intermedia agenda-setting research.

Sweetser, Golan, and Wanta (2008) examined correlations between candidate blogs, political advertisements, and television news networks

(ABC, CBS, NBC, CNN, Fox) during the 2004 election and found strong correlations between the two candidates' blogs and between candidate blogs and the media agenda. According to the authors, there is a reciprocal agenda-setting effect between candidate blogs and broadcast television news. Though the traditional media continue to have a more dominant role in influencing candidate blogs, the media agenda is somewhat shaped by the candidate blog agenda. Moreover, candidate blogs also set the agenda for one another.

The impact of website campaigning on traditional news media was documented in a study of the 2000 election. Ku, Kaid, and Pfau (2003) compared two months of newspaper and television coverage (the *New York Times*, the *Washington Post*, ABC, CBS, NBC) and news releases from the websites of George W. Bush and Al Gore. The analysis reveals several multidirectional relationships between Web campaign agenda and media agenda over time. Both candidate websites, particularly Bush's, were more likely to lead and influence newspaper and television agendas than the reverse.

Moving beyond the mainstream mass media, Tedesco (2005) investigated intercandidate agenda setting through online campaigning to study the possibility that candidates might set the issue agenda for their opponents. The findings suggest that candidate websites shared a common set of issues and each campaign's Web issue agenda was consistent. Intercandidate agenda-setting effects were detected in every month of the 2000 Democratic primary campaign, but no single candidate played the dominant role. Candidate websites were more likely to take turns in setting the issue agenda.

The Web technology allows political groups to join forces and advance their own agendas. Internet agenda can be shared broadly across platforms and channels or adopted narrowly within group media networks. Ragas and Kiousis (2010) tested both levels of intermedia agenda setting among online political advertisements generated by a political activist group (MoveOn.org), citizen activists ("Obama in 30 Seconds" contest), the Obama campaign (Obama's official YouTube channel), and partisan media coverage (*The Nation*). An analysis of online advertising content and news content during the 2008 elections provides some evidence of the transfer of issue and affective attribute salience within this political media network. While a modest first-level relationship was found among the online political advertisements, the robust correlation between the Obama negative attribute agenda and the MoveOn.org issue agenda demonstrates an impact of attribute salience on subject salience, thereby supporting the compelling-arguments hypothesis.

News sites. As the Internet spreads, mainstream media entities migrate online and niche news outlets (hyperlocal, alternative) mushroom in cyberspace. News sites have an advantage over traditional news media thanks to their capability to deliver news in a 24-hour cycle. Online agenda-setting research has examined the influence of news websites on media agendas.

Two publications explored the intermedia agenda-setting function of mainstream online media in South Korea. The first project (Lim, 2006) focused on the relationship between two online newspapers (Chosun, JoongAn) and an online wire service (Yonhap). According to the results, online newspapers were more likely to influence the issue agendas of the online wire service than the reverse. The leading online newspaper (Chosun) played a dominant role. The second study (Lim, 2011) looks at both levels of intermedia agenda setting among three South Korean major newspaper sites (Joins.com, Chosun.com, Donga.com) and Yonhap online news service. A similar pattern of effects emerged. Specifically, two primary online newspapers (Chosun and Donga) shaped the issue agenda of the online wire service. At the attribute level, these two sites led the secondary online newspaper (Joins) and the leading newspaper site (Chosun) impacted the online wire service.

Groshek (2008) compared the home pages of CNN and CNN International sites in a 30-day period and found a high correlation between their issue agendas. Though the coverage on the CNN site was less extensive, its topical choice was strikingly similar to the CNN International version. On the other hand, the two websites showed some differences in their attribute agendas. CNN homepages contained more American and violence frames than did CNN International.

The role of independent news sites in focusing attention on marginal issues was documented in a study of anti-U. S. sentiment during 2002 in South Korea (Song, 2007). Comparing the content of alternative online news services (OhmyNews, PRESSian) and that of national newspapers (conservative Chosun and JoongAng, progressive Hankyoreh), this research found a divide along the conservative and progressive lines. Specifically, online and progressive media pushed the agenda, thereby building reactions to the deaths of two schoolgirls by a U. S. military vehicle into massive anti-U. S. protests. In other words, independent news sites set agendas by allying and competing with different traditional media outlets in expanding the scope and intensity of a political issue.

Social media and bulletin boards. The ascendance of digital technology enables individuals to "take the media into their hands" and publish their own coverage (Sayre et al., 2010). User-generated content may fill in what is left out by professional media and even set off media coverage. In other words, alternatives to the mainstream media have become new players in the intermedia agenda-setting process.

Sayre et al. (2010) paired YouTube content with content specifically mentioning Proposition 8 (a controversial plan to ban same-sex marriage in California) in eight state newspapers and in the Google News archive during a 14-month timeframe. The study findings point to a reciprocal influence among social media, newspaper coverage, and search traffic. The relationships between social media and mainstream media shifted over time and were determined by trigger events (elections, court decisions, etc.) Social media content followed mainstream media coverage before the 2008 election, but social media led the way in the aftermath of the election and during the period surrounding the 2009 California Supreme Court decision.

Although early online agenda-setting research used Internet bulletin board discussion to measure public salience online (Roberts, Wanta, & Dzwo, 2002), recent studies examined this platform as part of intermedia agenda setting. An analysis of source use in bulletin board discussions, Internet portal news, newspapers, and government press releases in a Chinese food safety scandal shows mutual influence among these media agendas. Bulletin board discussions initially led source agenda of press releases, while portal news followed the newspaper source agenda. Over time, bulletin board and portal source use influenced the source use of newspapers. In terms of attribute agenda setting, the tone in bulletin board discussions impacted both newspaper coverage and portal news (Siu, 2008). A two-way relationship between Internet bulletin boards on traditional media was documented during the 2000 election in South Korea (Lee, Lancendorfer, & Lee, 2005). At the issue level, mainstream newspapers (Chosun, Donga) set the agenda for bulletin boards. On the other hand, bulletin board discussions affected the newspaper agenda of substantive attributes.

Public Agenda Setting and Consequences

Research tends to examine online media effects from two distinct perspectives: the medium/message and the receiver. The first approach focuses on effects of online media, while the second one addresses effects on online audiences. This also holds true for the studying of online agenda setting.

Effects of the Internet. "The medium is the message" is the notion that the characteristics of a medium, not the content alone, influence perception and response (McLuhan & Lapham, 1994). The breadth and speed of the Internet make it far different from traditional mass media. The nature of agenda setting might change with online media platforms. The question of whether the Internet induces, enhances, or undermines agenda-setting effects on public opinion has stimulated various empirical investigations.

Coleman and McCombs (2007) looked at Internet use as a moderator of public agenda setting. In a two-study project, these scholars demonstrate that greater use of online media among the youngest generation did not eliminate the impact of traditional media. Meanwhile, Internet use did factor in agenda-setting effects. Heavy Internet users' agenda correlated with traditional media agenda (print and television), but the correlation was lower compared to that of lighter Internet users.

Above and beyond the important role of newspaper and television coverage, candidate websites have success in influencing public opinion. Using polling data as a measure of public salience, Ku and associates (Ku, Kaid, & Pfau, 2003) examined the power of online campaigning in the 2000 elections. They found the impact of candidate site agenda over time was greater than that of mainstream agenda. Moreover, the agenda-setting function of candidate websites was more evident in the later phase of the campaign.

Experimental research accounts for differential effects of Internet news vis-à-vis traditional news in setting the public agenda. Such undertakings assess individual-level agendas to capture the subtle way in which new media platforms alter agenda-setting effects. Althaus and Tewksbury (2002), for example, explored whether readers of the print and online versions of the *New York Times* would think differently about the importance of political issues. An experiment was conducted to detect effects as a function of prolonged exposure to either form of news coverage. According to the results, the structure of online news provides individuals greater control over exposure, thereby leading online readers to focus on different types of information and to develop different perceptions of important problems than those developed by print readers. Another study compared the agenda-setting effects of online and television news (Conway & Patterson, 2008). The investigators exposed respondents to either a CNN newscast or the CNN news site and then probed their recall for top stories and ranking of issue importance. The influence of Internet news on users proved to be weaker than the impact of television news on viewers. It is conceivable that the format of Web pages does not always contain salience cues and, therefore, Internet users are less likely to be affected by online media agenda.

Effects on netizens. Each time a survey on Internet use is completed, new data would show a growing online population. Broadband access makes it easier for people to perform a wide range of activities on the Web, from news consumption to political learning and democratic participation. As the Internet has established itself as a social space for individuals to gather, discuss, and take collective action, it is important to understand how the online public is influenced by agenda-setting effects.

An early online agenda-setting study utilized AOL bulletin board discussions to measure public salience online (Roberts et al., 2002). This research paired bulletin board agenda with mainstream online media agenda (*The New York Times*, AP, Reuters, *Time*, CNN) during the 1996 political campaign. In three out of four issues studied (immigration, health care, taxes), effects on bulletin board discussion were observed. Basically, media coverage stimulated discussion of issues on the Internet. The *New York Times* remained the key player in setting the online public agenda.

Recent research gauged agenda-setting effects through online search terms. Hester and Gibson (2007) studied how public salience is shaped by national and local media coverage (newspaper and television) of a controversial issue—same-sex marriage. Tracking search patterns in two different locations (Atlanta and Chicago), the authors demonstrate that online public interest in the issue was predicted by media coverage during the synchronous and previous weeks. Moreover, traditional media elicited differential effects on search behavior, with local media having greater agenda-setting influence when the salient issue was both local and national. Taking a similar approach, Weeks and Southwell (2010) examined issue salience in aggregate search as a function of issue salience in newspaper and television news. This research compared search data and traditional media coverage of a rumor that Obama was Muslim during the 2008 presidential election. As shown in the results, intense media coverage, especially television, immediately affected spikes in online searches. Put another way, online public agenda is set by conventional media agenda.

Consequences of agenda setting are also seen beyond online information-seeking behavior. A study of charitable giving and media coverage following the 2004 Asian tsunami reveals an interesting correlation (Waters & Tindall, 2011). Specifically, U. S. broadcast news coverage exerted a significant impact on e-philanthropy through third-party outlets such as Amazon.com and eBay. This finding links agenda-setting effects to actual behaviors as manifested in the online environment.

Departing from the use of aggregate data to account for affective elements in agenda setting, an experiment conducted on Internet news users addresses priming effects in connection with opinion direction and opinion strength as well as the association between two levels of agenda setting and priming (Lee, 2010). Over a three-week period, the researcher assigned participants to three versions of a website with different mixtures of stories about global warming. The findings show that greater exposure to the global warming issue and its attributes led to stronger support for efforts to reduce global warming effects. In addition, attribute agenda setting was more closely connected to priming.

Time Span for Online Effects

The optimal time lag involved in the transfer of salience has long been a controversial topic in agenda-setting research. In the previous review of the extant literature, McCombs (2004) sets the range of four to eight weeks for agenda setting to take effect. The decay of agenda-setting influence is believed to be between eight and 26 weeks.

This benchmark is based on the accumulation of evidence from the studying of traditional media. The optimal effect span for elite newspapers to set the agenda is between four and six weeks (Winter & Eyal, 1981). Tracing the agenda-setting effects of five media, Wanta and Hu (1994) estimate time lags of one week for national network news, two weeks for local TV news, three weeks for regional newspapers, four weeks for local newspapers, and eight weeks for national news magazines.

Conceivably, the immediacy and speed of online communication have shortened the time span for examining agenda setting in the Internet environment. Online information is delivered, sought, received, processed, and exchanged constantly in a 24-hour cycle, often in real time. Reactions to information through Web-based platforms tend to occur almost instantly. The media landscape has changed, thereby altering the way an agenda is set.

In an attempt to determine the optimal effect span for online agenda setting, Roberts, Wanta, and Dzwo (2002) set the time lags from one to seven days. These scholars compared online media salience (mainstream news sites) and online public salience (bulletin board discussions) and found immediate effects after just one day. The agenda-setting influence also decayed rather rapidly, after only two or three days. In other cases, effects did not occur until day six or seven.

Subsequent studies adopt this approach and find a relatively short time frame for online agenda-setting effects (Hester & Gibson, 2007; Ku et al., 2003; Lee et al., 2005; Sweetser, Golan, & Wanta, 2008; Wallsten, 2007; Weeks & Southwell, 2010). Most recently, an analysis of television news effects on online search behavior over a two-week period indicates that the transmission of salience online may take place within the same day, begin to attenuate on the next day, prolong for three days, diminish on following days, and disappear after one week (Weeks & Southwell, 2010).

Summary

Research during the last 10 years demonstrates that the nature of agenda-setting effects has altered with the advancement of online media communication. Mainstream mass media entities still exert their agenda-setting power, but they are no longer a sole force. Nontraditional platforms such as political blogs, partisan and alternative online media, political websites, and social media become fire starters or shapers of media agendas in subtle ways. These external media forces may even bypass mainstream media to target and influence public opinion. Online agenda-setting studies reveal how Internet use may attenuate the impact of traditional media on the public agenda and why the unique characteristics of online media may inhibit agenda-setting effects. The extant literature also provides evidence regarding a shorter time span in the transmission of salience from the media agenda to the public agenda in the Web environment.

Methodology

Agenda-setting theory is premised on the assumption that certain elements emphasized by the media become the focal point of public attention. Over time, such emphases lead the public to perceive them as important. Empirical testing of this hypothesis involves three steps: a) monitoring media content; b) tracking public opinion; c) comparing media and public agendas. Research of traditional mass media tends to utilize content analysis, survey, and correlations to document agenda-setting effects. With the rise of the Internet as a new mass medium, the measurement of salience and salience transfer has evolved.

Salience

Salience is a central variable in agenda-setting scholarship. Conceptually, it can be understood as the importance or relevance of something. Salience is concerned with both the media and the public. At the operational level, salience has been defined in various ways (McCombs, 2005).

Media salience. There are three dimensions of media salience: attention, prominence, and valence (Kiousis, 2004). Both attention and prominence refer to overall coverage volume, with the former focusing on the frequency and the latter emphasizing elements such as placement, size, space, or time. Valence can be measured as tone or level of conflict.

Online agenda-setting methodology has developed new measures to gauge salience in Web content. Several studies used the frequency of issue, attribute, and source mentions as the surrogate for salience in blog posts (Meraz, 2011b; Messner & DiStaso, 2008; Schiffer, 2006; Sweetser et al., 2008), online political ads (Ragas & Kiousis, 2010), bulletin boards and portals (Lee et al., 2005; Siu, 2008), and online news (Song, 2007). The sum of keywords is another measure of salience on websites (Ku et al., 2003; Tedesco, 2005) and blogs (Wallsten, 2007). Some scholars counted the total number of unique blog posts, articles, hyperlinks, or YouTube clips as markers of issue salience (Lim, 2006, 2011; Meraz, 2009, 2011a; Sayre et al., 2010). These approaches largely focus on the attention dimension of media salience.

There also has been an attempt to measure the valence aspect of salience. In addition to frequency counts of issue-specific headlines and frames, Groshek (2008) devised a five-point scale to rate pictorial depictions of conflict on news websites. Departing from content analysis, Maier (2010) utilized secondary data provided by the Project for Excellence in Journalism to examine salience in new media agenda. The New Media Index is developed on top of data from two Web-tracking sites, Technorati and Icerocket, which monitor the way more than 100 million news blogs and 250 social media Web pages link to news articles. Percentage of links provides an index of news influence on the agenda of social media.

Public salience. Citing Evatt and Ghanem (2001), McCombs (2005) refers to at least two dimensions of public salience: social salience and personal salience. In traditional agenda-setting research, public salience is oftentimes gauged through surveys. The most conventional method is to ask the MIP (Most Important Problem) question. This instrument has been validated for years and is the foundation of agenda-setting methodology. On the other hand, it is a self-reported measure and susceptible to response bias.

Studying agenda setting in the context of the Web, scholars have come up with alternative methods to measure public salience online. Roberts et al. (2002), for example, analyzed Internet bulletin board discussion as an indicator of salience. According to the authors, the frequency of bulletin board discussions provides evidence of a behavioral effect. More recently, a new measure of public salience has been introduced: aggregate search volume. Hester and Gibson (2007) used the Yahoo! Buzz Index as the marker of public interest in a topic. Sayre et al. (2010) monitored search traffic through the number of hits on Google News. In a study of the 2008 U. S. elections, Weeks and Southwell (2010) adopted the Search Volume Index generated by Google Trends as a marker of information-seeking behavior.

The introduction of online search terms as indicators of public salience is an innovative approach to agenda-setting methodology. Data on search queries are rich and easily available. Search indicates a strong motivation to seek for information. Aggregate search volume appears to capture all dimensions of public salience. Intuitively, scholars praise this measure as an unobtrusive, natural index that is not influenced by survey wording or response bias (Granka, 2010; Hester & Gibson, 2007; Weeks & Southwell, 2010). Meanwhile, empirical efforts should be made to ensure the validity of search query data as a measure in agenda-setting methodology (Mellon, 2011).

Salience transfer and causality. At the heart of it, agenda setting is a theory of salience transmission. Inherently, the measurement of agenda relationships requires a time order and an element of causality. While several online studies still rely on cross-sectional, aggregate data and simple correlations to account for agenda-setting effects, steps have been taken toward more stringent approaches. Time series and experimental designs were used to draw inferences about directional and causal relationships.

In this sample of publications, time series research accounts for roughly 27% of online agenda-setting studies. The figure reported here (see Table 10.1) indicates a methodological progression toward sophistication in exploring time lags of effects. Time series analysis allows for testing across different time periods and for identifying the time order in relationships among variables (Hester & Gibson, 2007). Time series models control for a given time series' past values when predicting its present value. There are various techniques for time series analysis. Tracking the transfer of salience online, some scholars utilized Vector Autoregression (VAR) with Granger causality tests to provide statistical summaries for any significant time-order relationships (Meraz, 2011a; Sayre et al., 2010; Schiffer, 2006). Others relied on Autoregressive Integrated Moving Average (ARIMA) model cross-correlation analysis, which allows the data to be analyzed with several different time lags and alternative models for each time series (Roberts et al., 2002; Wallsten, 2007; Weeks & Southwell, 2010). Though time series analysis does not imply causality, it suggests some sort of directional influence. This is a powerful technique to draw inferences about the transfer of salience from one agenda to another.

A widely cited limitation of agenda-setting methodology is a violation of the "correlation-is-not-causation" principle and a lack of individual level explanations of the cognitive processes underlying agenda setting (Son & Weaver, 2006). Experimental design is an obvious methodological choice to attribute causality and to account for effects at the individual level. Experiments allow researchers to manipulate the presumed cause and to observe the outcome afterward. That helps detect the relatively direct stimuli–response relationships, while reducing the plausibility of alternative explanations for the observed effects. Scholars have begun to adopt experimental design as a method in studying online agenda setting. Conway & Patterson (2008), for example, conducted an experiment to demonstrate discrepancies in using free recall (MIP question) versus aided ranking instruments to measure public salience. Two

Table 10.1 Online agenda-setting research methodology (2002–2011)

Study	*Topic*	*Data*	*Analysis*
Meraz (2011a)	Intermedia agenda setting (political blogs/mainstream media)	Content analysis	Time series
Meraz (2011b)	Intermedia agenda setting (political blogs/mainstream media)	Content analysis	Correlations
Lim (2011)	Intermedia agenda setting (online newspapers/online wire service)	Content analysis	Correlations
Waters & Tindall (2011)	Consequences of agenda setting (broadcast news/e-philanthropy)	Content analysis/ fundraising totals	Correlations
Lee, G. (2010)	Public agenda setting/ consequences of agenda setting (online news)	Experiment	Regression/t-test/ correlations
Maier (2010)	Intermedia agenda setting (online media/legacy media)	Secondary data-PEJ media indices	Correlations/ t-test
Ragas & Kiousis (2010)	Intermedia agenda setting (online political ads/partisan media)	Content analysis	Correlations
Sayre et al. (2010)	Intermedia agenda setting (online/social media/traditional media)	Content analysis/ Google News hits	Time series
Siu (2010)	Source/intermedia agenda setting (portal/bulletin board/ government release/newspaper coverage)	Content analysis	Correlations
Weeks & Southwell (2010)	Public agenda setting (mainstream media/online search behavior)	Content analysis/ Google Trends	Time series
Meraz (2009)	Source/intermedia agenda setting (political blogs/mainstream media)	Content analysis	Analysis of variance
Conway & Patterson (2008)	Public agenda setting (online news/TV news)	Experiment	Correlations
Groshek (2008)	Intermedia agenda setting (online CNN networks)	Content analysis	Correlations/ analysis of variance/chi-square test

(Continued)

Table 10.1 (Continued)

Study	*Topic*	*Data*	*Analysis*
Messner & DiStaso (2008)	Source/intermedia agenda setting (blogs/mainstream media)	Content analysis	Chi-square test
Sweetser et al. (2008)	Intermedia agenda setting (political ads/candidate blogs/TV news)	Content analysis	Correlations
Coleman & McCombs (2007)	Public agenda setting (moderating role of Internet use)	Content analysis/ survey	Correlations
Hester & Gibson (2007)	Public agenda setting (traditional media/online search behavior)	Content analysis/ Yahoo! Buzz	Time series
Song (2007)	Intermedia agenda setting (online news service/mainstream newspapers)	Content analysis	Frequency distribution
Wallsten (2007)	Intermedia agenda setting (blogs/ mainstream media)	Content analysis	Time series
Lim (2006)	Intermedia agenda setting (online newspapers/online wire service)	Content analysis	Correlations
Schiffer (2006)	Intermedia agenda setting (political blogs/mainstream media)	Content analysis	Time series
Lee, B. et al. (2005)	Intermedia agenda setting (bulletin boards/newspapers)	Content analysis	Correlations
Tedesco (2005)	Intermedia/intercandidate agenda setting (candidate websites)	Content analysis	Correlations
Ku et al. (2003)	Intermedia/public agenda setting (candidate websites/mainstream media/public opinion)	Content analysis/ secondary data (polling results)	Correlations
Althaus & Tewksbury (2002)	Public agenda setting (online news/print news)	Experiment	Analysis of variance
Roberts et al. (2002)	Public agenda setting (mainstream media/bulletin boards)	Content analysis	Correlations

other studies (Althaus & Tewksbury, 2002; Lee, 2010) used a pretest-posttest design to capture the transmission of salience over time. The experiment on priming and two levels of agenda setting (Lee, 2010) has important implications for theorizing consequences of agenda setting. McCombs (2004) considers opinion strength and opinion direction two different things, aligning the former with first-level agenda setting and the latter with second-level agenda setting. This study indicates the two concepts are interdependent: opinion direction innately has opinion strength (Lee, 2010). Without experimental design, it is extremely difficult, even impossible, to develop these insights.

Summary

In collecting data for online agenda-setting research, scholars have developed various new methods to measure salience and transmission of salience in the Web environment. Media salience is most likely to be gauged through the attention dimension and operationalized as the volume of mentions in online content. Relying on search volume as aggregate data, more recent studies link public salience to online information-seeking behavior. The availability and richness of online data also enable scholars to take a time series approach to research design and data analysis. This helps capture the transmission of salience among agendas as a time-related process. Although content analysis remains the most common method in online agenda-setting research, experimental studies have been conducted to better explain causality and the mechanism underlying agenda-setting effects.

Looking Ahead

The swift emergence and spread of the Internet has brought both challenges and opportunities for agenda-setting research. On the one hand, the wide variety of online communication channels enables individuals to mix various agendas into their personal comfortable fit. Changes in the media landscape make it less likely for the mass media to tell people what to think about. On the other hand, the vast wealth of competing agendas and individuals' greater control over information selection allow researchers to test agenda-setting theory with the bottom-up approach, which takes into account the reciprocal relationships between the public, media, and social actors. The Internet as a mass medium provides a means to move research beyond basic agenda-setting effects.

Figure 10.1 maps existing Web-related studies against the theoretical model of the agenda-setting process as extrapolated from McCombs (2004) and Dearing and Rogers (1996). In the initial phase of exploration, online agenda-setting research centered on sources of the media agenda (illustrated as a thick circle line). The majority of published studies addressed questions pertaining to who sets the media agenda, interactions among the news media (online media vs. legacy media; alternative media vs. mainstream media), as well as interactions with sources and their agendas communicated online. Such a high degree of concentration on intermedia agenda-setting and source agenda-setting

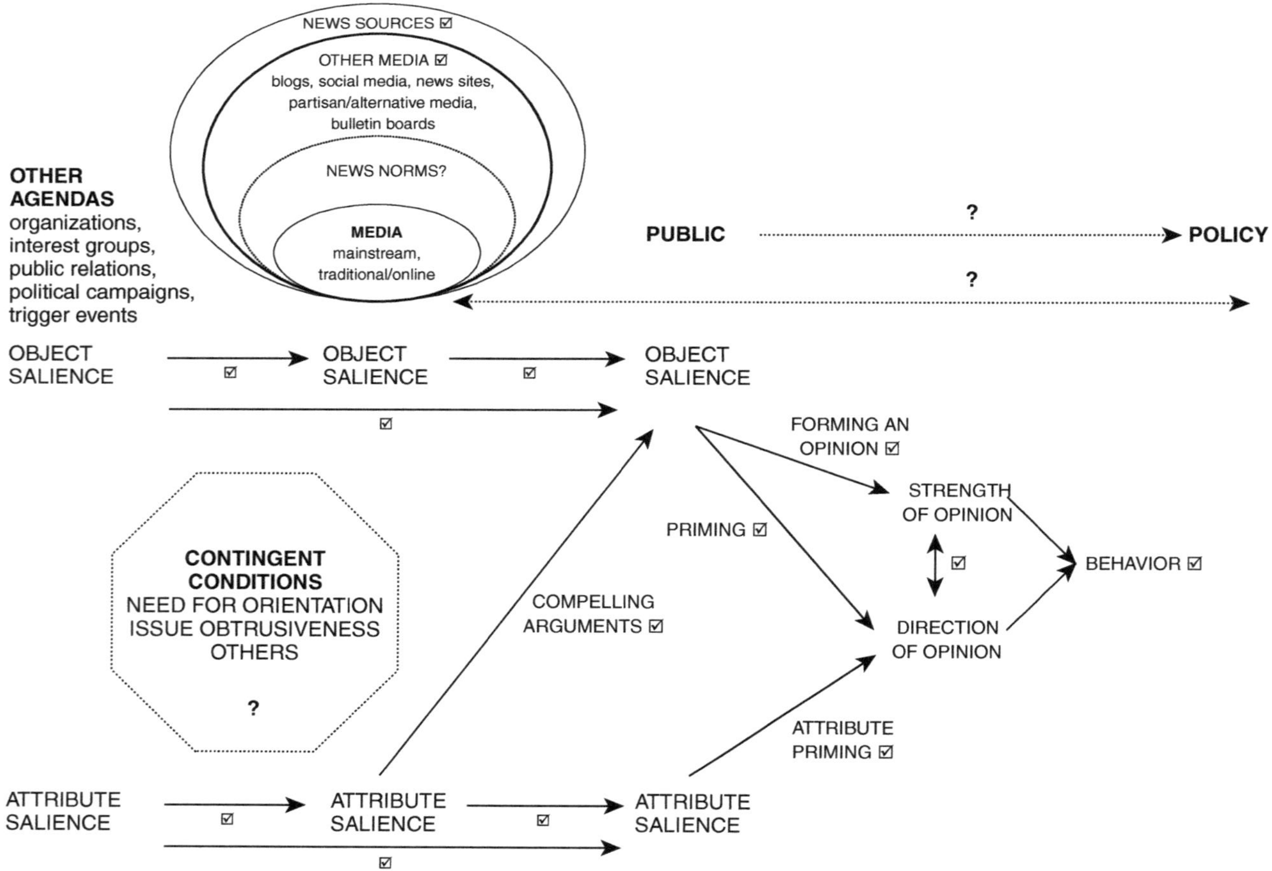

Figure 10.1 Mapping online agenda-setting research

effects may be due to the fact that early online work is most likely to focus on how the introduction of new information technologies have changed the media landscape.

To a lesser extent, relevant studies examine more closely the relationships between various media agendas (online, traditional, nonmainstream, and mainstream outlets) and the public agenda (general and online audiences). Evidence of public agenda setting has been documented at both object and attribute levels. More recent work helps capture the transmission of salience from the media to the public over time in the Web environment. In addition, salience transfer has been found among other agendas and the public agenda with social actors (political candidates, interest groups) seeking to bypass the mass media and exert a direct influence on public opinion.

There are some exploratory, though sporadic, attempts to extend online research to consequences of agenda setting. Experimental data, for example, provide a more nuanced understanding of how an opinion is formed as a function of agenda setting and priming. In another case, media salience is linked to real-world, online behaviors.

The studying of online agenda setting has several areas that are ripe for exploration. As shown in Figure 10.1, important research questions remain unanswered in other components and stages (illustrated as dotted lines) of the agenda-setting process. Some direct extensions of existing research are listed below.

1. Contingent conditions or mechanisms underlying public agenda-setting effects in the online environment. Individual differences, including need for orientation, issue obtrusiveness, and other moderators and mediators are important variables in agenda-setting research. Moreover, the Internet medium is a user-controlled system, which is characterized by its highly interactive features. Levels of media exposure/attention, media trust, interpersonal communication (including computer-mediated communication), and differences in information processing style and disposition may come into play. Our understanding of factors that enhance or inhibit agenda-setting effects is getting more and more complex (Chernov, Valenzuela, & McCombs, 2011). Need for orientation, for example, was traditionally defined by relevance and uncertainty (Weaver, 1980; Weaver, Shaw, & McCombs, 1977). More recently, this concept has become multi-dimensional with three distinct levels interacting with each other: need for orientation toward an issue; need for orientation toward attributes or frames; and need for orientation in regard to journalistic evaluations (Matthes, 2006; 2008). Three chapters in this book (Camaj, Chapter 5, this volume; Kim, Chapter 4, this volume; Valenzuela, Chapter 3, this volume) open new venues for research on various contingent conditions. Future studies should take up the task of further examining the psychology of agenda setting as it occurs in the Web environment.

2. Policy agenda setting and the symbiotic relationship between the media and policymakers and the transfer of salience from the public agenda to the policy agenda. The Web technology enables a form of constant, even real-time, communication between various entities. Online interactions are often described as a two-way, high-speed traffic flow. As news organizations and government agencies and organizations use the Web as a key channel to communicate information messages, it is important to analyze how salience is moved back and forth among the media, public, and policy agendas.

3. Media agenda setting and the role of news norms and online news production in the age of social media and user-generated content. The workflow in online newsrooms is a non-stop, 24/7 cycle. In many cases, breaking news comes from far-flung, remote places, where news organizations do not have immediate access to the events. There have been several instances in which the mainstream media utilize social media curation tools and crowdsourcing in their reporting. Examples include the coverage of Mumbai terrorist attacks, Iranian election protests, the death of Osama bin Laden, and the Arab spring movement (Veritas, 2011). On those occasions, the news media prioritize immediacy over verification of information (Zuckerman, 2010). Research should address how news norms have changed in the digital age and how groups lead and force coverage of certain issues through social media.

4. Reverse agenda setting or the influence of the public agenda on the media agenda. Anecdotal evidence shows that online media take cues from public interest when covering news (Keefe, 2009; Peters, 2010a, 2010b). Journalists are instructed to write SEO (search engine optimization)-friendly headlines and news in an attempt to boost online traffic. The Huffington Post, in particular, has assigned an editor to monitor search traffic and trends. Reverse agenda setting, or the power of the public to tell the media what they want to think about (Chaffee & Metzger, 2001), might be an interesting phenomenon to be empirically tested in the Internet environment.

5. Agenda setting and newer media platforms. According to 2012 statistics, 63% of U. S. adults access the Internet wirelessly, 88% have a cell phone, 57% have a laptop, 19% have a tablet (Zickuhr & Smith, 2012). This calls for investigations of agenda setting in conjunction with mobile devices and mobile audiences. In addition, future research should examine the differential agenda-setting effects of various online media platforms. News and information delivered through computers versus touchscreen tablets have structural differences, which may induce different cognitive responses.

6. Agenda setting as a social process. Since the initial study in Chapel Hill, agenda setting has evolved through a gradual theoretical refinement. Original research emphasized the influence of the media agenda on the focus of public opinion (McCombs & Shaw, 1972). Today, agenda setting refers to the transfer of salience from one agenda to another (McCombs, 2005). According to Shaw and colleagues (Shaw, McCombs, Weaver, & Hamm, 1999), agenda setting is part of a larger "ongoing social process through which individuals relate to larger groups by media connections" (p. 2). Individuals have a powerful impulse to affiliate with others in groups to remove social dissonance. They join groups by joining agendas. This is called agendamelding or the social learning process in which people are driven to seek information that is compatible with the group to which they want to belong.

As intriguing as it is, the concept of agendamelding warrants an empirical search for evidence. The Internet, with its diverse agendas and fragmented audiences choosing different mixes of media to form their views of public affairs, is an ideal testing ground for this new stage of theory development. Challenges and opportunities are ahead for scholars to explore.

Note

1 Data on published research in 2011 are incomplete at the time of this writing.

References

Althaus, S. L., & Tewksbury, D. (2002). Agenda setting and the "new" news: Patterns of issue importance among readers of the paper and online versions of the New York Times. *Communication Research*, *29*, 180-?. doi:10.1177/009365020202900200 4

Chaffee, S., & Metzger, M. (2001). The end of mass communication? *Mass Communication & Society*, *4*, 365–379. doi:10.1207/S15327825MCS0404_3

Chernov, G., Valenzuela, S., & McCombs, M. E. (2011). An experimental comparison of two perspectives on the concept of need for orientation in agenda-setting theory. *Journalism & Mass Communication Quarterly*, *88*(1), 142–155. doi: 10.1177/107769901108800108

Coleman, R., & McCombs, M. E. (2007). The young and agenda-less? Exploring age-related differences in agenda setting on the youngest generation, baby boomers, and the civic generation. *Journalism & Mass Communication Quarterly*, *84*(3), 495–508. doi:10.1177/107769900708400306

Conway, M., & Patterson, J. R. (2008). Today's top story? An agenda-setting and recall experiment involving television and Internet news. *Southwestern Mass Communication Journal*, *24*(1), 31–48.

Dearing, J. W., & Rogers, E. M. (1996). *Agenda-setting*. Thousand Oaks, CA: SAGE.

Evatt, D., & Ghanem, S. (2001, May). *A salience scale to enhance interpretation of public opinion*. Paper presented at the World Association for Public Opinion Research, Rome, Italy.

Granka, L. A. (2010, September). *Measuring agenda setting with online search traffic: Influences of online and traditional media*. Paper presented at the Annual Meeting of the American Political Science Association, Washington, DC.

Groshek, J. (2008). Homogenous agendas, disparate frames: CNN and CNN International coverage online. *Journal of Broadcasting & Electronic Media*, *52*, 52–68.

Hester, J. B., & Gibson, R. (2007). The agenda-setting function of national versus local media: A time-series analysis for the issue of same-sex marriage. *Mass Communication & Society*, *10*, 299–317.

Jansen, J. (2010). *The better-off online*. PEW Internet & American Life Project. Retrieved from http://pewresearch.org/pubs/1809/internet-usage-higher-income-americans

Kamhawi, R., & Weaver, D. H. (2003). Mass communication research trends from 1980 to 1999. *Journalism & Mass Communication Quarterly*, *80*, 7–27.

Keefe, J. (Producer). (2009, October 9). *On the media* [radio broadcast transcript]. NewYork:WNYC Radio.

Kiousis, S. (2004). Explicating media salience: A factor analysis of New York Times issue coverage during the 2000 U. S. Presidential Election. *Journal of Communication*, *54*, 71–87.

Kiousis, S., Mitrook, M., Wu, X., & Seltzer, T. (2006). First- and second-level agenda-building and agenda-setting effects: Exploring the linkages among candidate news releases, media coverage, and public opinion during the 2002 Florida gubernatorial election. *Journal of Public Relations Research*, *18*, 265–285.

Ku, G., Kaid, L. L., & Pfau, M. (2003). The impact of web site campaigning on traditional news media and public information processing. *Journalism & Mass Communication Quarterly*, *80*(3), 528–547.

Lee, B., Lancendorfer, K., & Lee, K. J. (2005). Agenda-setting and the Internet: The intermedia influence of Internet bulletin boards on newspaper coverage of the 2000 general election in South Korea. *Asian Journal of Communication*, *15*(1), 57–71.

Lee, G. (2010). Who let priming out? Analysis of first- and second-level agenda setting effects on priming. *International Communication Gazette*, *72*, 759–776.

Lim, J. (2006). A cross-lagged analysis of agenda setting among online news media. *Journalism & Mass Communication Quarterly*, *83*, 298–312.

Lim, J. (2011). First-level and second-level intermedia agenda-setting among major news websites. *Asian Journal of Communication*, *21*, 167–185. doi:10.1080/01292986.2010.539300

Maier, S. (2010). All the news fit to post? Comparing news content on the Web to newspapers, television, and radio. *Journalism & Mass Communication Quarterly*, *87*, 548–562.

Matthes, J. (2006). The need for orientation towards news media: Revising and validating a classic concept. *International Journal of Public Opinion Research*, *18*, 422–444.

Matthes, J. (2008). Need for orientation as a predictor of agenda-setting effects: Causal evidence from a two-wave panel study. *International Journal of Public Opinion Research*, *20*, 440–453.

McCombs, M. E. (2004). *Setting the agenda: The mass media and public opinion*. Cambridge: Polity.

McCombs, M. E. (2005). A look at agenda-setting: Past, present and future. *Journalism Studies*, *6*, 543–557.

McCombs, M., & Shaw, D. (1972). The agenda-setting function of the media. *PublicOpinion Quarterly*, *36*, 176–187.

McLuhan, M., & Lapham, L. H. (1994). *Understanding media: The extensions of man* (1st ed.). Cambridge, MA: The MIT Press.

Mellon, J. (2011). Search indices and issue salience: The properties of Google Trends as a measure of issue salience (Paper No. 2011–01). Sociology Working. University of Oxford. Retrieved from www.sociology.ox.ac.uk/documents/working-papers/2011/swp1101.pdf

Meraz, S. (2009). Is there an elite hold? Traditional media to social media agenda setting influence in blog networks. *Journal of Computer-Mediated Communication*, *14*(3), 682–707.

Meraz, S. (2011a). Using time series analysis to measure intermedia agenda-setting influence in traditional media and political blog networks. *Journalism & Mass Communication Quarterly*, *88*(1), 176–194.

Meraz, S. (2011b). The fight for "how to think": Traditional media, social networks, and issue interpretation. *Journalism*, *12*, 107–127.

Messner, M., & DiStaso, M. W. (2008). The source cycle: How traditional media and weblogs use each other as sources. *Journalism Studies*, *9*, 447–463.

Peters, J. W. (2010a, July 4). At Yahoo, using searches to steer news coverage. *New York Times*. Retrieved from http://www.nytimes.com/2010/07/05/ business/ media/05yahoo.html

Peters, J. W. (2010b, September 5). Some newspapers, tracking readers online, shift coverage. *New York Times*. Retrieved from http://www.nytimes.com/2010/09/06/ business/ media/06track.html

Ragas, M. W., & Kiousis, S. (2010). Intermedia agenda-setting and political activism: MoveOn.org and the 2008 presidential election. *Mass Communication & Society*, *13*,560–583.

Rainie, L., & Horrigan, J. (2002). *Counting on the Internet: Most find the information they seek, expect*. PEW Internet & American Life Project. Retrieved from http://www.pewinternet.org/Reports/2002/Counting-on-the-Internet-Most-find-the-information-they-seek-expect/1-Summary-of-Findings.aspx

Roberts, M., Wanta, W., & Dzwo, T.-H. (Dustin). (2002). Agenda setting and issue salience online. *Communication Research*, *29*, 452–465.

Sayre, B., Bode, L., Shah, D., Wilcox, D., & Shah, C. (2010). Agenda setting in a digital age: Tracking attention to California proposition 8 in social media, online news and conventional news. *Policy & Internet*, *2*, 7-32.

Schiffer, A. J. (2006). Blogswarms and press norms: News coverage of the Downing Street memo controversy. *Journalism & Mass Communication Quarterly, 83*(3), 494–510.

Shaw, D. L., Hamm, B., & Knott, D. (2001). Technological change, agenda challenge and social melding: Mass media studies and the four ages of place, class, mass and space. *Journalism Studies, 1*, 57–79.

Shaw, D. L., McCombs, M. E., Weaver, D. H., & Hamm, B. J. (1999). Individuals, groups, and agenda melding: A theory of social dissonance. *International Journal of Public Opinion Research, 11*(1), 2–24.

Siu, W. (2008). Source agenda setting: Intermedia influence in the food safety scandal. *Journal of Communications Research, 1*(4), 355–381.

Son, Y. J., & Weaver, D. H. (2006). Another look at what moves public opinion: Media agenda setting and polls in the 2000 U. S. election. *International Journal of Public Opinion Research, 18*, 174–197.

Song, Y. (2007). Internet news media and issue development: A case study on the roles of independent online news services as agenda-builders for anti-US protests in South Korea. *New Media & Society, 9*, 71–92.

Strömback, J., & Kiousis, S. (2010). A new look at agenda-setting effects—Comparing the predictive power of overall political news consumption and specific news media consumption across different media channels and media types. *Journal of Communication, 60*, 271–292.

Sweetser, K. D., Golan, G. J., & Wanta, W. (2008). Intermedia agenda setting in television, advertising, and blogs during the 2004 Election. *Mass Communication & Society, 11*, 197–216.

Tedesco, J. C. (2005). Intercandidate agenda setting in the 2004 Democratic presidential primary. *American Behavioral Scientist, 49*, 92–113.

Tomaszeski, M., Proffitt, J. M., & McClung, S. (2009). Exploring the political blogosphere: Perceptions of political bloggers about their sphere. *Atlantic Journal of Communication, 17*, 72–87.

Veritas, J. (2011). Social media: The next frontier for journalism. Technorati. Retrieved from http://technorati.com/technology/article/social-media-the-next-frontier-for/

Wallsten, K. (2007). Agenda setting and the blogosphere: An analysis of the relationship between mainstream media and political blogs. *Review of Policy Research, 24*, 567–587. doi:10.1111/j.1541–1338.2007.00300.x

Wanta, W., & Hu, Y.-W. (1994). Time-lag differences in the agenda-setting process: An examination of five news media. *International Journal of Public Opinion Research, 6*, 225–240. doi:10.1093/ijpor/6.3.225

Waters, R. D., & Tindall, N. T. J. (2011). Exploring the impact of American news coverage on crisis fundraising: Using media theory to explicate a new model of fundraising communication. *Journal of Nonprofit & Public Sector Marketing, 23*, 20–40. doi:10.1080/10495142.2010.494875

Weaver, D. H. (1980). Audience need for orientation and media effects. *Communication Research, 7*, 361–373. doi:10.1177/009365028000700305

Weaver, D. H., Shaw, D. L., & McCombs, M. E. (1977). Political issues and voter need for orientation. *The Emergence of American Political Issues*. St. Paul, MN: West.

Weeks, B., & Southwell, B. (2010). The symbiosis of news coverage and aggregate online search behavior: Obama, rumors, and presidential politics. *Mass Communication & Society*, *13*, 341–360. doi:10.1080/15205430903470532

Winter, J. P., & Eyal, C. H. (1981). Agenda setting for the civil rights issue. *Public Opinion Quarterly*, *45*, 376–383. doi:10.1086/268671

Zickuhr, A., & Smith, A. (2010). *Digital differences*. Retrieved from PEW Internet & American Life Project website: http://www.pewinternet.org/Reports/2012/Digital-differences/Main-Report/The-power-of-mobile.aspx

Zuckerman, E. (2010). International reporting in the age of participatory media. *Dædalus, 139*, 66–75. doi: 10.1162/daed.2010.139.2.66

Chapter 11

Consensus-Building Function of Agenda Setting in Times of Crisis

Substantive and Affective Dimensions

Vanessa de Macedo Higgins Joyce

At its essence, the theory of agenda setting deals with the exchange of priorities within a social system: an exchange between groups and media, media and other media, or individuals and media. It deals with the effects that this exchange can have on what we see in the world and how we see the world, how we think about issues, problems, and people. If indeed we are "a reflection of the agendas that we hold" (Shaw, McCombs, Weaver, & Hamm, 1999, p. 7), it is important that we understand the exchange of agendas, why we are drawn to them, where the agendas originate from, the different dimensions of agendas, and what the consequences are of this exchange.

The news media, with its presentation of a limited set of issues, images, and frames, have been shown to be an important institution influencing the audience's adoption of priorities, as hundreds of agenda-setting studies have found support for such theses. One of the consequences of this exchange of priorities is the potential to provide common ground on which divergent groups of society may deliberate and discuss (McCombs, 1997; Shaw & Martin, 1992). In other words, the news media have the potential to build consensus among groups of society through the transference of issues and attributes of the issues. This chapter focuses on this consequence of agenda setting, specifically its consensus-building function.

Many communication and social science studies focus on differences, on conflict of power and interests, and indeed so do the news media. Although differences exist and should not be ignored, there is much need to understand how groups can come together on aspects of life that impact these groups. Consensus, or general agreement, is essential for a constitutional democratic government (McClosky, 1964). Without it, there is no set of issues and problems set to be solved, no laws or provisions passed. Indeed, it is not difficult to see, in a polarized government, how rare it is to have anything being approved. Consensus is an important democratic value, yet few social science and communication studies focus on it. Consensus is essential for government and it is also important for the social system, allowing for enough commonality for groups to work together. As the world becomes increasingly complex and interconnected, it intensifies the importance of understanding consensus, and how the agenda-setting role of the news media can act in building consensus. This study is interested in particular in understanding how divergent groups can come together, through the news media, in times of crisis.

In times of crisis, the need for access of information and orientation about such events through news media is heightened. At these times, the news media are especially equipped to fulfill such roles, and audiences respond with massive and loyal viewership of crisis coverage (Graber, 2010). News media are needed for vital information that affects how audiences lead their daily lives, from flights cancellations, to evacuation and safety mechanisms. And, during such times, news media operate within their usual process of selection and emphasis, affecting how people think about and feel about the crisis, its causes and consequences. Therefore, the development of a crisis can be related to both the agenda-setting roles and framing processes of media (Ma, 2005). High need for orientation (high relevance and uncertainty) is also a necessary contingent condition for increased agenda-setting influence (Kim, Chapter 4, this volume; McCombs, 2004). Understanding consensus building during times of crisis may provide a helpful perspective in analyzing the effects of news media at such important times. This chapter focuses on the case of European Union audience reaction to the events of September 11, 2001, in the United States to test the consensus-building consequence of agenda setting.

In the early morning of September 11, 2001, four civilian airplanes hijacked by the extremist group al-Qaeda struck The World Trade Center, the Pentagon, and rural Pennsylvania, killing almost 3,000 people. Across the Atlantic, in Europe, people closely followed the developing events through the lenses of news media. In Europe, it was already afternoon, and media attention was fast to re-focus to this event. Many TV stations stopped their regular programming to air breaking news about the unfolding events (Pludowski, 2007). Europeans watched, read, listened to information about the attacks and their consequences for their countries and themselves. Although 9/11 happened in the United States, the unusual nature of the events brought worldwide attention, as did the consequence of the strikes to one of the world's biggest superpowers. The focus of this study is on the attributes of terrorism and their transference from a national media agenda to the agenda of an audience in the European Union after 9/11. It understands that the news media play an important function in times of crisis, and that building consensus in such times may be an important consequence of their use.

News media in Europe is certainly no unified block, with countries holding their own media system and culture, from opinionated and political to detached observer and objective focused. Looking at consensus building in the European Union then only makes sense as a cross-national comparison of the influence of their news media.

This chapter also differentiates between attributes of the issue of terrorism in general, the substantive dimension of attribute agenda setting, and attributes of Muslim and Arab community, and the affective dimension of attribute agenda setting. This study seeks to compare the news media potential of building consensus on how people in Europe thought about the terrorism issue (a substantive dimension) and how people felt about the Muslim and Arabic communities (an affective dimension) within the context of 9–11 as a crisis. This study conducts a second-level agenda-setting analysis and asks, specifically, what is the role of the news media in bridging differences among opposing demographic groups?

Consensus Building in Agenda Setting

Consensus building as a consequence of agenda setting has its roots in Lasswell's functions of mass media. Lasswell (1960) identified surveillance of the environment, transmission of social heritage, and correlation of societal groups in responding to the environment as the basic functions of the communication process performed by mass media. Within this last function, correlation, mass media potentially function to aid mobilization and prevent threats to social stability by connecting different groups of society in response to the environment. Correlation may also provide a dysfunction by increasing social conformism and preventing social change if there is a lack of opportunities for diverging perspectives to be present. (Wright, 1960). The news media then, by presenting their audience with a limited set of issues and a particular representation of these issues, function to provide agreement on important issues and a general perception of how to understand those issues. These presentations by the news media provide at least a common ground on which divergent groups of society may deliberate.

The term *consensus* means a general group agreement in feeling or beliefs (Lijphart, 1998). In political science, the term has been described as a cultural concept, as a group's shared pattern of "political beliefs, feelings and judgments" (Lijphart, 1998, p. 100). In a consensual democratic system, instead of the rule of the majority of the people, one would find the rule of the largest possible number of people (Anderson & Guillory, 1997). Consensus can more easily be reached on certain issues and values than others. Studies have shown that issues that are more general and abstract tend to reach consensus at a higher level than issues that connote direct applications (McClosky, 1964). Greater consensus has been found, for example, on broad principles of democracy than on specific propositions, and opinion varies systematically by socio-demographic groups on these specific propositions (Prothro & Grigg, 1960). These authors suggested that general issues and values might have a greater connection with culture, as a shared principle (Prothro & Grigg, 1960). One possible explanation of why specific propositions come closer to dissent is that these issues and values might lead to distinct interpretations and appeal more to individualistic experiences, desires, and necessities.

Consensus is used in the present chapter to describe general agreement on certain issues and attributes of those issues. The argument here is not that consensus would be achieved at all times, or that it is completely universal. It is also not arguing that people would have the same reasons for achieving such consensus. It is a thesis of this study, though, that the news media can function as a bridge between different segments of societies to provide common ground for deliberation on the issues of the time.

These concentrations can be occurring in different levels of localities: regional, national, and even transnational. This present chapter focuses on consensus building at the national level.

The news media may then be reducing differences between various segments of society, emphasizing consensus on perceptions of issues. By correlating different segments of society in terms of the priorities that need to be addressed, the news media may be providing an important function in concentrating the community's and government's efforts toward social action (López-Escobar, Llamas, & McCombs, 1998). The consensus-building function of agenda setting focuses on how the news media can reduce group differences and bring together opposing segments of society, bridging communities and helping "the community at large achieve consensus on the priority issues of the day" (McCombs, 1997). Given this potential, the news media have a responsibility to include in their discourse different issues and different positions within those issues that are pertinent to all ends of the community at large.

Scholars have found consensus, or general agreement, to be one of the consequences of news media agenda setting. Individuals with different characteristics (such as race, age, and gender, for example) that are exposed to the issues and attributes of those issues presented by the media's agenda are pulled toward a social consensus. This collective influence of mass media, as presented through the consensus-building consequences of agenda setting, is a consequence at the societal level. The differences in attitudes and opinions between diverse groups in society (such as young versus old, low education versus high education), often highlighted in studies of public opinion, diminish as these individuals increase their news media exposure, which in turn increases the agreement between agendas (McCombs, 2004). In other words, the consensus-building consequence of agenda setting assumes that these diverse demographic groups in society are different and look for circumstances, such as television and press news use, where these differences are reduced.

Shaw and Martin (1992) found that the very use of the news media can build consensus among opposing demographic groups, enhancing agreement about important public issues and reducing the gap within segments of society. This was the first study conducted on this specific consequence of agenda setting. Using data from a statewide survey in North Carolina, as well as content analysis data from newspapers in the state, they found that as individuals increased their newspaper reading (also replicated for television), the level of agreement on the important issues of the day increased within different gender, racial, age, and education groups (Shaw & Martin, 1992). The news media may bring a dialogue between otherwise disparate groups by reducing the gap within these different social groups regarding current important issues.

Ching-Yi Chiang (1995) replicated Shaw and Martin's original study on consensus building. Chiang expanded the concept to an international setting with a study of Taiwan's 1992 legislative election. Chiang differed in the demographic subgroups analyzed, having included the geographic groups of Taiwanese and Mainlanders, and she excluded race and age. These findings generally supported the hypothesis that increasing exposure to the news media leads to an increased consensus about the most important issues perceived within each social group (Chiang, 1995).

López-Escobar, Llamas, and McCombs (1998) conducted a third study on agenda setting and community consensus building in Spain. The first phase of this study explored consensus building of prioritized social issues within the first level of an agenda-setting framework, within the context of a political contest. The second phase of this study expanded the research to consensus building in relation to the attributes of political candidates, that is, within the second level of agenda-setting framework. Second-level agenda setting, also known as attribute agenda setting, deals with the transference of salience of an object's attributes from the media agenda to the public's agenda. For the two phases of this study, the authors used survey data from Navarra, Spain's parliamentary elections and an analysis of local newspaper and television coverage of the issues and their attributes. The demographic variables analyzed in this study were age, gender, education, and ideology (or political affiliation), excluding income and race. Overall, the findings supported their hypotheses that increased exposure to the news media corresponded to an increased consensus among the different demographic groups of the public agenda of issues and the public agenda of the candidates' attributes (López-Escobar et al., 1998).

This study seeks to replicate the three previous consensus-building studies mentioned earlier (Chiang, 1995; López-Escobar et al., 1998; Shaw & Martin, 1992). All three of these studies focused on the first level of agenda setting, and the study in Spain also expanded to include the second level of agenda setting. This current study focuses strictly on consensus building at the second level of agenda setting. At the second level, the consensus-building consequence of agenda setting suggests an important function of the news media in creating common ground for the deliberation about the characteristics of issues and objects, aiding mobilization on what aspects of an issue need consideration.

Substantive and Affective Dimensions of Attribute Agenda Setting

This chapter also seeks to replicate the distinction made in the consensus-building consequence of agenda setting conducted in Spain between the substantive and affective dimensions of attributes. This study seeks to expand our understanding of consensus building by conducting an extensive cross-national replication of consensus building at the second level of agenda setting, but applying it within the context of a crisis situation. It analyzes the influences of national television and press in Europe in building consensus on the substantive and affective dimensions of attributes of the events of 9/11 in the United States and its consequences in Europe.

In this study, attributes of the issue of terrorism and attributes of the Muslim and Arab communities are conceptualized as being different in terms of their attribute dimensions. Ghanem (1997) delineated four major dimensions of media frames and attributes: topics, presentation, cognitive attributes, and affective attributes. She defined the cognitive dimension as dealing with "general cognitive categories that might shed a light on whether the media and the audience are thinking about the problem in the same way" (Ghanem, 1997, p. 13), and the affective dimension as dealing with "the public's emotional response that may result from media coverage" (Ghanem, 1997, p. 12). McCombs identified the dimensions of attributes in terms of substantive (distinct characteristics such as ideology and personality) and affective (such as positive and negative). He said that media effects might differ when the attributes are explored within these different dimensions (McCombs, 2004). The substantive dimension closely relates to the cognitive dimension and is reflected in the beliefs about aspects of an object, such as changes in awareness and knowledge. The affective dimension relates to feelings about an object.

In the consensus-building study of López-Escobar et al. (1998), these two dimensions were examined. They described the substantive dimension as descriptive categories, in their case relating to attributes of political candidates, and the affective dimension as an appraisal of those political candidates, reflecting media's tone when presenting these candidates (López-Escobar et al., 1998). The authors found greater support for consensus building within the second level of agenda setting for the affective dimension, as opposed to the substantive dimension, although both provided support for their hypotheses.

Researchers have focused their studies of the substantive and affective dimensions with different objects (or issues). Some researchers have conducted experiments to test agenda-setting effects on substantive and affective attributes of candidates (Kiousis, Bantimaroudis, & Ban, 1999). Others focused their study on one dimension, such as the affective dimension, proposing different objects, such as nation as the focus (Wanta, Golan, & Lee, 2004).

Attributes themselves can be differently understood when the analysis focuses on the object as an issue versus on the object as a person, such as a political candidate, or a community. Craft and Wanta (2004) stated that, when dealing with an issue, "attributes are consequences, and not the type of characteristics—such as intelligence or experience—that would be the focus of second level agenda setting regarding a person" (p. 460). With the intention of having an in-depth understanding of consensus building at the second level of agenda setting, this study proposes to test these different types of attributes through two measurements. The first one is the analysis of consensus building as a consequence of attribute agenda setting regarding an issue, specifically, the attacks of 9/11 and terrorism, a substantive dimension. The second is the analysis of consensus building as a consequence of attribute agenda setting regarding attitudes toward the Muslim and Arab community, an affective dimension. In the case of this study, the context is a crisis, the events of 9/11, and its resulting implications for a European Union audience.

This study is thus an extensive replication of Shaw and Martin's consensus-building consequence of agenda setting, analyzing the effects of national television and press news in building consensus in 15 European Union countries. The focus, however, is on the second level of agenda setting, with analysis for both substantive and affective dimensions of attributes. It builds on previous studies by analyzing both consensus building on attributes of an issue (substantive dimension) and attributes of a community (affective dimension) within the context of a crisis. Therefore, based on the previous studies of the consensus-building consequence of agenda setting (Shaw & Martin, 1992; Chiang, 1995) focusing on the second level of agenda setting (López-Escobar et al., 1998) and making the distinction between the substantive and affective dimensions, this study tests the following hypotheses and research questions:

H1: Increased use of national news media corresponds to increased consensus among demographic subgroups regarding the agenda of substantive attributes

H2: Increased use of national news media corresponds to increased consensus among demographic subgroups regarding the agenda of affective attributes

RQ1: Are there differences between the substantive and affective attributes on the magnitude of consensus increase?

RQ2: Are there differences among national television and national press in how they pull the different demographic subgroups closer together in consensus on the substantive and affective attributes?

Methodology

This study employed a secondary data analysis to test the consensus-building function of agenda setting at the second level. Specifically, this study tests the influence of national television and press in building consensus among the education and age demographic subgroups on how they thought about the terrorism issue (substantive dimension) and how people felt about the Muslim communities (affective dimensions). Data for this analysis originated from Flash Eurobarometer 114, a topical telephone-based survey conducted by EOS Gallup, commissioned by the Directorate General of the European Union. The study was done among the European Union population in the immediate aftermath of the 9/11 attacks in the United States (Flash Eurobarometer, 2001).

A total of 14,986 interviews were conducted between November 13 and 23, 2001, in the 15 countries comprising the European Union at the time (Belgium, Denmark, Germany, Greece, Spain, France, Ireland, Italy, Luxemburg, the Netherlands, Austria, Portugal, Finland, Sweden, and United Kingdom), approximately 1,000 interviews per country.

Calculations of consensus for this study are based only on the public opinion data. This study follows the example of Shaw and Martin's (1992) analysis of television influence on consensus building, by comparing agreement of attributes of issues and community between those who reported using national media and those reporting not using those national media. This allows the researcher to analyze the level of agreement between demographic subgroups, which is the core relationship modeled in the three previous studies on consensus building (Chiang, 1995; López-Escobar et al., 1998; Shaw & Martin, 1992).

Three main groups of variables are necessary for the hypotheses and research questions of this study: a set of dichotomized demographics or social groups; media exposure or use; and public opinion or attribute salience. It is also necessary to have a location variable, specifically the 15 European Union member countries, which were controlled for this cross-national analysis.

The dichotomization of the demographic variables is needed to determine key group agreement, following the base of the triangular comparison presented in the seminal consensus-building study by Shaw and Martin (1992). This would mean comparing men and women, younger and older population, and so on. The demographic groups used in this study, such as age and education, had to have the subgroups collapsed a posteriori into categories traditionally used in the social science, such as "younger" and "older" populations and "lower" and "higher" income populations. Such categories have traditionally shown discord in how they think about issues, therefore making an interesting case for media's potential in building consensus.

For the set of dichotomized demographics, this study replicates the choices of age and education from previous studies of consensus building. For the demographic age, respondents were asked to state their exact age. These responses were collapsed a posteriori into the variables "younger" and "older" subgroups. The operational definition of "younger" is those respondents who reported their age as being 15 through 42 years old, roughly 49% of the respondents. The operational definition of "older" is respondents who reported their age as being 43 through 98 years old, roughly 51% of the respondents. Thirty-seven respondents refused to state their age. For education, respondents were asked to state their age when their full-time education was finished, which was dichotomized a posteriori into "lower" and "higher" education subgroups. The operational definition of "lower" education is those reporting having finished their full-time education between "never having gone to full time school" and "having been up to 18 years old when they finished full time education" (roughly 54% of the respondents) The operational definition of "higher" education is those reporting "being 19 years old and older when they finished their full time education" (roughly 46% of the respondents).

Media exposure is measured in this study through a bivariate question that asked the respondents about their use of various media for information on the events in Afghanistan, Pakistan, and the United States in the aftermath of 9/11 and their impact in Europe. Respondents were asked to state, from a presented list, which media they used and deemed reliable. This closed-ended question asked the respondents about their use of 10 different sources of information. Of those sources, this study is mainly concerned with two specific sources of information, national television and national press.

This study analyzes national press and television news, in replication of the previous consensus-building studies, within each of the 15 European Union member countries. Overall, 65% of the respondents stated they use the national press news for information, meaning press from their present country, and 82% of the respondents stated they use national TV news for information, meaning TV news from their present country.

The third set of variables needed for measuring the consensus-building function of agenda setting is opinion salience. This study is measuring consensus at the second level of agenda setting at two instances: the substantive dimension, on attribute of an issue (**H1**), and the affective dimension, on attributes of a community (**H2**).

The operational definition of attribute salience regarding an issue, the substantive dimension, is the hierarchy of importance of the attributes related to measures that respondents deemed the European Union should take in response to the attacks of September 11th. Respondents were asked, in a closed-ended itemized question, to express their opinion regarding the importance of each item on a four-point scale. The value "essential" has been used in this analysis to rank the hierarchy of importance of the attributes of the issue. Of the respondents, 44% stated that it would be essential to conduct multiple actions to prevent extension of the conflict to other countries (item a); 46% of the respondents stated it would be essential to increase humanitarian assistance to the victims (item b); 42% stated it would be essential to support restoration of democracy in Afghanistan (item c); 24% stated it would be essential to massively finance reconstruction in Afghanistan (item d); 52% stated it would be essential to launch a specific action plan for air safety (item e); 42% said it would be essential to take measures to improve cooperation between police services and judicial bodies in Europe (item f); 31% said it would be essential to encourage political and cultural relations between Europe and Arabic countries (item g); and 34% stated it would be essential to improve the quality of relationships between Muslim communities and local populations within the European territory (item h).

The operational definition of attribute salience regarding a community, the affective dimension, is the hierarchy of agreement of attributes related to the Muslim and Arab community, as a consequence of the terrorist attacks of September 11th. Respondents were asked, in a closed-ended itemized question, to express their agreement for each of five statements on a four-point scale. The values "totally agree" and "agree" have been collapsed to determine an aggregated agreement regarding the attributes of the community. The collapsed category "totally agree and agree" represents agreement on the attributes of a community and was used to rank order the importance of the attributes of a community. Of the respondents, 55% stated they totally agreed/agreed that the Muslim community has sufficiently condemned the attacks of September 11 (A); 69% stated they totally agreed/agreed that one can understand some distrust toward the Muslim community in Europe (B); 86% said they totally agreed/agreed that one can't judge the Arab world based on terrorist acts of a few individuals (C); 71% said they totally agreed/agreed that terrorists do not act like real Muslims (D); and 60% said they totally agreed/agreed that the real motivation of the terrorists was not religion (E).

Agreement between each of the demographic subgroups on the attributes was measured with rank-order correlation, using Spearman's rho. Correlation calculations could not be done within specific countries in certain segments of the analysis when the base of the cross-correlation was too small for any meaningful comparison.

Findings

Studies focusing on the consensus-building function of agenda setting attempt to test whether within-group agreement on a certain issue improves with increased use of a certain media. It calculates, for example, if there is an increased correlation in how younger people and older people perceive issues (or attributes) with the use of national television. Figure 11.1 illustrates this concept, using age and national media use from this study as an application for three out of the 15 countries being analyzed here.

This study finds strong support that increased use of national news media corresponds to increased consensus among demographic

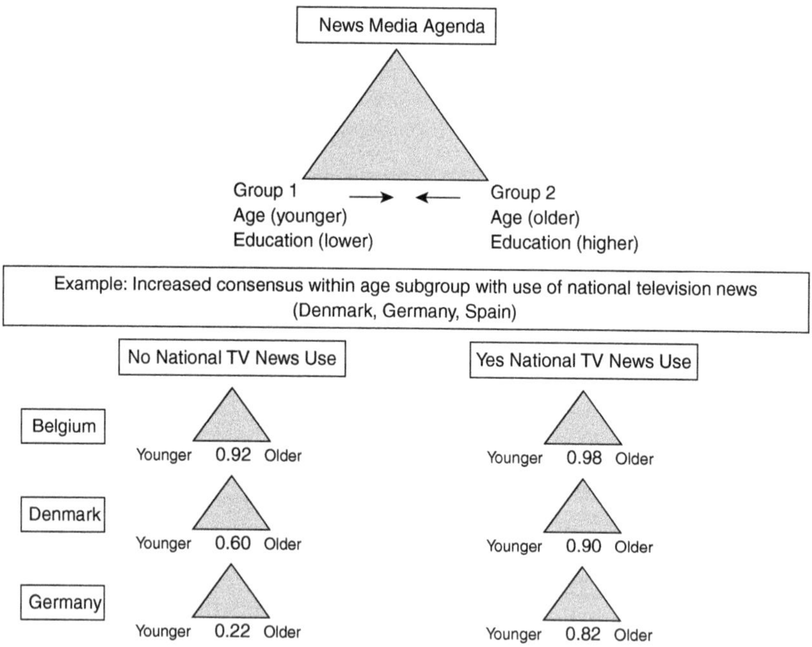

Figure 11.1 Increased consensus among demographic subgroups on attributes of issue with use of media

subgroups regarding the agenda of substantive attributes (**H1**). It found that in 72% of the 58 calculated correlations, younger and older population and lower and higher education population, when considering both national television and national press, followed the expected pattern of increased consensus on the attribute of terrorism-related issue (substantive dimension). Some differences were seen in the pull toward consensus according to what media they used, as Figure 11.2 indicates.

The study indicates that these two demographic subgroups were pulled toward social consensus on the substantive dimension with the use of national television, with 82% of the calculated correlations indicating increased agreement. In 13 of the 15 calculated countries, younger and older populations were drawn closer together in how they thought about the issue of terrorism with the use of national TV news. These countries were Belgium, Denmark, Germany, Greece, Spain, Italy, Luxemburg, the Netherlands, Austria, Portugal, Finland, Sweden, and the UK. In Germany, for example, younger and older populations indicated a much higher correlation on attributes of an issue for those using national TV news (0.82, a very high positive correlation) than for those not using national TV news (0.22, a low positive correlation). The average correlation among younger and older population on the attributes of an issue for those not using national TV was 0.69, while the average correlation among those using national TV was 0.91.

In 10 out of the 13 calculated countries, lower and higher education populations were drawn closer together on the substantive dimension

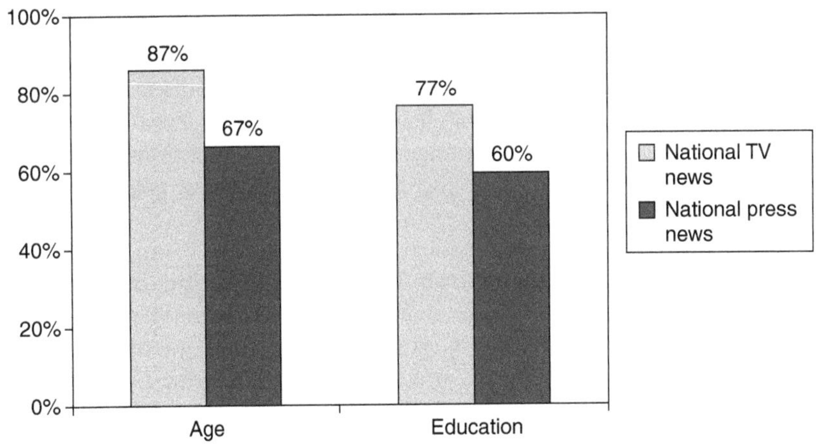

Figure 11.2 Increased consensus on the substantive dimension of attributes among age and education demographic subgroups by national TV news and national press news

of attributes with the use of national TV news. These countries were Belgium, Germany, Spain, France, Ireland, the Netherlands, Austria, Portugal, Sweden and the UK. Once again, Germany presents quite a high increase, where lower and higher education not using national TV news presented a -0.12 correlation (a slight negative correlation) and those using national TV indicated a 0.90 correlation on attributes of an issue (a very high positive correlation). The average correlation among lower and higher education for all 13 countries for those not using national TV news was 0.63, while the average correlation for those using national TV news was 0.88.

This indicates that use of national television was found to be an indicator that differences among the younger and older population and lower and higher education populations on the substantive dimension were reduced. And indeed, in the vast majority of the calculated correlations, these groups were brought closer together in social consensus. Belgium, Germany, Spain, the Netherlands, Austria, Portugal, Sweden, and the UK presented an increased consensus in both demographic subgroups on the substantive dimension of attributes with the use of their national television news.

When looking at the influence of national press news in pulling these two demographic subgroups closer together in consensus on the substantive dimension of attributes, this study found that over half of the correlations (63%) indicated increased agreement. In 10 of the 15 calculated countries, younger and older populations were drawn closer together in how they thought about the issue of terrorism with the use of the national press. These countries were Belgium, Denmark, Germany, Greece, Ireland, Luxemburg, the Netherlands, Austria, Portugal, and Finland. In the Netherlands, younger and older populations not using their national press news presented a 0.38 correlation, whereas those using their national press news presented a 0.72 correlation on attributes of an issue, a clear increase. The average correlation for all 15 countries among younger and older populations not using national press was 0.84, while the correlation for those using national press was 0.89.

Among lower and higher education populations, 9 out of the 15 calculated countries presented an increased correlation on the substantive dimension of attributes with the use of national press news. These countries were Belgium, Germany, Greece, Italy, Luxemburg, the Netherlands, Austria, Portugal, and Finland. Again, in the Netherlands, lower and higher education populations presented quite a steep increase of consensus, where those not using national press news indicated a slight correlation of 0.10 whereas those using national press news indicated a very high correlation (0.83) on attributes of an issue. The average correlation for all countries among lower and higher education population not using national press was 0.80, while the average correlation for those using national press was 0.85.

This indicates that younger and older populations and lower and higher education populations were also drawn closer together in consensus on attributes of an issue with the use of national press news, although fewer countries indicated this trend than with the national TV news. Belgium, Germany, Greece, Luxemburg, the Netherlands, Austria, Portugal, and Finland presented an increased consensus in both demographic subgroups on the substantive dimension of attributes with the use of their national press news.

This study also found support for the hypothesis that increased use of national news media corresponds to increased consensus among demographic subgroups regarding the affective agenda of attributes (**H2**). In over half of the 56 calculated correlations (57%), younger and older populations and lower and higher education populations were brought closer together in consensus on the attributes of the Muslim community, the affective dimension of attributes. It is important to note, however, that these demographic subgroups were more likely to be drawn closer together in the substantive dimension than in the affective dimension. This study found, however, that in a substantial number of countries, these two demographics were drawn closer together in consensus in how they thought about the attributes of the Muslim community with the use of national media. This study also found some differences in consensus building on the affective dimension for those using national TV and those using national press news, as Figure 11.3 indicates.

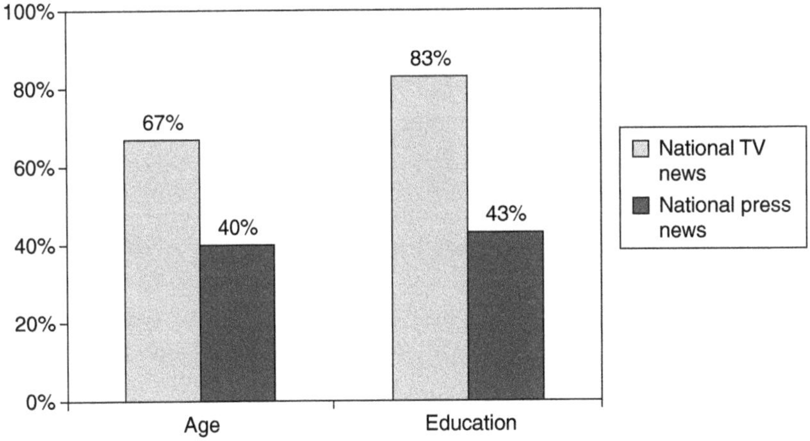

Figure 11.3 Increased consensus on the affective dimension of attribute among age and education demographic subgroups by national TV news and national press news

The study indicates that these two demographic subgroups were pulled towards social consensus on the affective dimension with the use of national television, with almost three-quarters of the calculated correlations indicating increased agreement. In 10 of the 15 calculated countries, younger and older populations were drawn closer together in how they thought about the Muslim community with the use of national TV news. In Belgium, Denmark, Germany, Greece, Spain, France, the Netherlands, Portugal, Sweden, and the UK, younger and older populations presented increased correlations on the affective dimension with the use of national TV news. In Germany, younger and older populations not using national TV news presented a 0.68 correlation on attributes of a community, whereas those indicating they used national TV news presented a 0.90 correlation on that attribute. The average correlation for all 15 countries among younger and older populations on the affective dimension with those not using national TV news was 0.69,whereas the correlation for those using national TV news was 0.87.

Among lower and higher education populations, 10 out of 12 calculated countries presented an increased consensus on the attributes of the Muslim community with the use of national TV news. In Belgium, Germany, Greece, Spain, France, Ireland, the Netherlands, Portugal, Sweden, and the UK, lower and higher education populations presented increased correlations on the affective dimension with the use of national TV news. In Germany, lower and higher education populations not using national TV news presented a 0.50 correlation on the affective dimension of attributes, while those using national TV news indicated a 0.90 correlation. The average correlation for those 12 calculated countries among lower and higher education population for those not using national TV news was 0.67, while the correlation for those using national TV news was 0.87.

Increased correlations on attributes of a community for those using national TV news were found, for both age and education subgroups, in Belgium, Germany, Greece, Spain, France, the Netherlands, Portugal, Sweden, and the UK.

When looking at the influence of national press news in pulling these two demographic subgroups closer together in consensus on the affective dimension of attributes, this study found that 41% of the correlations indicated increased agreement. In 6 of the 15 calculated countries, younger and older populations were drawn closer together in how they thought about the attributes of the Muslim community, the affective dimension. These countries were Belgium, Denmark, Germany, Greece, Italy, and the Netherlands. In the Netherlands, younger and older populations not using national press indicated a 0.78 correlation on the attributes of the Muslim community, while those using national press news indicated a 0.90 correlation. The average correlation on the affective dimension for the 15 calculated countries among younger and older population for those not using national press was 0.82, while the average correlation for those using national press news was 0.87.

In 6 out of the 14 calculated countries, lower and higher education subgroups were drawn closer together in how they thought about the Muslim community with the use of national press. In Germany, Greece, France, Italy, Luxemburg, and Austria, this demographic subgroup was brought closer together on the affective dimension with the use of national press. In Italy, this demographic subgroup indicated a 70 correlation among those who did not use national press news, while those using national press news indicated a 90 correlation on the attributes of the Muslim community. The average correlation between lower and higher education subgroups in the 14 calculated countries was 0.78 for those not using national press news and 0.81 for those using national press news.

In Germany, Greece, and Italy, increased correlations on the affective dimension for those using their national press news were found for both age and education subgroups.

When looking at how much correlations on attributes of an issue and a community increased between the demographic subgroups with the use of media, or the magnitude of increase of consensus, interesting differences were found (RQ1). It is important to note that, as a whole, correlations for both those using and not using the media are high, which stresses the strength of the findings of increased consensus. For example, the average correlation among younger and older population on the substantive dimension for those not using national TV news is 0.69, while the average correlation among younger and older population on the substantive dimension for those using national TV news is 0.91.

Focusing on the magnitude of the increase, the average increase of correlation on the substantive dimension among younger and older population for those using national TV news is 0.22, while the average increase of correlation on the affective dimension among younger and older populations for those using national TV news is 0.17. Graph 3 indicates the increase of consensus within age demographics on the substantive and affective dimension with the use of national TV news.

The average increase of correlation on the substantive dimension among lower and higher education population for those using national TV news is 0.25, while the average increase of correlation on the affective dimension among lower and higher education population for those using national TV news is 0.17. Figure 11.5 indicates the increase of consensus within education demographics on the substantive and affective dimension with the use of national TV news.

The average increase of correlation on the substantive and affective dimensions between the age demographic subgroups was small. Both

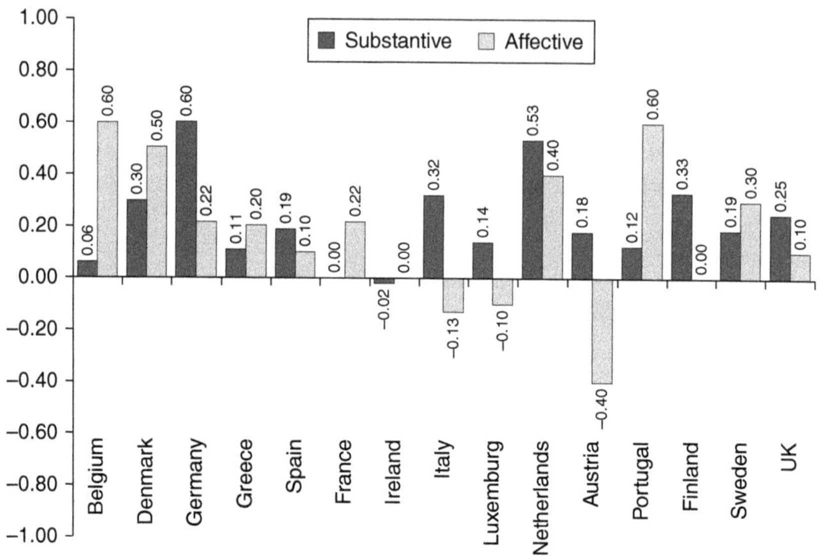

Figure 11.4 Increase of correlation among younger and older populations using national TV news by substantive and affective dimensions of attribute per country

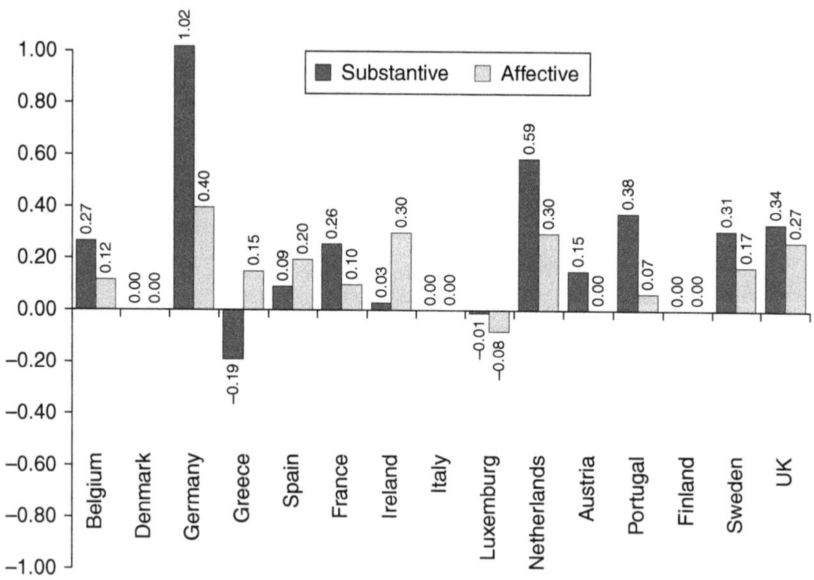

Figure 11.5 Increase of correlation among lower and higher education populations using national TV news by substantive and affective dimensions of attribute per country

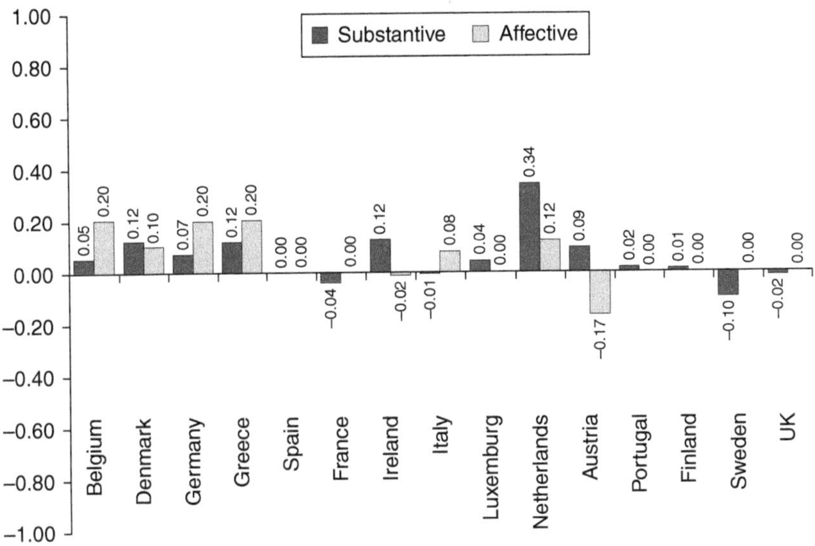

Figure 11.6 Increase of correlation among younger and older populations using national press news by substantive and affective dimensions of attribute per country

averages were 0.05 for those using national press news. Figure 11.6 indicates the increase of consensus within age demographics on the substantive and affective dimensions with the use of national press news.

The average increase of correlation on the substantive dimension lower and higher education for those using national press news is 0.05, while the average increase of correlation on the affective dimension among lower and higher education for those using national press news is 0.03. Figure 11.7 indicates the increase of consensus within education demographics on the substantive and affective dimensions with the use of national press news.

Differences also arose according to the potential of national television news and national press news in how they pulled the two demographic subgroups closer together in consensus (**RQ2**). More than three-quarters of the calculated correlations (43 out of 55 correlations) followed the expected pattern of increased consensus with the use of national television news, considering both the substantive and affective dimensions of attributes and both demographics. For those using national press news, 53% of the 59 calculated correlations indicated that pattern of increase, also considering both the substantive and affective dimensions of attributes and both demographics. This indicates that those using national TV were more likely to be brought closer in consensus, but there is

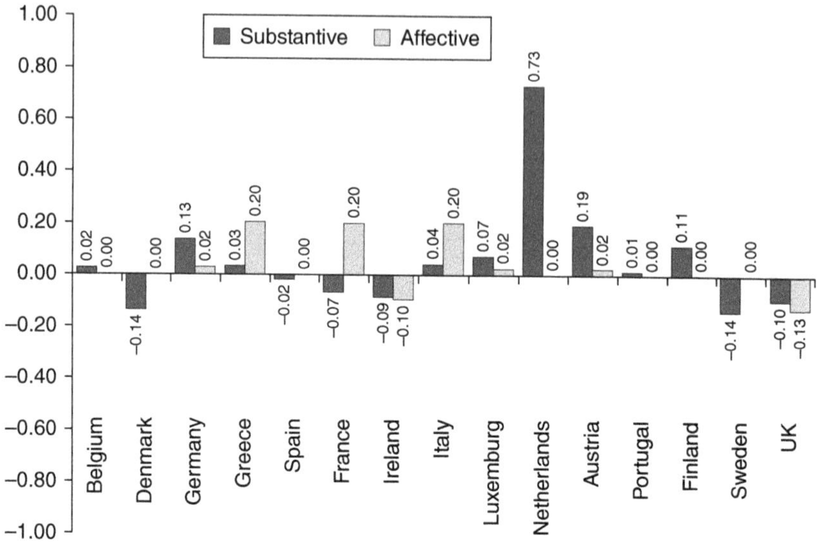

Figure 11.7 Increase of correlation among lower and higher education populations using national press news by substantive and affective dimensions of attribute per country

still strong evidence that national press news functioned to bring these opposing demographic subgroups closer together.

In a closer look at the total correlations of each of the demographic subgroups, this study has found that both age and education demographic subgroups were, in general, as likely to be brought closer together in consensus on the substantive and affective attributes with the use of national TV and national press news combined. In 65% of the total calculated correlations for the age as well as for the education demographic subgroups, opposing groups were brought closer together in agreement of the attributes.

Discussion

Shaw and Martin (1992) developed and tested a research design grounded in agenda-setting theory, whose central proposition is that the agendas of demographically diverse groups in society will come into closer agreement with increased exposure to the news media. This research design understands that younger and older population subgroups, lower and higher education subgroups, for instance, are different, and looks for instances, such as news media use, when these differences are reduced. It thus states that the news media, through its selections of issues and representation of these issues, function to provide a correlation of society, building consensus that could aid mobilization and prevent social instability, as proposed by Harold Lasswell (1960).

In a broad and systematic replication and expansion of Shaw and Martin's original research, as well as the previous studies in Taiwan and Spain, this chapter tests the consensus-building proposition in 15 Western European countries, examining two demographic groups (age and education), two national news media (television and press), and two dimensions of attributes (substantive and affective). Altogether, this is a total of 114 tests of Shaw and Martin's consensus-building proposition. And as a whole, this study found strong support for such a proposition: 66% of all 114 correlations indicated an increased consensus on attributes with the use of national media.

Among its findings, this study identifies national television as the news media more likely to be associated with consensus building. Overall, 78% of the 55 correlations indicated that when younger and older populations and lower and higher education populations accessed their national TV news, they were more likely to be brought closer together in how they thought about the issue of terrorism and how they felt about the Muslim and Arab populations. Those accessing national press were also brought closer together in consensus on the substantive and affective dimension (and 53% of the correlations indicated so), just not at the same rate. Not only were those accessing national TV news more likely to be drawn closer together in consensus than those accessing national press news, they were also more likely to have a higher increase of consensus. This indicates that national TV indeed fulfills Lasswell's function of correlating society in times of crisis.

Straubhaar (2007) stated that television, specifically national television broadcasts, is a dominant framework for cultural forums. The topic of this study, the attacks of 9/11 in the United States and their repercussions among the European Union population, was very unique in terms of the issue itself and of television broadcasting in Europe. The nature of the attacks was highly unusual and the reaction of European journalists to the crisis did not initially follow the traditional scripts. They were more emotional and, even in countries traditionally more critical of the United States, such as France, coverage was mainly sympathetic and supportive of the United States (Pludowski, 2007).

The nature of television, highly visual and easily transmitting the emotions of those experiencing the events, helps explain the greater influence of television here, compared to press. Coleman and Wu (2006), in an analysis of the nonverbal communication cues transmitted by American TV journalists during their coverage of 9/11, found that broadcasters exhibited more non-neutral expressions than neutral expressions during the immediate coverage of this crisis. And perhaps the power of such expressions, coupled with the catastrophic nature of the images presented, helps explain such strong influence of consensus building by national TV news. Many television stations throughout the European Union nations rescheduled their programs, and such images were repeatedly presented. The particularly tragic images and ideas presented by national TV news had a stronger influence in the European Union population in building consensus than did the national press news.

This study also replicated agenda setting's theoretical distinction between the substantive and affective dimensions at the second level of agenda setting, within the context of the consensus-building proposition. In the case of this study, the substantive dimension is defined by the attributes of the terrorism issue, as seen by the European Union population. The affective dimension is defined by the attributes of the Muslim and Arab communities, as seen by the European Union population. This study found greater support for increased consensus on the substantive dimension (72% of the comparisons) for those who used national TV and press, compared to 57% for the affective dimension among those who used national TV and press news. This study also found that not only were the different demographic subgroups more likely to be brought closer together in the substantive dimension than the affective dimension, but also that correlations increased at a higher rate in the substantive dimension than the affective dimension. It is possible to say, then, that in times of crisis, the news media may be more successful in building consensus on how people think about the issue (or crisis) than on how they feel about a certain group. It is important to say, however, that this study found support for both the substantive and affective dimensions. This study thus suggests the importance of the news media in crisis communication, in influencing how people react to the crisis, but also how the transmission of such events may bring people together in trying to identify solutions to potential issues arising from the crisis situation.

López-Escobar et al. (1998) found greater support for the affective dimension of second-level agenda setting than for the substantive dimension. In the Spanish study, the affective dimension related to appraisal of political candidates and the substantive dimension related to candidate's characteristics. The issues from the Spanish study and this current one are very different in nature. In the Spanish study, the attributes are connected to a political context, while in this present study the context is related to a crisis. This is not to say that a political issue cannot be a crisis—indeed it can be. If one defines crisis as an event posing uncertainty and threatening the lives or property of a large amount of people, then certainly a political issue could be a crisis situation. But the nature of a political context, focus of the Spanish study and many other political communication studies, is not a crisis situation in nature. Although both have the potential to affect many, and do, the directedness of such effect is different. Future studies should focus not only on what types of issues we can expect news media to act as a consensus builder but also on the different dimensions of these issues.

The contingent conditions from agenda-setting theory of relevance and obtrusiveness of issues may help explain why the different demographic subgroups were more likely to be brought closer together on the substantive dimension than on the affective dimension within this specific crisis situation. The substantive dimension was associated with the attributes of the terrorism issue, as it related to the events of 9/11. Since this study deals with the European Union population's reactions to the events of 9/11, it is highly unlikely that the respondents of this study had any personal experience with the issue. Although some of the countries within the European Union have had terrorist attacks, the attacks perpetrated before the duration of this survey had very different contexts. Some of these countries have had terrorists attacks occur within contexts more parallel to the events of 9/11, but those occurred after this survey was conducted. However, the past terrorist attacks experienced in this region bring relevance to the terrorism issues. High relevance and unobtrusive issues make it more likely that they will be transferred from the media's agenda to the public's agenda (Kim, in press; McCombs. 2004). The high need for orientation on the issues related to 9/11 and its repercussion in Europe is a contingent condition for the agenda-setting effects, and helps explain some of the impacts of media in this particular event, but not all.

The affective dimension was associated with the attributes of the Muslim and Arab communities, or how the European Union audience felt about that community in the wake of 9/11. The European Union as a region has experienced an increased flow of Muslim and Arab immigrants in recent decades. Although these groups are still minorities in all of the European Union countries analyzed in this study and their presence varies by country, their increased presence makes it more likely that the participants of this study already have some personal experience with the Muslim and Arab communities. It is possible that this more personal experience makes this topic a more obtrusive topic, hence less likely that people would rely on news media agenda. On the other hand, this topic is highly relevant in the region, with a recent history of difficult integration, and sometimes violent relations, with the Muslim and Arab communities. It is unfortunate that no question in this study probed respondents further in their religious affiliation and interpersonal interactions with this group.

A rough estimate of the relation of the percentage of Muslim population per country and the consensus-building influence of their national media in the affective dimension brings some interesting connections. Table 11.1 indicates the percentage of the population that is Muslim in each of these 15 countries and the estimate of the total Muslim population as of 2009 (The Pew Forum on Religion and Public Life, 2009) and whether they had their opposing demographic subgroups brought closer together in the attributes of the Muslim community with the use of national media.

Table 11.1 shows that the six countries with the highest percentage of Muslim population (France, Netherlands, Germany, Austria, Belgium, and Greece) were also more likely to have their opposing demographic subgroups closer together in consensus on the affective dimension, with the exception of Austria. Countries with lower percentages of Muslim

Table 11.1 Percentage of population identified as Muslim and Muslim population by country; indication of increased consensus on affective dimension by country by national media and demographic groups

COUNTRY	*Percentage of population Muslim**	*Estimate 2009 Muslim population**	*Television News*		*Press News*	
			Age	*Education*	*Age*	*Education*
France	6.0%	3,554,000	Yes	Yes	No	Yes
Netherlands	5.7%	946,000	Yes	Yes	Yes	No
Germany	5.0%	4,026,000	Yes	Yes	Yes	Yes
Austria	4.2%	353,000	No	–	No	Yes
Belgium	3.0%	281,000	Yes	Yes	Yes	No
Greece	3.0%	310,000	Yes	Yes	Yes	Yes
Luxemburg	3.0%	13,000	No	No	No	Yes
UK	2.7%	1,647,000	Yes	Yes	No	No
Denmark	2.0%	88,000	Yes	–	Yes	–
Sweden	2.0%	149,000	Yes	Yes	No	No
Spain	1.0%	650,000	Yes	Yes	No	No
Italy	<1%	36,000	No	No	Yes	Yes
Finland	0.5%	24,000	No	–	No	No
Ireland	0.5%	22,000	No	Yes	No	No
Portugal	0.1%	15,000	Yes	Yes	No	No

*Data from "Mapping the Global Muslim Population," a 2009 report by The Pew Forum on Religion and Public Life.

*Yes indicates increased consensus with media use, No indicates had not increased consensus with media use, - indicates calculation could not be conducted.

population seem to be less likely to have their population subgroups brought closer together in consensus. Perhaps that increased presence does reinforce the relevance of the issue.

When looking at the two demographics, age and education, this study found that they were just as likely to be brought closer together in consensus. This finding reinforces the proposition set by Shaw and Martin (1992) that the news media has the potential to bring different demographic subgroups closer together in consensus. This study focused on age and education. Previous studies have also looked into gender, race, income, and localities as demographic subgroups in consensus-building studies. Future studies should focus on other social demographic groups and social indicators, such as political and religious affiliation and social customs.

This study also limited the analysis to the influence of national TV and national press in building consensus. Future studies should also focus on the increasingly important Internet news (Tran, Chapter 10, this volume) and other media for news that might be building consensus. With the increased availability and use of transnational media it is also important to focus on where the influence of consensus might be originating from, and what that means to the social system.

This study indicates that the nature of the issue itself, whether within a political contest context or a crisis context, may be an important factor in whether consensus is more likely to be achieved within an affective or substantive dimension. It would be important to have consensus-building studies replicated within different contextual issues. For example, one could explore the contextual issue of crime, more specifically drug trafficking: Can national and transnational news media act as a bridge to different social groups in how they perceive and understand the issue of international drug trafficking? If so, what are the potential consequences of that?

This extensive cross-national replication of the consensus-building proposition finds strong support that the news media has the potential to bring opposing groups closer together and provide a common ground for deliberation of important issues that society faces. It furthers our understanding that, within different contexts, agenda setting has among its consequences, building consensus. It finds that the use of national television and press (to a lesser extent) can bring people together in how they think about issues in society and how they feel about people and communities. It therefore reinforces the powerful role of news media, as an institution, in influencing our social system (not universal, nor at all times, but a powerful role nonetheless). In a time of much divergence, understanding consensus, and the role of news media in bridging communities, might provide important insights.

References

Anderson, C. J., & Guillory, C. A. (1997). Political institutions and satisfaction with democracy: A Cross-national analysis of consensus and majoritarian systems. *American Political Science Review*, *91*(1), 66–81.

Chiang, C. Y. (1995, May). *Bridging and closing the gap of our society: Social function of media agenda-setting. A Case study in Taiwan*. (unpublished master thesis). The University of Texas at Austin, Texas.

Coleman, R., & Wu, D. (2006). More than words alone: Incorporating broadcasters' nonverbal communication into the stages of crisis coverage theory—Evidence from September 11th. *Journal of Broadcasting & Electronic Media*, 50(1), 1–17.

Craft, S., & Wanta, W. (2004). U. S. public concerns in the aftermath of 9–11: A Test of second level agenda-setting. *International Journal of Public Opinion Research*, **16**(4), 456–463.

Flash Eurobarometer. (2001). *International crisis, flash eurobarometer 114*. EOS Gallup Europe. Wavre, Belgium: German Social Science Infrastructure Service/ EOS Gallup Europe.

Ghanem, S. (1997). Filling in the tapestry: The second level of agenda setting. In M. McCombs, D. L. Shaw, & D. Weaver (Eds.), *Communication and democracy: exploring the intellectual frontiers in agenda-setting theory* (pp. 3–14). Mahwah, NJ: Erlbaum Associates.

Graber, D. (2010). *Mass media and American politic* (8th ed.)Washington, D.C: CQ Press.

Kiousis, S., Bantimaroudis, P., & Ban, H. (1999). Candidate image attributes: Experiments on the substantive dimension of second level agenda setting. *Communication Research 28*(4), 414–428.

Lasswell, H. (1960). The structure and function of communication in society. In W. Schramm, & W. Schramm (Eds.), *Mass Communications* (2nd ed., pp. 117–130). Urbana, IL: University of Illinois Press.

Lijphart, A. (1998). Consensus and consensus democracy: Cultural, structural, functional, and rational-choice explanations. *Scandinavian Political Studies*, 21(2), 99–108.

López-Escobar, E., Llamas, J. P., & McCombs, M. (1998). Agenda setting and community consensus: First and second level effects. *International Journal of Public Opinion Research*, 10(4), 335–346.

Ma, R. (2005). Media, crisis, and SARS: An Introduction. *Asian Journal of Communication*, *15*(3), 241–246.

McClosky, H. (1964). Consensus and ideology in American politics. *The American Political Science Review*, *58*(2), 361–382.

McCombs, M. (1997). Building consensus: The news media's agenda-setting roles. *Political Communication*, 14, 433–443.

McCombs, M. (2004). *Setting the agenda: The mass media and public opinion*. Malden, MA: Blackwell Publishing Inc.

Pludowski, P. (Ed.) (2007). *How the world's news media reacted to 9/11: Essays from around the globe*. Spokane, WA: Marquette Books.

Prothro, J., & Grigg, C. M. (1960). Fundamental principles of democracy: Bases of agreement and disagreement. *The Journal of Politics*, 22(2), 276–294.

Shaw, D., & Martin, S. E. (1992). The function of mass media agenda setting. *Journalism Quarterly*, **69** (4-Winter), 902–920.

Shaw, D., McCombs, M., Weaver, D., & Hamm, B. (1999). Individuals, groups, and agenda melding: A theory of social dissonance. *International Journal of Public Opinion, 11*(1), 2–24.

Straubhaar, J. D. (2007). *World television: From Global to local.* Los Angeles: Sage.

The Pew Forum on Religion and Public Life. (2009, October). *Mapping the global Muslim population: A Report on the size and distribution of the world's Muslim population.* Retrieved October 20, 2009, from The Pew Forum on Religion and Public Life: http://pewforum.org/docs/?DocID = 465

Wanta, W., Golan, G., & Lee, C (2004). Agenda setting and international news: Media influence on public perceptions of foreign nations. *Journalism & Mass Communication Quarterly, 81*(2), 364–377.

Wright, C. R. (1960). Functional analysis and mass communication. *Public Opinion Quarterly, 24*(4), 605–620.

Chapter 12

Agenda Setting in the Corporate Sphere

Synthesizing Findings and Identifying New Opportunities in this Growing Domain

Matthew W. Ragas

While agenda-setting theory will always have its roots in public issues and political elections (for example, see Martin, in press), in recent decades, the agenda-setting proposition has been tested with a diverse range of new actors, settings, topics, and methods (Coleman, McCombs, Shaw, & Weaver, 2009; McCombs & Reynolds, 2009). A particularly active new area of agenda-setting inquiry has been devoted to business news and its effect on public opinion toward companies and the economy (Berger, 2001; Carroll, 2011; Carroll & McCombs, 2003).

Research into agenda setting in the corporate sphere not only helps test the predictive power, scope, and generality of this theoretical metaphor (Shoemaker, Tankard, & Lasorsa, 2004), but is also of growing societal relevance, given the sharp growth over the past 20 years in media coverage dedicated to corporations, the economy, and the stock market (Henriques, 2000; Roush, 2006, 2011). This includes the launch of business-only media outlets such as cable channels CNBC and FOX Business, global business newswire Bloomberg News, and a multitude of business news websites. In a sign of the general rise of business topics on the media and public agendas, *The Wall Street Journal* has surpassed *USA Today* as the largest circulation daily newspaper in the United States. This growth in business news has been matched by a similar increase in the resources that companies dedicate to communication functions that directly or indirectly interface with the media and the public, including media relations, crisis communication, public affairs, and issues management (Goodman & Hirsch, 2010; van Riel & Fombrun, 2007).

With nearly half of U. S. households indirectly or directly owning stocks or bonds today, making sense of the swirl of business news is an increasingly relevant undertaking for many Americans (Investment Company Institute & Securities Industry and Financial Markets Association, 2008). As an easily accessible window into "the world outside," how the corporate sphere is covered by business journalists carries important ethical responsibilities. For example, much of what the public learned—and did not learn—about the recent financial crisis, the taxpayer-funded bailouts, and the debt ceiling debate was acquired second-hand through the lens of the media's reporting on these topics. Given this backdrop, exploring *what effect* the financial press has on shaping public perceptions of corporations, corporate figures, the stock market, and business issues, as well as assessing *who* builds the media agenda within this domain, *how*, and to *what purpose*, is crucial to informing the public's understanding of these various influences.

This chapter provides one of the first detailed reviews of agenda-setting research conducted within the corporate sphere. This chapter will review corporate agenda-setting research conducted by journalism, mass communication, and public relations scholars, all traditional homes of agenda-setting scholarship, as well as the growing body of business media research conducted in adjacent disciplines, such as management, finance, and accounting. Even when agenda-setting theory and terminology has not been explicitly invoked within research in these adjacent fields, the concept of media salience has often been implicit, with business scholars assessing how the *quantity* and *quality* of media attention toward corporate objects impacts public perceptions—and often market behaviors—toward these same objects. This chapter provides an important opportunity to better integrate these two complementary, but at times disconnected, schools of research so that future research can better leverage both schools.

The corporate agenda-setting literature reviewed in this chapter is based on a search of three major databases (EBSCOhost's Academic Search Complete, Communication & Mass Media Complete, and Business Source Complete) for relevant journal articles, as well as a search of Google Books for books and book chapters on this topic. This chapter will briefly describe the genesis of corporate agenda-setting research before moving on to a review and synthesis of the three major lines of research that have emerged in this domain, followed by the identification of new opportunities among this expanded theoretical map, and then brief concluding thoughts.

The Genesis of Corporate Agenda Setting

Though most agenda-setting explorations into the corporate sphere and business news have taken place over the past decade, it is important to note that the genesis of research in this area goes back more than 25 years. A pioneering study by Kieffer (1983) of the agenda-setting consequences of business media coverage found that the stocks of companies featured in *Fortune* magazine over a three-year period outperformed the stock market during this span. In what would now be referred to as second-level agenda setting (affective attribute salience), the stocks receiving *favorable* coverage in *Fortune* increased the most in value, but even stocks receiving *neutral* or *negative* coverage outperformed the overall stock market (Kieffer, 1983).

Around the same time as the *Fortune* study, Behr and Iyengar (1985) were explicating the applicability of agenda setting specifically to economic issues. While agenda-setting research has often included a generic business-oriented topic such as the economy as one of the issues analyzed on the media and public agendas, Behr and Iyengar (1985) were two of the first scholars to hone in on this area. Their times-series study demonstrated that agenda setting for two economic issues—energy and inflation—was *unidirectional* with shifts in TV news coverage prompting changes in the level of public concern. For the issue of unemployment, public concern appeared to be linked with actual real-world economic conditions.

Exploration of the agenda-setting process in a business or economic context then lay relatively dormant until research by Berger (2001) investigated attempts by the Business Roundtable, an elite business policy organization that represents corporate interests, to influence the federal policy agenda. This study found that the association's use of information subsidies, such as lobbying and testimonies, *did* exert some influence over time on several issues on the policy agenda, with this influence preceding media and public attention (Berger, 2001). With the publication of this study, corporate actors had begun to be acknowledged as agenda-setters.

A subsequent review of the literature by Berger, Hertog, and Park (2002) concluded that agenda-setting theory was a potentially useful framework for assessing influence in the corporate sphere. These researchers argued that the "construction or framing of issues and issue images by business political actors is a crucial area of research opportunity" that had been largely neglected (Berger et al., 2002, p. 189). Shortly thereafter, Carroll and McCombs (2003) further explicated the concept of corporate agenda setting, specifically defining the corporate agenda as "the aggregate of discourse by a firm including attempts through public relations strategies to influence the media with a variety of information subsidies" (Carroll & McCombs, 2003, p. 42).

Collectively, these initial studies (Behr & Iyengar, 1985; Berger, 2001; Berger et al., 2002; Carroll & McCombs, 1983; Kieffer, 1983) helped to lay the empirical groundwork for the growing stream of corporate agenda-setting scholarship that has appeared in recent years.

Three Focus Areas for Corporate Agenda Setting

Having looked at the genesis of this theoretical extension, we will now more closely examine the accumulated research in this growing domain.

The corporate agenda-setting research conducted to date generally falls under one of three broad areas of focus:

1. *Agenda setting and corporate reputation*: This line of research examines the role of the news media in influencing perceptions of corporate reputation and issues associated with corporations among the general public and specific stakeholder groups. Research in this area also examines agenda-building efforts by firms to shape media coverage.
2. *Agenda setting and the economy*: This line of research examines the interplay among media coverage of the economy, public perceptions of the economy (e.g., consumer confidence), real-world economic indicators (e.g., GDP, CPI, employment rate, energy prices), and, in some cases, presidential approval ratings and information subsidies.
3. *Agenda setting and the stock market*: This line of research examines the consequences of media coverage on the stock market performance of publiclyheld corporations, including stock returns, turnover, underpricing, and other behavioral measures. This research area also assesses how corporate agenda-building efforts impact this process.

Although these three lines of research may seem on the surface to have somewhat limited overlap, they all explore in a corporate context elements of the same underlying core proposition, *the transfer of salience between the media agenda and the public agenda*—thereby providing theoretical consistency (McCombs, 2005, 2006). The following sections of this chapter will explore each of these three focus area for corporate agenda-setting scholarship more in-depth.

Agenda Setting and Corporate Reputation

Mass Media and Corporate Reputation

There is fairly widespread agreement among academics and practitioners that corporate reputation is an intangible asset (Lev, 2004, 2005) that provides companies with a variety of valuable tangible benefits and competitive advantages, including superior financial performance (Dowling, 2001; Hutton, Goodman, Alexander, & Genest, 2001; Kim, 2001). Therefore, an ongoing area of interest to corporate reputation scholars is identifying and assessing the various factors that impact the impressions that the public hold of companies. Early reputation research (e.g., Fombrun & Shanley, 1990) identified mass media content, including news media coverage of companies and company advertising, as one such factor. As explained by Fombrun (1996), given that the public generally has limited access to the inner workings of companies and lacks direct experience with many firms, news media accounts often help the public "make sense of companies' complex activities—and so affect company reputations" (p. 139).

In their seminal research into the forces influencing corporate reputation, Fombrun and Shanley (1990) found an unexpected *inverse* relationship between the amount of news media attention companies received and their perceived corporate reputation, based on *Fortune*'s Most Admired Companies rankings. That is, the more media coverage a company received, as measured by the frequency of company mentions in the Business Periodical Index, the lower its reputation score. Also, an expected positive relationship between the favorable tone of media coverage and reputation scores did not transpire. Only for the subset of firms classified as "high diversity" firms, meaning these companies had complex business operations, were favorable media tenor toward firms linked with higher reputation rankings. Even for these "high diversity" firms, the volume of coverage was negatively associated with the reputation rankings.

At least part of the explanation for the unexpected findings of early studies (e.g., Fombrun & Shanley, 1990; Wartick, 1992) is that they assessed the influence of media reports on reputation in its form as a *global variable*, rather than with its *specific reputational dimensions* (Carroll, 2009, 2011). Top-level reputation measures, such as *Fortune*'s Most Admired Companies and the Harris Reputation Quotient (RQ), are based on respondents' ratings on batteries of items measuring different reputational dimensions (Fombrun, Gardberg, & Sever, 2000). At least three distinct reputational dimensions have been identified in recent years: 1) *organizational prominence*—the degree to which a firm receives widespread collective recognition and top-of-mind awareness; 2) *organizational public esteem*—the degree to which the public likes, trusts, admires, and respects a firm; and 3) perceptions of the *organizational attributes of quality* associated with a firm, such as its products and services, workplace environment, social responsibility, executive performance, and financial performance (Carroll, 2009, 2011; Rindova, Williamson, Petkova, & Sever, 2005). Research that assesses the influence of first- and second-level media salience in relation to specific reputational dimensions, rather than the global variable, has often detected stronger effects (Carroll, 2011).

Reputational Dimension: Organizational Prominence

At its most basic level, the amount of news media attention a company receives has been shown to affect the level of organizational prominence that company achieves on the public agenda. In general, the greater the volume of news reports about a company, the higher a company's level of brand recognition and the greater its top-of-mind awareness. Using data from the nomination phase of the 2000 U. S. Harris RQ survey, which is based on a random sample of the public, Carroll (2004, 2011) found that the more news stories written about RQ-ranked companies in *The New York Times*, the greater the number of nominations (i. e., level of organizational prominence) the firms received. This relationship persisted even after controlling for the effects of firm-issued news releases, firms' advertising expenditures, and the firms' previous levels of prominence (based on the 1999 RQ data). Similar linkages have been found in a variety of international settings, including Denmark (Kjaer & Morsing, 2011), Finland (Luoma-aho, Uskali, Heinonen, & Ainamo, 2011), Norway (Ihlen & Bronn, 2011), Spain (Arrese & Baigorri, 2011), and Korea (Cha & Yang, 2011).

Reputational Dimension: Organizational Esteem

The news media has been found to impact more than just *which* companies the public "think about" (first-level agenda setting), but also *how* the public thinks about companies (second-level agenda setting). The reputational dimension of organizational esteem—essentially how the public feels about a company—is shaped in part by affective attribute media salience, the tone of the overall media coverage of a firm (Carroll, 2009, 2011; Dalpiaz & Ravasi, 2011) or the tone toward a specific cognitive attribute associated with a firm (Einwiller, Carroll, & Korn, 2010). Carroll's analysis (2009) of media coverage in *The New York Times*, paired with the three-item emotional appeal dimension of the 2000 Harris RQ instrument, found a positive association between media tenor toward companies and these firms' level of public esteem.

A replication and extension of the Carroll studies (2009, 2011) in Italy using the same measure of emotional appeal to gauge organizational public esteem, only in this case based on 2002 Harris RQ survey data, found a significant positive linkage as well (Dalpiaz & Ravasi, 2011). This positive association persisted even after controlling for the firms' age and level of advertising spending. Finally, a German study found that the media's positive tone toward a company's 'social and environmental responsibility' attribute influenced respondents' evaluation of the firm's level of emotional appeal (Einwiller et al., 2010). The tone of coverage toward this specific attribute was also linked with respondents' evaluation of the company on its "product and services" and "management and corporate strategy" cognitive attributes as well. These findings intersect with recent political agenda-setting research (Coleman & Wu, 2010), which indicates that affective agendas have two dimensions: emotions and trait evaluations.

Reputational Dimension: Attributes of Quality

There is mixed evidence that the news media's use of specific cognitive attributes of quality to define and describe companies in coverage shapes which attributes the public perceive as being most salient and associate cognitively with companies when forming impressions of them. Carroll's (2004) analysis of the 2000 U. S. RQ data and coverage in *The New York Times* found there was a positive association between the amount of media attention dedicated to the firm attributes "executive performance" and "workplace environment" and the usage of these two attributes for assessing companies' reputations. However, this linkage was not found for the attributes of "financial performance," "products and services," or "social responsibility." A subsequent investigation based on 2005 U. S. RQ data paired with content analyses of media coverage found only minimal associations regarding cognitive attribute salience (Kiousis, Popescu, & Mitrook, 2007). Drawing from media dependency theory, a possible explanation for these findings is that the public attends more closely to media information on company attributes that are more difficult for the public to experience or learn about directly (Einwiller et al., 2010).

Corporate issues can also be thought of as cognitive attributes of companies. Often companies are in the news in relation to their handling of certain issues. On some corporate issues, such as charitable giving and environmental sustainability initiatives, companies are generally viewed favorably by the public, while on other issues, such as corporate scandals, major layoffs, and bankruptcy talks, companies are generally viewed unfavorably. A detailed multi-year study of issue news and reputation among large Dutch organizations found that the more news there was about a certain issue in relation to an organization, the more likely that issue was associated with the organization by the public (Meijer & Kleinnijenhuis, 2006a). Overall, respondents who associated an organization with a positive issue assigned a higher reputation score to the organization, whereas respondents who associated the organization with a negative issue gave a lower reputation mark. These findings suggest evidence of "attribute priming" (Kim & McCombs, 2007; Kim, Scheufele, & Shanahan, 2002) in which the attribute emphases found in media coverage help set the criteria the public use to form judgments.

Information Subsidies and Corporate Reputation

The business news agenda is, in part, the result of the interactions between sources and journalists. The media have traditionally placed an emphasis on gathering newsworthy information from official and authoritative sources (Gans, 2003; Iyengar & McGrady, 2007). Within the corporate sphere, corporations often serve as a primary source for the raw materials—known as information subsidies (Gandy, 1982)—that journalists use to cost-effectively and efficiently produce the news. While business journalists are hardly stenographers for corporate interests, this backdrop places companies, particularly large corporations (which are generally viewed as being more newsworthy), in an advantageous position to shape how they are portrayed in the news (Davis, 2002, 2006). The accumulating evidence (e.g., Carroll, 2010; Fombrun & Shanley, 1990; Kiousis et al., 2007) suggests that the corporate communication function, particularly public relations through the use of information subsidies, plays a meaningful role in managing corporate reputation through the news media, and in turn, affects public opinion of firms. While firm actions and policies are paramount in building reputation, firm impression management activities matter as well (Deephouse, 2000; van Riel & Fombrun, 2007).

Kiousis et al. (2007) conducted one of the most detailed studies to date of the role of public relations in shaping media coverage and corporate reputation. This study found strong evidence of first-level agenda building, specifically that large U. S. firms that devoted more resources to disseminating a greater number of news releases received more stories about them in *The New York Times* and *The Wall Street Journal*. Turning to second-level agenda building, positive linkages were generally detected between the tone taken toward the companies in releases and in the news, and between the attributes emphasized by firms in releases and the amount of attention these attributes received in the news. This set of findings suggests that large firms can influence not just *how much* media attention they receive, but the *type* of attention.

Research indicates that working through the news media—the traditional focus of media relations activities—rather than taking company messages directly to the public, such as through wire-issued news releases or advertising, appears to be a more effective strategy for reputation management (Carroll, 2010). Multiple studies over the years that have simultaneously examined the relative influence of media coverage and company ad spending on corporate reputation indicate that the news media exert a stronger influence (Carroll, 2004, 2011; Dalpiaz & Ravasi, 2011; Fombrun & Shanley, 1990; Fombrun & van Riel, 2004; Meijer & Kleinnijenhuis, 2006b). Though trust in the news media has declined, corporate reputations still seem to be more credibly and cost-effectively built through earned media coverage rather than with ad campaigns.

Agenda Setting and the Economy

Though not as robust an area of inquiry as corporate reputation, a steady stream of agenda-setting research over the years has focused on the effect of news about the economy on public perceptions of the economy, often expressed as consumer confidence, and the perceived salience of economic issues on the public agenda. There is general agreement in this literature that the news media—both in the United States and internationally—tends to frame economic news in negative terms and that public opinion is most responsive to negative news (e.g., Harrington, 1989; Hester & Gibson, 2003; Ju, 2008; Soroka, 2006; Wattenberg, 1984). In support of the causal proposition underlying agenda setting, time-series analyses suggest that shifts in economic news coverage generally lead to changes in public perception of the economy rather than vice versa (e.g., Behr & Iyengar, 1985; Blood & Phillips, 1997; Goidel & Langley, 1995), although there are exceptions (Stevenson, Gonzenbach, & Prabu, 1991). Where there is inconsistency is whether media coverage of the economy and economic issues or actual real-world economic indicators are more influential in shaping public opinion on the economy and economic issues.

Economic News, Public Opinion, and Real-World Cues

Early longitudinal research in this area by Behr and Iyengar (1985) found that for two of three U. S. economic issues—energy and inflation (but not unemployment)—increases in news coverage boosted public concern, while media attention was, for the most part, unaffected by public opinion. The issue of unemployment seemed to respond exclusively to actual real-world economic conditions, specifically the level of unemployment and the average duration of unemployment. An important contribution of this study is that it added objective real-world indictors of the economy (i. e., "the world outside") as independent variables alongside media coverage to more rigorously test the influence of the news on the "pictures in our heads" in this setting. Behr and Iyengar (1985) concluded that economic news and public opinion were at least partially determined by actual economic conditions, therefore "analyses of media agenda setting that ignore real world conditions will arrive at severely inflated estimates of media influence" (p. 53). Virtually all subsequent economic-oriented agenda-setting research has followed this model of examining the news, public opinion, and real-world indicators in tandem.

Contrary to Behr and Iyengar's (1985) findings on unemployment, a multi-year study by Shaw and Slater (1988) set in a local community (Richmond, Virginia) found that the public's perceived importance of this issue was correlated with both national unemployment figures *and* national TV news coverage of unemployment. Cross-lag correlations suggest that the real-world national state of unemployment was the most closely associated with local opinion on this topic, but that national TV news, and to a lesser extent, local newspaper coverage, were linked as well.

An extensive series of time-series analyses by Blood and Phillips (1997) indicate that coverage in *The New York Times* undirectionally influenced consumer sentiment toward the U. S. economy over more than a decade (1980–1993). A rising number of positive economic headlines boosted consumer sentiment, while a growing number of negative headlines dampened sentiment. Analyses using various time lags found that shifts in media coverage were generally associated with subsequent changes in the leading economic indicator index, a measure of real-world economic behavior. Evidence of economic headlines reacting to changes in the index was more limited. This finding of media effects extending beyond the cognitive level to actual behavior serves as an example of how "perception becomes reality," and in this case, how economic news coverage can potentially adversely impact subsequent economic performance.

Subsequent work by Hester and Gibson (2003) suggests that the news has a meaningful impact on shaping the public's perception of the U. S. economy, but that real-world economic cues play at least as important a role. Content analyses of economic news appearing in *The New York Times* and ABC World News over a four-year period (1998–2002) paired with secondary economic and public opinion data found that unfavorably framed news served as a significant predictor of *future* economic performance expectations. However, negative news did not impact *current* consumer assessments of the economy. In separate regression models predicting current and future consumer expectations regarding the economy, multiple real-world indicators, specifically the CPI (a measure of inflation), and the unemployment rate, had a significant influence. Hester and Gibson (2003) conclude that the public's feelings about the *current* economy may be based largely on personal observations and experiences, whereas the public may be more reliant on media reports for forming *future* judgments about the economy.

International Evidence of Economic Agenda Setting

Recent scholarship has provided much-needed tests of economic agenda setting across additional times and places, exploring the interplay among economic news, public opinion, and the real-world economy in several international settings. An analysis by Soroka (2006) set in the United Kingdom from 1986 to 2000 found that public opinion responded more to negative economic media coverage in the elite newspaper *The Times* (London) than to positive coverage. Patterns of economic news coverage were not found to affect the public's perception of unemployment as a Most Important Problem. Only an increase in the unemployment rate was a significant predictor of MIP responses on this issue. However, negative news stories, as well as negative changes in the actual economy, were found to significantly dampen the public's future outlook regarding the economy. Though the media played a role in shaping public perceptions of the economy, "much of the asymmetry in opinion is in fact a function of individuals' reactions to the economy itself, rather than the effect of media coverage" (Soroka, 2006, p. 381).

International research continues the inconsistency of results about the relative influence of real-world events and the news media on the public's opinion of the economy. Ju's (2008) investigation in South Korea found that the news media, rather than real-world economic cues, drove public opinion on the economy during the 1999–2004 timeframe. As has been the case in the United States, the Korean news media, specifically news stories appearing on the front pages of the elite newspapers *Chosun-Ilbo* and *Donga-Ilbo*, displayed a negative news bias in regard to economic coverage. Both negatively and positively valenced economic news, as well as a real-world economic indicator (a composite index of economic performance), were significantly associated with the public's feelings about the economy. Out of these three variables, negative news coverage was most closely linked with these feelings. Ju (2008) argues that these findings demonstrate that "people form a general perception of the economy more by reading and listening to economic news than by looking for the statistics released by the governmental office or directly feeling the economic changes of unemployment, consumer price, national products, etc. that are indicated by these statistics" (p. 245).

Finally, an economic agenda-setting study set in Japan found relatively little evidence that the media shaped public opinion regarding the economy (Wu, McCracken, & Saito, 2004). These scholars suggest that the decade-long recession in Japan may have reduced the media's influence in this area. In sum, research conducted in both the United States and abroad demonstrates that the media typically plays a meaningful role in influencing public perceptions of the economy, although the magnitude of this effect can vary considerably and the influence of "the world outside" on both the media and public agendas in this setting should not be underestimated.

The Role of the White House in Economic Agenda Setting

The role of the president—arguably the top agenda setter in the land—and the White House in influencing the salience of topics on the media and public agendas has been well documented (Kiousis & Strömbäck, 2010; Miller & Wanta, 1996; Strömbäck & Kiousis, 2011; Wanta & Foote, 1994). Several investigations of agenda setting and the economy have incorporated the actions and statements of the president and public approval of the president into this process. The early work of Behr and Iyengar (1985) found that presidential attention to specific economic issues, as measured by the number of speeches given on a topic, generally induced the public to assign greater importance to these issues. The approval rating of the president has also been found to impact future economic headlines, with a more popular president reducing the number of future negative headlines (Blood & Phillips, 1997).

These findings raise the question of whether other official and authoritative government actors, besides the White House and the president, may also play a role in shaping the salience of economic issues in the news and among the public. Federal government agencies, such as the U. S. Federal Reserve, Treasury, Commerce, and Labor Departments, all maintain sophisticated public relations operations that devote significant resources to producing and disseminating information subsidies, which communicate their priorities to the media and the public (Gans, 2003). Future research into agenda setting and the economy should examine the influence of these other actors in the creation of economic news content and the shaping of public opinion.

Agenda Setting and the Stock Market

Contemporary agenda-setting scholarship demonstrates that the news media at times do more than shape people's *cognitions*, such as impressions of corporate reputations or perceptions of the economy; the media may also impact people's *behaviors* toward these objects (McCombs, 2005, 2006). The stock market provides a rich backdrop for exploring the potential *consequences* of agenda setting on the behaviors of investors toward the securities of publicly held companies. Echoing Lippmann, noted business journalist Michael Lewis has argued that "even people as hardboiled as the typical Wall Street trader gets his understanding of the world from some place" such as a "really well written piece of journalism about Wall Street" (Gladstone, 2008, para. 30). Lewis claims that financial reporting "does shape the thing's view of itself" (Gladstone, 2008, para. 30). While research in this area is still emerging, the accumulated empirical findings to date provide solid support for these assertions: More often than not, business media coverage *does* seem to have some effect on how the stock market treats companies, as well as how the market perceives and reacts to firms during transformative events, such as initial public offerings (IPOs), and corporate proxy contests and takeover battles.

Business News and Stock Market Performance

A variety of studies have examined whether the professional stock recommendations featured in prominent financial news media outlets outperform the performance of the overall stock market over various time frames. These studies have examined a wide range of outlets including *The Wall Street Journal*'s "Heard on the Street" column (Huth & Maris, 1992), *Business Week* magazine's "Inside Wall Street" column (Sant & Zaman, 1996), the pages of *Fortune* magazine (Kieffer, 1983), the PBS television program "Wall $treet Week with Louis Rukeyser" (Ferreira & Smith, 2003), and CNBC's "Mad Money with Jim Cramer" program (Engelberg, Sasseville, & Williams, in press). Overall, these studies indicate that the stocks featured by these prominent media outlets do tend to generate statistically significant, albeit economically small returns compared to the market—at least over the short term.

What is unclear from these studies is how much of the resulting outperformance is due to the stock picking prowess of the featured investment professionals and how much may be due to the media visibility the recommendations received. Huth and Maris (1992) posit that "apart from the value of security analysts' recommendations, it is possible that investors will react to information in the column simply because it appears in print" (p. 28). In other words, did the recommendation made by the expert influence the stock's price or was it simply the fact that the recommendation was legitimized and amplified by the notable media attention it received?

Recent research by Pfarrer, Pollock, and Rindova (2010) provides a cleaner test of the role of accumulated media presentations in influencing investor reactions to companies' securities. Publicly held firms are evaluated by investors in part on their ability to meet or exceed the quarterly earnings estimates made by Wall Street analysts. Exploring a concept called "firm celebrity"—essentially an interaction of first- and second-level agenda-setting effects—Pfarrer et al. (2010) found that celebrity firms were better rewarded by investors for generating positive quarterly earnings surprises, and were less penalized for negative surprises. A firm was deemed to have attained "celebrity" only if it was in the top quartile for both: (1) the amount of media coverage it had received in *Business Week* (first-level media salience), and (2) how positive the tone of this coverage was toward the firm (second-level affective salience). In short, celebrity firms saw their stock prices increase more in value on average than non-celebrity firms following better-than-expected earnings reports. Celebrity status also provided a valuable buffer of sorts to negative earnings reports, with the stock prices of these firms declining less on average.

Business News and Initial Public Offerings

An initial public offering (IPO) occurs when a previously private company raises capital by selling stock to the public market, thereby becoming a public company. The IPO market provides an ideal setting for exploring news media influence since it is an area in which investors tend to look toward information intermediaries, such as the news media, to help them form impressions of relatively young companies. A rigorous multivariate analysis of the 1992 U. S. IPO market by Pollock and Rindova (2003) found that the amount of media coverage firms received in the year prior to their IPO was associated with the performance of their stock on the first day of trading. The tone of coverage had a marginally significant effect as well. In short, even in the face of rigorous controls, the media was found to impact both the amount of capital firms raised from their IPOs and the ease with which investors could buy and sell their stock.

More recent research into the IPO market yields similar findings. Scholarship by Pollock, Rindova, and Maggitti (2008), set during the 60-day period *after an IPO*, finds that sustained media attention can lead to positive effects on market performance. Based on these results, Pollock et al. (2008) advise managers to "consider developing a 'media strategy' that enables them to attract ongoing media attention, thereby influencing their firms' early market performance" (p. 353). While agenda building and information subsidies were not the focus of these IPO market studies (Pollock & Rindova, 2003; Pollock et al., 2008), it is worth noting that in both cases, bivariate linkages were found between firms' news releases and media attention.

Business News, Proxy Contests, and Hostile Takeover Battles

Research into the role of the mass media during corporate proxy contests and hostile takeovers, two transformative events for public companies and their shareholders, more directly gets at the impact of corporate agenda-building efforts on media coverage, and the outcomes of agenda setting (e.g., Ohl, Pincus, Rimmer, & Harrison, 1995; Ragas, 2010; Ragas, Kim, & Kiousis, 2011). Both of these events tend to garner high levels of media coverage given that they feature conflict and impact many different stakeholders, including employees and customers.

When a shareholder group becomes sufficiently dissatisfied with the board of directors and management of a public company, they may nominate their own alternate slate of board of directors nominees. The ensuing political-style election campaign, in which each side solicits the support of shareholders, is known as a proxy contest. A detailed analysis of the 25 largest U. S. proxy contests over a five-year period (2005–2009) reveals evidence of issue agenda-building linkages between campaign information subsidies (news releases and shareholder letters) and financial media coverage of the contests. Cross-lag correlation analyses indicate that reciprocal influence between the corporate and media issue agendas was the norm during the contests. Building issue salience in media coverage may also be consequential, as the side in the contest that enjoyed the stronger agenda-building bond with the media ultimately ended up winning the proxy fight vote in nearly 60% (37/62) of possible comparisons (Ragas, 2010).

Related to proxy contests are corporate takeover battles in which one company or investor group attempts to acquire another company *without the support* of that firm's board or management. The two sides then take their case directly to shareholders in which they try to convince investors to either support or reject the offer. An analysis of a takeover battle in the United States found that the newspapers covering the conflict often adopted the key message points and tone provided in the dueling corporate news releases (Ohl et al., 1995). The side that was more accessible and responsive to journalists, in this case the firm resisting the takeover attempt, received more extensive coverage in favor of its points of view. The hostile acquirer ultimately abandoned its bid. These studies (Ohl et al., 1995; Ragas, 2010) suggest that corporate actors, through the use of information subsidies, have some influence in shaping business news during financial-oriented events. Further, the media may even affect how shareholders vote at times.

Expanding the Playing Field

Corporate reputation, the economy, and the stock market have emerged as the three core focus areas for corporate agenda-setting research to date, but there are additional aspects of each of these areas yet to explore, as well as other areas in the corporate sphere in need of exploration. Figure 12.1 provides a theoretical map for agenda setting in the corporate sphere. This figure represents an attempt to "expand the playing field" in regard to the kinds of sources, news media,and information intermediaries, manifestations of public opinion, and consequences of agenda setting that fall under the corporate sphere and may be studied from this perspective.

Starting with sources, most research to date has looked at the role of corporations, and to a lesser extent the White House, in building media agendas and influencing public opinion. As previously mentioned, federal government agencies, such as the Federal Reserve, Treasury, Commerce, and Labor Departments, are also potential agenda builders on business issues (Gans, 2003). Trade and professional organizations, such as the U. S. Chamber of Commerce and the Business Roundtable (Berger, 2001), also actively attempt to influence the media and policy agendas. Regulators, such as the Securities and Exchange Commission (SEC), also exert influence on corporate issues. Finally, on virtually every major business issue—ranging from ethanol subsidies to foreign worker visas—there are interest groups on both sides of the debate

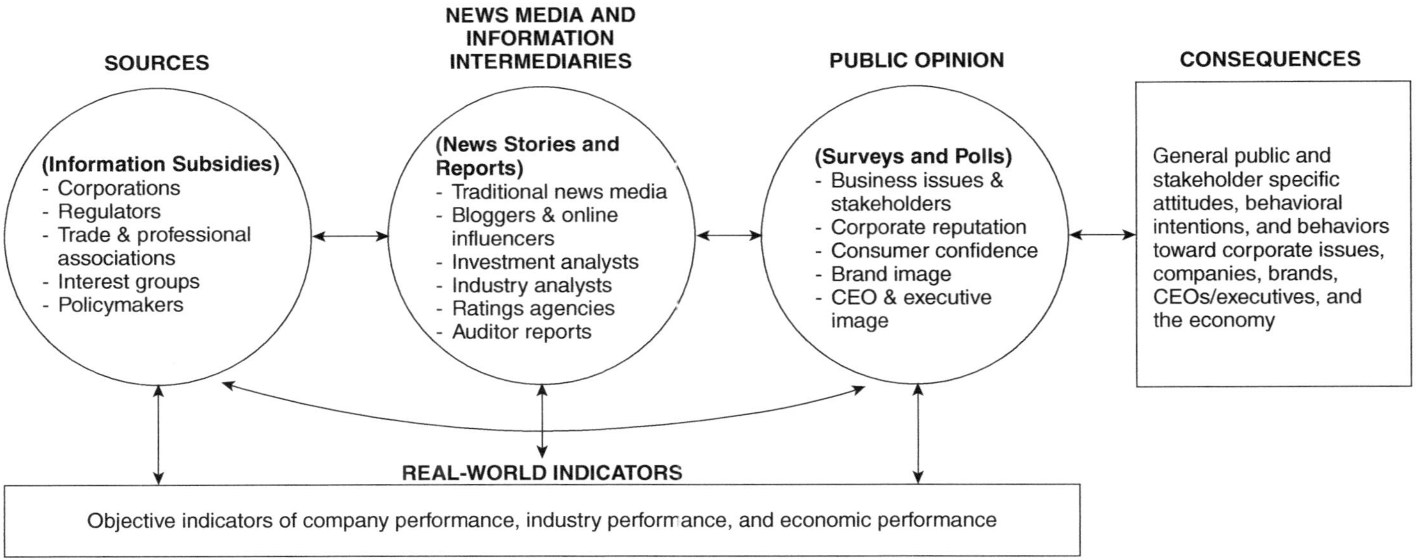

Figure 12.1 Theoretical map of agenda setting in the corporate sphere

attempting to sway public opinion directly and by working through the news media.

Within the corporate sphere, the traditional news media remains an important information intermediary, but the press is not the only expert, objective provider of information on companies, stocks, business issues, and related topics. The Web and social media has helped usher in a new breed of information intermediaries, such as bloggers and other online influencers (Meraz, Chapter 1, this volume; Tran, Chapter 10, this volume). Financial stakeholders also rely on reports issued by Wall Street and industry analysts, accounting firms, and credit ratings agencies for forming impressions of firms and business issues (Fombrun, 1996). The news media is far from the only game in town.

Public opinion, generally the dependent variable in agenda-setting research, has often been operationalized in the corporate sphere based on perceptions of business issues and stakeholders, consumer confidence and the economy, or corporate reputations. Other manifestations of public opinion toward corporate objects include the CEO image (Park & Berger, 2004) and the brand image (Ragas & Roberts, 2009). Companies invest millions of dollars each year on corporate communication efforts that attempt to manage how their CEOs and brands are presented by intermediaries, such as the media, and thought about by the public.

Moving on to outcomes, most corporate agenda-setting research to date has focused on the consequences of agenda setting in regard to the attitudes and behaviors of the *general public* or *financial-oriented stakeholders*. Far less attention has been paid to the attitudinal and behavioral outcomes of agenda setting for other stakeholder groups, including employees, suppliers, customers, the government, and the local community (Freeman, 1984).

Finally, corporate agenda-setting researchers have done a fairly good job of incorporating real-world indicators into their analyses, perhaps because financial and economic time-series are readily available. It is important that this tradition is continued as incorporating the "real world" as a potential third variable enhances the validity of findings.

Identifying New Opportunities

There remain substantial opportunities for building on, and extending, research into mass media influence in the corporate sphere. Based on this review of the corporate agenda-setting literature, the following additional recommendations for future research are offered:

- *Beyond issue agenda setting*: Most first-level agendas in corporate agenda-setting research to date have been operationalized based on the salience of business issues or corporations. However, more than issues and corporations receive varying levels of attention in business-oriented media messages. Other sets of objects that compete for attention, and therefore fit the agenda-setting metaphor, include stakeholder groups (Ragas, 2010), brands (Ragas & Roberts, 2009), CEOs and other business leaders (Park & Berger, 2004).

- *Intermedia agenda setting among business news*: There is an assumption in some corporate agenda-setting research (e.g., Carroll, 2009, 2010; Kiousis et al., 2007) that *The New York Times* and *The Wall Street Journal* set the overall business news agenda, although this has not been explicitly tested. Intermedia agenda-setting research is needed by media type (newspapers, newswires, magazines, websites, cable channels, etc.) and media outlet to establish if there is indeed a convergence in coverage patterns, and a topic- and situation-specific hierarchy of influence among the business news outlets.
- *Working through or bypassing the media filter*: In the age of social media and the mobile Web (Meraz, Chapter 1, this volume; Tran, Chapter 10, this volume), organizations now have the ability to cost-effectively and broadly disseminate their messages directly to their stakeholders and bypass the media filter (Sullivan, 2011). More research is needed that compares the relative influence and effectiveness of direct organization-controlled communication efforts via the Web versus working through the news media on shaping public opinion in the corporate sphere.
- *Social media and the corporate sphere:* Building on the previous recommendation, firms increasingly devote resources to monitoring and participating in conversations on social networks (Nassauer, 2010; Steel, 2009). Journalists also use social media for research and as sources for stories. More concentrated research is needed to understand how the priorities expressed by active, vocal publics on social networks may influence, as well as be influenced by, the agendas of corporate and government actors, business news, and the general public.
- *Diversity of information subsidy types:* Firms use much more than news releases to articulate their agendas. Other information subsidies at their disposal include ads, websites, social media updates, mobile alerts, videos, letters, testimonies, speeches, and meetings. Political agenda-setting research has demonstrated that the effects of different subsidy types on media coverage (Kiousis, Kim, McDevitt, & Ostrowski, 2009) and public opinion (Kiousis & Strömbäck, 2010) are not uniform. Further, not all subsidy types may be positively associated with coverage. This line of inquiry should be extended to corporate studies.
- *Methodological integration:* The strengths of one research method generally offset the limitations of another method and vice versa

(Campbell & Fiske, 1959; Iyengar & Kinder, 1987). For the most part, the existing body of corporate agenda-setting findings has been based on a mix of content analyses and public opinion surveys, but little in the way of experimental and qualitatively derived scholarship. Greater diversity in the methods deployed—what Carl Hovland (1959) called methodological "integration" (p. 14)—would provide a stronger foundation for this body of knowledge and help uncover new insights.

- *Consequential effects*: Over the past four decades, much has been learned about the relationship between media salience and people's cognitions, where agenda-setting theory first started. Much less is known about the potential behavioral consequences of agenda setting. The stock market and the economy, with their readily accessible data series, provide compelling backdrops for exploring the outcomes of the agenda-setting process on the public's behaviors. Consequences are a logical area to explore in this data-rich setting.

Conclusion

Business news has always had a place on the media agenda, but it has risen in prominence in recent years as news organizations respond more to advertiser and audience interest (Shoemaker & Reese, 1996) in coverage of the stock market, the economy, and corporate issues. The accumulated evidence to date suggests that the patterns of news coverage in each of these areas generally have had at least a modest agenda-setting influence on both "*what* the public thinks about" and "*how* the public thinks about" objects under the media spotlight.Emerging research into agenda-setting consequences in the corporate sphere, particularly the stock market, suggest that the mass media may even impact in some cases "what the public thinks—*and does*." Of course, as has been demonstrated in the theory's original domain of public affairs, agenda-setting effects in the corporate sphere are not universal or all-powerful, and should not be overstated (Dearing & Rogers, 1996; McCombs, 2006). The public learns from a variety of information sources besides the mass media, and they are far from automatons.

In conclusion, understanding how and which sources and source materials influence the media agenda and the effect of business news decisions and content on shaping audience cognitions, attitudes, and behaviors toward the business arena is a crucial area of scholarly inquiry. The stakes are high and the cascading impact is not limited to corporate titans. The results of these interactions have broad societal repercussions that cut across the cross-sections of society, including which companies gain and lose resources and jobs; which and how economic issues are addressed by policymakers; how voters respond to perceptions of the economy and corporate America; and which companies the public decides to buy from and work for.

While this line of inquiry is off to a promising start, scholars have much to learn. Given the media's demonstrated influence on business issues, one thing that can be said with confidence is that financial reporting carries incredible ethical responsibilities. The choices business journalists make are far from trivial; they impact the lives of the ordinary citizens.

References

Arrese, A., & Baigorri, M. (2011). Corporate reputation and the news media in Spain. In C. E. Carroll (Ed.), *Corporate reputation and the news media: Agenda-setting within business news coverage in developed, emerging, and frontier markets* (pp. 168–191). New York: Routledge.

Behr, R. L., & Iyengar, S. (1985). Television news, real-world cues, and changes in the public agenda. *Public Opinion Quarterly*, *49*(1), 38–57.

Berger, B. K. (2001). Private issues and public policy: Locating the corporate agenda in agenda-setting theory. *Journal of Public Relations Research*, *13*(2), 91–126.

Berger, B. K., Hertog, J. K., & Park, D. (2002). The political role and influence of business organizations: A communication perspective. In W. B. Gudykunst (Ed.), *Communication yearbook 26* (pp. 160–200). Mahwah, NJ: Erlbaum.

Blood, D. J., & Phillips, P. C. B. (1997). Economic headline news on the agenda: New approaches to understanding causes and effects. In M. McCombs, D. L. Shaw, & D. Weaver (Eds.), *Communication and democracy: Exploring the intellectual frontiers in agenda-setting theory* (pp. 97–113). Mahwah, NJ: Erlbaum.

Campbell, D. T., & Fiske, D. W. (1959). Convergent and discriminant validation by the multitrait-multimethod matrix. *Psychological Bulletin*, *56*, 81–105.

Carroll, C. E. (2009). The relationship between firms' media favorability and public esteem. *Public Relations Journal*, *3*(4).

Carroll, C. E. (2010). Should firms circumvent or work through the news media? *Public Relations Review*, *36*(3), 278–280.

Carroll, C. E. (Ed.). (2011). *Corporate reputation and the news media: Agenda-setting within business news coverage in developed, emerging, and frontier markets*. New York: Routledge.

Carroll, C. E., & McCombs, M. (2003). Agenda-setting effects of business news on the public's images and opinions about major corporations. *Corporate Reputation Review*, *6*(1), 36–46.

Cha, H., & Yang, S. (2011). Corporate reputation and the news media in South Korea. In C. E. Carroll (Ed.), *Corporate reputation and the news media: Agenda-setting within business news coverage in developed, emerging, and frontier markets* (pp. 340–262). New York: Routledge.

Coleman, R., McCombs, M., Shaw, D., & Weaver, D. (2009). Agenda setting. In K. Wahl-Jorgensen, & T. Hanitzsch (Eds.), *The handbook of journalism studies* (pp. 147–160). New York: Routledge.

Coleman, R., & Wu, H. D. (2010). Proposing emotion as a dimension of affective agenda setting: Separating affect into two components and comparing their second-level effects. *Journalism & Mass Communication Quarterly, 87*(2), 315–327.

Dalpiaz, E., & Ravasi, D. (2011). Corporate reputation and the news media in Italy. In C. E. Carroll (Ed.), *Corporate reputation and the news media: Agenda-setting within business news coverage in developed, emerging, and frontier markets* (pp. 105–128). New York: Routledge.

Davis, A. (2002). *Public relations democracy: Public relations, politics and the mass media in Britain*. Manchester, UK: Manchester University Press.

Davis, A. (2006). The role of the mass media in investor relations. *Journal of Communication Management, 10*(1), 7–17.

Dearing, J. W., & Rogers, E. M. (1996). *Agenda-setting*. Thousand Oaks, CA: Sage.

Deephouse, D. L. (2000). Media reputation as a strategic resource: An integration of mass communication and resource based theories. *Journal of Management, 26*(6), 1091–1112.

Dowling, G. R. (2001). *Creating corporate reputations: Identity, image, and performance*. New York: Oxford University Press.

Einwiller, S. A., Carroll, C. E., & Korn, K. (2010). Under what conditions do the news media influence corporate reputation? The roles of media dependency and need for orientation. *Corporate Reputation Review, 12*(4), 299–315.

Engelberg, J., Sasseville, C., & Williams, J. (in press). Market madness? The case of *Mad Money. Management Science*.

Ferreira, E. J., & Smith, S. D. (2003). "Wall $treet Week": Information or entertainment? *Financial Analysts Journal, 59*(1), 45–53.

Fombrun, C. J. (1996). *Reputation: Realizing value from the corporate image*. Boston: Harvard Business School Press.

Fombrun, C. J., Gardberg, N. A., & Sever, J. M. (2000). The Reputation Quotient[sm]: A multi-stakeholder measure of corporate reputation. *Journal of Brand Management, 7*(4), 241–255.

Fombrun, C., & Shanley, M. (1990). What's in a name? Reputation building and corporate strategy. *Academy of Management Journal, 33*(2), 233–258.

Fombrun, C., & van Riel, C. B. M. (2004). *Fame & fortune: How successful companies build winning reputations*. Upper Saddle River, NJ: Prentice Hall.

Freeman, R. E. (1984). A stakeholder theory of the modern corporation. In R. E. Freeman (Ed.), *Stakeholder theory and organizational ethics* (pp. 38–48). Boston: Pitman.

Gandy, O. H. (1982). *Beyond agenda setting: Information subsidies and public policy*. Norwood,NJ: Ablex.

Gans, H. J. (2003). *Democracy and the news*. New York: Oxford University Press.

Gladstone, B. (Narrator). (2008, November 21). Risk and return [Radio broadcast]. In K. Rogers (Senior Producer), *On the Media*. New York: National Public Radio. Retrieved July 11, 2011, from http://www.onthemedia.org/2008/nov/21/risk-and-return/transcript/

Goidel, R. K., & Langley, R. E. (1995). Media coverage of the economy and aggregate economic evaluations: Uncovering evidence of indirect media effects. *Political Research Quarterly, 48*, 313–328.

Goodman, M. B., & Hirsch, P. B. (2010). *Corporate communication: Strategic adoption for global practice.* New York: Peter Lang.
Harrington, D. E. (1989). Economic news on television: The determinants of coverage. *Public Opinion Quarterly, 53,* 17–40.
Henriques, D. M. (2000, November-December). Business reporting: Behind the curve. *Columbia Journalism Review, 39*(4), 18–26.
Hester, J. B., & Gibson, R. (2003). The economy and second-level agenda setting: A time-series analysis of economic news and public opinion about the economy. *Journalism & Mass Communication Quarterly, 80*(1), 73–90.
Hovland, C. I. (1959). Reconciling conflicting results derived from experimental and survey studies of attitude change. *American Psychologist, 14*(1), 8–17.
Huth, W. L., & Maris, B. A. (1992). Large and small firm stock price response to "Heard on the Street" recommendations. *Journal of Accounting, Auditing & Finance, 7*(1), 27–44.
Hutton, J. G., Goodman, M. B., Alexander, J. B., &, Genest, C. M. (2001). Reputation management: The new face of corporate public relations? *Public Relations Review, 27*(3), 247–261.
Ihlen, O., & Bronn, P. S. (2011). Corporate reputation and the news media in Norway. In C. E. Carroll (Ed.), *Corporate reputation and the news media: Agenda-setting within business news coverage in developed, emerging, and frontier markets* (pp. 153–167). New York: Routledge.
Investment Company Institute & Securities Industry and Financial Markets Association. (2008). *Equity and bond ownership in America, 2008.* Retrieved July 5, 2011, from http://www.ici.org/pdf/rpt_08_equity_owners.pdf
Iyengar, S., & Kinder, D. R. (1987). *News that matters: Television and American opinion.* Chicago: University of Chicago Press.
Iyengar, S., & McGrady, J. A. (2007). *Media politics: A citizen's guide.* New York: Norton.
Ju, Y. (2008). The asymmetry in economic news coverage and its impact on public perception in South Korea. *International Journal of Public Opinion Research, 20*(2), 237–248.
Kieffer, N. (1983). *Agenda-setting and corporate communication issues: Can the mass media influence corporate stock prices?* Unpublished master's thesis, Syracuse University, Syracuse, NY.
Kim, Y. (2001). Measuring the economic value of public relations. *Journal of Public Relations Research, 13*(1), 3–26.
Kim, K., & McCombs, M. E. (2007). News story descriptions and the public's opinions of political candidates. *Journalism & Mass Communication Quarterly, 77*(2), 273–291.
Kim, S. H., Scheufele, D. A., & Shanahan, J. (2002). Think about it this way: Attribute agenda-setting function of the press and the public's evaluation of a local issue. *Journalism & Mass Communication Quarterly, 79*(1), 7–25.
Kiousis, S., Kim, S., McDevitt, M., & Ostrowski, A. (2009). Competing for attention: Information subsidy influence in agenda building during election campaigns. *Journalism & Mass Communication Quarterly, 86*(3), 545–562.
Kiousis, S., Popescu C., & Mitrook, M. (2007). Understanding influence on corporate reputation: An examination of public relations efforts, media coverage, public opinion, and financial performance from an agenda-building and agenda-setting perspective. *Journal of Public Relations Research, 19*(2), 147–165.

Kiousis, S., & Strömbäck, J. (2010). The White House and public relations: Examining the linkages between presidential communications and public opinion. *Public Relations Review, 36*, 7–14.

Kjaer, P., & Morsing, M. (2011). Corporate reputation and the news media in Denmark. In C. E. Carroll (Ed.), *Corporate reputation and the news media: Agenda-setting within business news coverage in developed, emerging, and frontier markets* (pp. 17–35). New York: Routledge.

Lev, B. (2004). Sharpening the intangibles edge. *Harvard Business Review, 82*(6), 109–116.

Lev, B. (2005). Intangible assets: Concepts and measurements. *Encyclopedia of SocialMeasurement, 2*, 299–305.

Luoma-aho, V., Uskali, T., Heinonen, J., & Ainamo, A. (2011). Corporate reputation and the news media in Finland. In C. E. Carroll (Ed.), *Corporate reputation and the news media: Agenda-setting within business news coverage in developed, emerging, and frontier markets* (pp. 36–61). New York: Routledge.

Martin, J. (in press). Agenda-setting in election campaigns: News and candidate websites. In T. J. Johnson (Ed.), *Agenda setting in a 2.0 world: A tribute to Maxwell McCombs*. New York: Routledge.

McCombs, M. (2005). A look at agenda-setting: past, present and future. *Journalism Studies, 6*(4), 543–557.

McCombs, M. (2006). *Setting the agenda: The mass media and public opinion.* Malden, MA: Polity Press.

Meijer, M., & Kleinnijenhuis, J. (2006a). Issue news and corporate reputation: Applying the theories of agenda setting and issue ownership in the field of business communication. *Journal of Communication, 56*(3), 543–559.

Meijer, M., & Kleinnijenhuis, J. (2006b). News and corporate reputation: Empirical findings from the Netherlands. *Public Relations Review, 32*(4), 341–348.

Miller, R. E., & Wanta, W. (1996). Sources of the public agenda: The president—press—public relationship. *International Journal of Public Opinion Research, 8*(4), 390–402.

Nassauer, S. (2010, June 24). 'I hate my room,' the traveler tweeted. Ka-boom! An upgrade! *The Wall Street Journal.* Retrieved July 12, 2011, from http://online.wsj.com/article/SB10001424052748704256304575320730977161348.html

Ohl, C. M., Pincus, J. D., Rimmer, T., & Harrison, D. (1995). Agenda-building role of news releases in corporate takeovers. *Public Relations Review, 21*(2), 89–101.

Park, D., & Berger, B. K. (2004). The presentation of CEOs in the press, 1990–2000: Increasing salience, positive valence, and a focus on competency and personal dimensions of image. *Journal of Public Relations Research, 16*(1), 93–125.

Pfarrer, M. D., Pollock, T. G., & Rindova, V. P. (2010). A tale of two assets: The effects of firm reputation and celebrity on earnings surprises and investors' reactions. *Academy of Management Journal, 53*(5), 1131–1152.

Pollock, T. G., & Rindova, V. P. (2003). Media legitimation effects in the market for initial public offerings. *Academy of Management Journal, 46*(5), 631–642.

Pollock, T. G., Rindova, V. P., & Maggitti, P. G. (2008). Market watch: Information and availability cascades among the media and investors in the U. S. IPO market. *Academy of Management Journal, 51*(2), 335–358.

Ragas, M. W. (2010). *Agenda-building and agenda-setting in corporate proxy contests: Exploring influence among public relations efforts, financial media coverage and investor opinion.* Unpublished doctoral dissertation, University of Florida.

Ragas, M. W., Kim, J., & Kiousis, S. (2011). Agenda-building in the corporate sphere: Analyzing influence in the 2008 Yahoo!-Icahn proxy contest. *Public Relations Review, 37*(3), 257–265.

Ragas, M. W., & Roberts, M. S. (2009). Agenda setting and agenda melding in an age of horizontal and vertical media: A new theoretical lens for virtual brand communities. *Journalism & Mass Communication Quarterly, 86*(1), 45–64.

Reynolds, A., & McCombs, M. (2009). How the news shapes our civic agenda. In J. Bryant & M. B. Oliver (Eds.), *Media effects: Advances in theory and research* (3rd ed., pp. 1–16). New York: Routledge.

Rindova, V., P., Williamson, I. O., Petkova, A. P., & Sever, J. M. (2005). Being good or being known: An empirical examination of the antecedents, and consequences of organizational reputation. *Academy of Management Journal, 48*(6), 1033–1049.

Roush, C. (2006). *Profits and losses: Business journalism and its role in society.* Oak Park, IL: Marion Street Press.

Roush, C. (2011). *Show me the money: Writing business and economic stories for mass communication* (2nd ed.). New York: Routledge.

Sant, R., & Zaman, M. A. (1996). Market reaction to Business Week 'Inside Wall Street' column: A self-fulfilling prophecy. *Journal of Banking & Finance, 20*(4), 617–643.

Shaw, D. L., & Slater, J. W. (1988). Press puts unemployment on agenda: Richmond community opinion, 1981–1984. *Journalism Quarterly, 65*(2), 407–411.

Shoemaker, P., & Reese, S. (1996). *Mediating the message.* White Plains, NY: Longman.

Shoemaker, P. J., Tankard, J. W., & Lasorsa, D. L. (2004). *How to build social science theories: Theories of influence on mass media content.* Thousand Oaks, CA: Sage.

Soroka, S. N. (2006). Good news and bad news: Asymmetric responses to economic information. *The Journal of Politics, 68*(2), 372–385.

Steel, E. (2009, November 23). Marketers find web chat can be inspiring. *The Wall Street Journal.* Retrieved July 12, 2011, from http://online.wsj.com/article/SB10001424052748703819904574551562382557556.html

Stevenson, R., Gonzenbach, W. J., & Prabu, D. (1991, August). *Economic recession and the news.* Paper presented to the Association for Education in Journalism and Mass Communication, Boston.

Strömbäck, J., & S. Kiousis (Eds.). (2011). *Political public relations: Principles and applications.* New York: Routledge.

Sullivan, J. (2011, May/June). True enough: The second age of PR. *Columbia Journalism Review*. Retrieved July 12, 2011, from http://www.cjr.org/feature/true_enough.php

van Riel, C. B. M., & Fombrun, C. J. (2007). *Essentials of corporate communication: Implementing practices for effective reputation management*. New York: Routledge.

Wanta, W., & Foote, J. (1994). The president-news media relationship: A time series analysis of agenda-setting. *Journal of Broadcasting & Electronic Media*, *38*(4), 437–449.

Wartick, S. L. (1992). The relationship between intense media exposure and change in corporate reputation. *Business and Society*, *31*(1), 33–49.

Wattenberg, B. (1984). *The good news is the bad news is wrong*. New York: Simon and Schuster.

Wu, H. D., McCracken, M. W., & Saito, S. (2004). Economic communication in the 'lostdecade': News coverage and the Japanese recession. *International Communication Gazette*, *66*, 133–149.

Index